VOLUME ONE HUNDRED AND SIXTEEN

Advances in
COMPUTERS

VOLUME ONE HUNDRED AND SIXTEEN

ADVANCES IN COMPUTERS

Edited by

ALI R. HURSON

Missouri University of Science and Technology, Rolla, MO, United States

VELJKO MILUTINOVIĆ

University of Indiana at Bloomington, Bloomington, IN, United States

ELSEVIER

ACADEMIC PRESS

An imprint of Elsevier

Academic Press is an imprint of Elsevier
50 Hampshire Street, 5th Floor, Cambridge, MA 02139, United States
525 B Street, Suite 1650, San Diego, CA 92101, United States
The Boulevard, Langford Lane, Kidlington, Oxford OX5 1GB, United Kingdom
125 London Wall, London, EC2Y 5AS, United Kingdom

First edition 2020

Notices
Knowledge and best practice in this field are constantly changing. As new research and experience
broaden our understanding, changes in research methods, professional practices, or medical
treatment may become necessary.

Practitioners and researchers must always rely on their own experience and knowledge in evaluating
and using any information, methods, compounds, or experiments described herein. In using such
information or methods they should be mindful of their own safety and the safety of others, including
parties for whom they have a professional responsibility.

To the fullest extent of the law, neither the Publisher nor the authors, contributors, or editors, assume
any liability for any injury and/or damage to persons or property as a matter of products liability,
negligence or otherwise, or from any use or operation of any methods, products, instructions, or ideas
contained in the material herein.

ISBN: 978-0-12-820196-1
ISSN: 0065-2458

For information on all Academic Press publications
visit our website at https://www.elsevier.com/books-and-journals

Publisher: Zoe Kruze
Acquisition Editor: Zoe Kruze
Editorial Project Manager: Peter Llewellyn
Production Project Manager: James Selvam
Cover Designer: Alan Studholme

Typeset by SPi Global, India

Working together
to grow libraries in
developing countries

www.elsevier.com • www.bookaid.org

Contents

6. Using clickstream data to enhance reverse engineering of Web applications

Preface

Traditionally, *Advances in Computers*, the oldest series to chronicle of the rapid evolution of computing, annually publishes several volumes, each one typically comprised of four to eight chapters, describing new developments in the theory and applications of computing. The 116th volume is an eclectic volume inspired by several issues of interest in research and development in computer science, ranging from how to review a research article and respond to a review to enhancing reverse engineering of web applications. The volume is a collection of six chapters that were solicited from authorities in the field, each of whom brings to bear a unique perspective on the topic.

Chapter 1 entitled "Teaching Graduate Students How to Review Research Articles and Respond to Reviewer Comments" should be of great interest to graduate students and young researchers specially those seeking academic positions. The chapter is a guidelines for how to properly do research, how to properly report the outcome of a research, and how to react to the view of others. It should be noted that computer science/computer engineering are the underlying disciplines behind this article. However, the guidelines is generic and can be used by researchers in other disciplines.

Chapter 2, "ALGATOR—An Automatic Algorithm Evaluation System," by Tomaž Dobravec introduces ALGATOR developed to facilitate and automate the algorithm-evaluation process. ALGATOR creates an environment in which the time complexity and other project-specific indicators can be measured, and provides tools for the automatic and semiautomatic analyses of the results of the algorithms' execution. The article articulates that such a platform allows a better and more realistic analysis of the solutions during the development stage, hence reducing the practical surprises. The article presents implementation and functionalities of the ALGATOR system and gives several practical examples of its use in the evaluation of real world algorithms.

In Chapter 3, "Graph Grammar Induction," Luka Fürst et al. propose an approach that finds a concise graph grammar whose language, i.e., the set of graphs which can be generated using the grammar's productions, contains all of the given "positive" graphs and none of the given "negative" graphs. The article argues that in contrast, most of the existing induction methods couple the induction process with a graph grammar parser, thus inducing grammars from both positive and negative graphs. The proposed approach has been applied to several nontrivial graph sets and obtained graph grammars which can be regarded as a concise and meaningful generalization of the input graphs.

In Chapter 4, "Asymmetric Windows in Digital Signal Processing," Robert Rozman is looking at symmetric windows and their potential shortcomings and hence proposes asymmetric windows for better signal representation, performance, and shorter time delay. A comprehensive overview of the past and current work in the field of asymmetric windows is presented, summarized, and analyzed. Several examples of interesting effects of asymmetric windows are presented, followed by empirical evaluations in the fields of pitch modification, shorter time delay audio processing, frequency analysis, speech processing, and FIR filter design.

In Chapter 5, "Intelligent Agents in Games: Review With an Open-Source Tool," Matej Vitek and Peter Peer articulate that the concept of intelligent agents provides a much needed theoretical background for the comparison of different approaches to intelligent and rational behavior of computer-controlled characters in games. The chapter introduces various types of agents that are used in games and show how to implement meaningful, reasonable limitations to agent capabilities into the game world, and provide a freely available, open-source application for the comparison of such agents. The chapter also demonstrates that even the simplest agents can succeed in their tasks in certain environments, while, as expected, more difficult task environments often require a more sophisticated agent architecture.

Finally, in Chapter 6, "Using Clickstream Data to Enhance Reverse Engineering of Web Applications," Marko Poženel and Boštjan Slivnik are interested in web applications and their evolution. Factors such as user sessions, results of clickstream analysis, and session reconstruction are employed in order to gain a quick insight into web application's source code for the sake of automatic reverse engineering. The proposed method was tested against the results obtained by an expert. As noted, the proposed method can also be used for verifying the structure obtained by manual reverse engineering of the application's source code.

We hope that readers find these articles of interest and useful for teaching, research, and other professional activities. We welcome feedback on the volume, as well as suggestions for topics of future volumes.

ALI R. HURSON

Missouri University of Science and Technology, Rolla, MO, USA

VELJKO MILUTINOVIĆ

University of Indiana at Bloomington, Bloomington, IN, USA

CHAPTER ONE

Teaching graduate students how to review research articles and respond to reviewer comments

Milan Banković[a], Vladimir Filipović[a], Jelena Graovac[a], Jelena Hadži-Purić[a], Ali R. Hurson[d], Aleksandar Kartelj[a], Jovana Kovačević[a], Nenad Korolija[b], Miloš Kotlar[b], Nenad B. Krdžavac[a,b,c], Filip Marić[a], Saša Malkov[a], Veljko Milutinović[e], Nenad Mitić[a], Stefan Mišković[a], Mladen Nikolić[a], Gordana Pavlović-Lažetić[a], Danijela Simić[a], Sana Stojanović Djurdjević[a], Staša Vujičić Stanković[a], Milena Vujošević Janičić[a], Miodrag Živković[a]

[a]School of Mathematics, University of Belgrade, Belgrade, Serbia
[b]School of Electrical Engineering, University of Belgrade, Belgrade, Serbia
[c]Department of Microbiology and Molecular Genetics, Michigan State University, East Lansing, MI, United States
[d]Missouri University of Science and Technology, Rolla, MO, United States
[e]Department of Computer Science, University of Indiana at Bloomington, Bloomington, IN, United States

Contents

Advances in Computers, Volume 116
ISSN 0065-2458
https://doi.org/10.1016/bs.adcom.2019.07.001

1

Abstract

This article deals with teaching students how to review research articles of other authors and how to respond to reviewer comments related to their own articles. By teaching these skills, one educates students how to do something which is undoubtedly much more important for their professional development: how to properly do their own research and how to properly write their own research articles. The guidelines given are primarily intended for students of computer science. Nevertheless, the notions presented here are applicable to areas other than computing. Since there are so many branches within computing, each one characterized by its own particular features, this article also includes a section on variations of presented methodology, which are of interest for each particular branch of computer science covered by the authors of this article. In general, the review process consists of two main parts: formal review and quality review. Led by the lack of results in the available literature, an analysis based on readability index evaluates average article acceptance rates, review times, and delay period for publishing accepted articles depending on the following: article types, willingness of reviewers to do reviews, journal impact factor, and the scientific field. Quality review must be taken into account for all types of articles. Appendix A describes specific experiences from various scientific fields. Appendix B gives specific guidelines for a number of fields covered by the authors of this article.

1. Introduction

Publication of scientific research results is a process which comprises a number of important stages. Each stage includes specific procedures and deals with specific notions, so PhD students, as one of the most important driving forces of publication, have to be properly educated before getting involved with it. First of all, a student should be aware of all the important stages of the publication process, and must know how particular tasks in each of the stages are done. An experienced researcher might take this part for granted; however, for a young student who is just making the first steps in their research career, this process may involve numerous unknowns.

Second, it is important that the student knows how a typical article is reviewed by an anonymous reviewer. This is important for at least two reasons. First, knowing how an article will be reviewed might help authors improve their work and its presentation before the article is submitted. If one knows what the reviewer would typically expect, they might notice potential flaws in their work and address such issues before they are raised by reviewers. Another reason is that a student might actually be asked to be a reviewer. This is usually done by the student's advisor who sometimes needs the student's help due to time constraints, or the advisor might want the student to become experienced in this important activity in the research process.

Further, it is important for the student to know how to respond to reviewers' comments and requests. This is usually the case when an article is conditionally accepted for publication, and a revised version of the article needs to be submitted. This revised version must be accompanied by a cover letter in which the authors of the article must provide information on how each of the reviewers' requests has been addressed in the article. The quality of this response may greatly influence reviewers' opinions in the next iteration. This type of knowledge is also useful if one is asked to do a review, since they will be more skilled when it comes to treating and evaluating authors' responses to their comments.

Finally, it is important that the student is aware of some statistics related to the reviewing process in order to make the right decisions when submitting articles.

Therefore this article has the following structure: in Section 2 we give a short overview of related work; in Section 3 we briefly describe the main stages of the publication process; in Section 4 we discuss writing good reviews and good articles; in Section 5 we consider writing a response to reviewers' requests and comments; Section 6 presents some statistics related to the reviewing process; finally, in Section 7 we provide conclusions.

The article has two appendices. Appendix A contains details about field variations concerning the reviewing process, especially in the context of a paradigm shift in selected research fields. Appendix B presents some setups for homework assignments which may help students adopt the skills needed for a successful research career.

2. Related works

The process of publishing scientific results consists of many different steps explained in a variety of papers. These can be divided into several categories: guidelines for finding a research topic and presenting research results, guidelines for peer reviews, guidelines for responding to reviewers' comments, discussions about journal options, and problems in publishing scientific articles.

2.1 Topic selection and presentation of results

PhD students have a short period of time within which they are expected to graduate, which means not too much time to develop research ideas and get results. This inspired the effort to help PhD students in computer science and engineering to generate good original ideas for their PhD research [1]. This paper by Blagojevic presents a proposal of 10 methods which can be implemented to derive new ideas, based on the existing body of knowledge in the research field. It also provides a case study based on the examples of PhD research of the authors of this paper, and shows how they fit into the proposed classification.

Some papers offer guidelines for organization of oral or written research presentations [2] with the aim of making them as easy and simple enough to understand. There are many examples when good research was presented so poorly that results remained unrecognizable. After such presentation, it was impossible even to understand either the essence of the contribution or specific research details. The paper by Milutinovic paper gives advice on title selection, abstract structure, the structure of figures and tables and their captions, the syntax of references, the structure of the written paper and the corresponding oral presentation using slides.

Some papers underline the importance of writing articles interesting to a reader (or reviewer) [3]. In such articles the tendency is to focus on data presentation rather than presentation of problem solving principles.

2.2 Guidelines for the peer review process

Most articles which deal with the peer review process propose guidelines for the process and give an overview of steps for successful peer reviewing and for reviewing revised manuscripts [4]. Some of them give descriptions of tools and resources which can improve the reviewing process [5].

Some papers offer a checklist for the full review process [6]. The first two chapters of the paper by Hames explain the review process. They are followed by technical explanations how submit a paper online, what reviewers do and what kind of work is required from authors. To help the novice in the review process, it also gives detail explanation how deal with the scope of manuscripts, research topic objectives, research design, and soundness of obtained results. Finally, the last chapter analyzes unprofessionalism in research and publication.

The second group of papers give guidelines for reviewing in the form of a flowchart [7]. The paper by Kimberley presents the goals of peer review: to ensure accuracy and improve the quality of reviewed (and possibly published) material through constructive criticism, pointing out shortcomings, and listing suggestions for their elimination. The author advises the reader to consider a request to serve as a reviewer. Questions are whether the author's field of expertise matches editors' expectations, whether reviewing can be unbiased, and whether a review can be completed before a set deadline. The reviewer should first read the entire paper and evaluate whether it meets the prescribed conditions (aims and scope, format, etc.) to be publishable in principle. In the second interaction, the peer reviewer should read the paper in detail, and note the main points of the review, making remarks and comments concrete and pointing to disputed places in the text.

Other papers give reviewing guidelines in the form of a template [8]. The provided template provided by Provenzale includes a list of directions which the reviewer should perform during the review process. Directions are provided in the form of guidelines on what each paper section (abstract, introduction, previous work, material, method, results, etc.) should contain. The suggested list can be used as a guideline for young reviewers when they prepare their own manuscripts.

Ethical guidelines for peer reviewers are also very important [9]. Depending on peer review models (double-blind, single-blind or open), reviewers are either anonymous or not. The paper gives a detailed description on how to perform a peer review including general and ethical guidelines, how to work with editors, and how to review revised manuscripts.

2.3 Guidelines for answering to reviewers' comments

Articles which cover response to reviewers mostly contain specific rules to be followed when writing a response [10]. It is usually advised to answer completely (i.e., address all points raised by reviewers), answer politely, and answer with evidence (e.g., add literature or experiments in order to support claims) [11]. The process of responding to reviewers can be delayed for unexpected reasons, which is why it is a good idea to start working on the answer as soon as possible.

A well-designed answer to the reviewer's document is the key part of the response. A response letter will typically have two parts: for the editor and for the reviewers. Letters must be polite both to the editor and to the pointers, regardless of whether the author believes that some of the reviewers made a mistake and did not understand the content of the work. The letter to the editor includes a summary of changes, the description of given answers to reviewers' comments, as well as new analysis preformed upon the request of reviewers. Letters to reviewers must include concrete answers to received comments, again with the description of possible new analysis preformed. If some reviewers did not understand some assertions or larger parts of the paper, authors should try to make it clearer or support their own claims with additional proofs or evidence.

Some papers propose ways in which journals could help with responding to reviews [12]. One way is to organize reviewers' comments and response to reviews in the form of a table which corresponds to appropriate structures in the online system of submitting and responding to reviews. Many journals also allow authors to ask for deadline extension for revision if they are unable to complete it by the initially set deadline. Sometimes they give templates [13] for responding to reviewers' comments which some researchers find useful when preparing an answer. It is suggested to categorize responses to comments into major or minor ones before starting to draft a response letter.

2.4 Journal characteristics and other problems

Some papers [14] deal with the interesting topic of open access (OA) journals versus subscription journals. The scientific impact of journals was compared based on the average number of citations which the journal received over a 2-year period (impact factors) in Scopus/Web of Science. Research also compared the speed of publication in OA and subscription journals. The material included 610 OA journals and 7609 subscription journals indexed in Web of Science, and 1327 OA journals versus 11,124 subscription journals indexed in Scopus. Results show that a paper can be published faster in OA journals than in subscription journals, while the number of citations between both groups was similar. The authors also found that OA journals with article processing charges (APCs) had a higher number of citations than non-APC OA journals.

Delays in article publishing are not the same across different fields of science [15]. In this study the average publishing delay is analyzed for

2700 papers from different disciplines published in 135 journals indexed in the Scopus citation index. Journals from chemistry, engineering and bio-medicine had the shortest delays as opposed to arts/humanities, and business/economics. The delay for business/economics was approximately twice as long (18 months) compared to chemistry (9 month average delay).

Further research confirms that there are delays in publishing. The publication process is becoming longer and longer, and to find out why, Nature examined some recent analyses of publication time [16]. The delay period varies among disciplines, with social sciences being much slower than others. Reasons are diverse: reviewers requesting more data, revisions and new experiments, a large volume of paper production and lack of qualified reviewers, the fact that a large number of editors and almost all reviewers (even in high impact journals) work for free so they do not have any benefits from speeding up the entire process.

An interesting topic in the review process is recognizing falsification or fabrication in articles (i.e., plagiarism and self-plagiarism) [17]. Using different methods, it was found out that on average about 2% of scientists admitted to have fabricated, falsified, or modified data or results at least once, while 34% admit that they have been involved with other questionable research practices. Based on these results it can be stated that actual percentages are probably higher.

3. Publication process stages

The process of publication of a scientific article consists of several important stages, summarized in Fig. 1. Prior to writing an article, it may take several months or even years to obtain results of research relevant enough to be published [1]. This may look as the most demanding part of the process (and it usually is the case), but what follows should also be done with much care in order to increase the chance of successful publication. The next stage is writing the article, where researchers must describe their research, motivation and goals, used methodology and obtained results. Fair and detailed comparison with related work in the field is also required. Researchers must take care of the structure of the text, presentation style, and readability of the article.

Once the article has been written, the next stage is choosing an adequate journal. This is often not a simple task, since one should consider several important aspects such as the aims and scope of the journal, its impact factor, publishing fee (if there is one), expected time for the reviewing process, the option of open access, and so on.

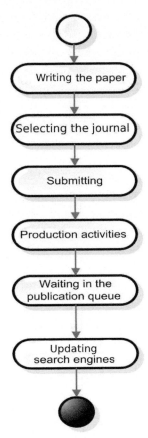

Fig. 1 The process of article submission.

The next stage is the submission of the article to the chosen journal, which is usually done by e-mail or through an online application provided by the selected journal. If the article is accepted for publication, it is further passed to a technical editor who is in charge of production activities. These activities are done in cooperation with authors, who must approve the final version of the article, after all technical corrections have been made. The article is then usually first published online, while it may stay in the publication queue for several months before it is assigned a journal number and issue, when its printed version becomes available.

More details about submission and production activities are shown in Fig. 2. After the article is initially submitted, it may take several months before the first response is received from anonymous reviewers. If the article is rejected, authors may try with another journal. If the article is conditionally accepted, a revision of the article is required before the article is

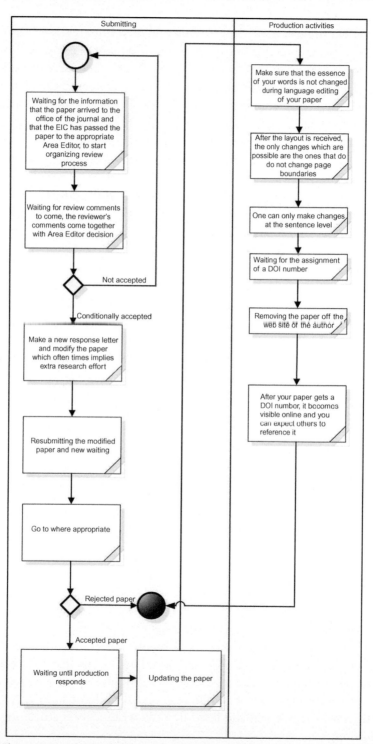

Fig. 2 The process of submission and production activities (*IEC*, editor in chief; *DOI*, digital object identifier).

resubmitted. In both cases, the feedback obtained from reviewers is crucial, since remarks and suggestions from reviews tell authors how to improve both their research and presentation of results. In the case of conditional acceptance, it is especially important to attend to each remark from reviews, since the revised version of the article will likely be read in the next iteration by the same reviewers who will expect that the article is improved according to their suggestions. There may be several iterations (i.e., article revisions) until final acceptance. However, there are usually three: initial submission, one major revision, and one minor revision.

After the article is accepted for publication, production activities start. The assigned technical editor will try to format the text according to the predefined style of the journal. It is of crucial importance to avoid changing the essence of what is written (or drawn) in the article during this process. Technical editors are usually not an expert in the research field of the article, so they may accidentally change some words or modify some figures or tables in the way which significantly alters their meaning. For this reason authors must monitor the whole process and intervene if something is not right.

Usually, at the end of the process, authors have a chance to take the final look at the article, when they may examine the article in detail before approving its publication.

When the final version is approved by authors, a DOI number is obtained and the article is usually published online on the web page of the journal. Copyright terms usually require that authors remove the draft versions of the article from their web pages, but sometimes keeping the draft version is allowed, provided that there is a copyright notice and inserted link to the final version.

4. How to write a review

In this section we consider an important activity in the publication process—writing a review. This activity is done by experienced reviewers who are experts in the research field which the article belongs to. However, article authors should also know how reviews are written and which particular properties and requirements are usually considered by reviewers, in order to attend to all potential flaws of the article in advance and increase the chance of receiving positive reviews. This means that this section is not just about how to write a good review but also about how to write a good article. In other words, while we consider important facts regarding writing a review, we will also learn how to write a good article for initial submission.

There are several types of scientific articles, each of which has its own specific traits which influence both writing an article and writing a review.

The types of articles which we consider here are:

1. survey article,
2. idea article (or a position article, or a theoretical article, or a methodology-explaining article),
3. simulation-based comparison (SBC) article,
4. research article (or a short communication article).

A survey article provides a detailed presentation, classification and analysis of existing results and their relations within a research field. An idea article provides some initial ideas which have the potential necessary in order to start a new research cycle in the field. An SBC article provides a simulation-based comparison between different solutions and approaches. Finally, a research article is an article in which a new approach, solution, methodology, statistical study, and experimental or theoretical results are presented in detail, supplied with a full analysis of obtained results and comparison to previous work. The activity diagram for the above-mentioned types of articles is shown in Fig. 3. As we can see from the figure, activities for an idea article and an SBC article may go in parallel, if the SBC article considers simulation-based comparison between the approach proposed in the idea article and other existing approaches.

Note that not all journals accept all types of articles. Survey articles get accepted by only about 10% of journals while SBC articles are welcomed by about 50% of journals. Idea articles are taken into consideration by only about 5% of journals. Research articles, being most the common category, are published by over 95%, but not by 100% of journals.

The review process includes two major phases: formal review and quality review. Formal review checks if all formal requirements of the journal are met. Quality review checks if an article is of good enough quality to justify publication in the journal it is submitted to. It is also advisable to do an internal review, prior to submitting an article, which is elaborated in more detail in Section 4.3.

4.1 Formal review

Formal review includes writing an official reviewer's report. It includes answering the questions in the form issued by the journal. Formal reviews can be divided into several parts: comments addressed to the editor, comments to authors, and a brief opinion about the article.

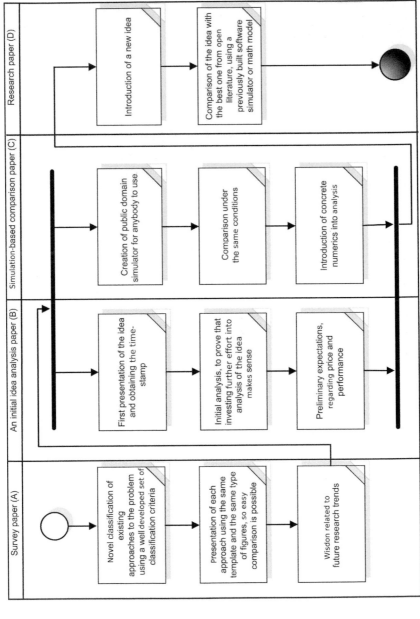

Fig. 3 An activity diagram for survey articles, initial idea analysis articles, SBC, and research articles. Solid lines indicate that the process may finish at that point.

Comments to authors usually begin with a brief summary of the article under review. The reviewer must then answer many questions which are concerned with various aspects of the article:

1. questions about technical soundness (i.e., whether statements and claims in the article are correct);
2. questions about originality and contribution (i.e., whether the work presented in the article is original);
3. questions about the relevance of the article in the scope of the journal (i.e., whether the work presented in the article has a significant impact on the research field which the journal covers and whether it is applicable in practice);
4. questions about readability, organization and presentation (i.e., whether the article is well organized and easy to follow and understand);
5. questions about reference to previous work (i.e., whether relevant related work is properly referenced in the article).

As the final step, the reviewer must state her/his decision about the article, which is usually one of the following: accept, accept with minor revision, conditionally accept (major revision), or reject.

4.2 Quality review

Quality review includes the reviewer's overall opinion about the article, as well as their particular judgments, remarks and suggestions concerning various statements in the article. It is important that each judgment is clearly stated and supported by adequate justification or reference. No matter which article type is under review, quality analysis includes, but is not limited to, the following elements:

1. quality of original contribution,
2. quality of analysis,
3. quality of presentation of results (of the analysis),
4. readability of the article.

These elements are obligatory, but are not the only issues which one should look at. Specific topics and viewpoints may require more elements, each of which will also need careful consideration.

4.2.1 Quality of original contribution

Quality of original contribution checks if the work presented in the article is significantly different from the work previously done in the field, as well as if the presented work can in some sense be considered as advancement in the field. In general, this means:

1. The contribution must be original (i.e. not plagiarized).
2. The contribution must in some sense be better compared to the most relevant approaches from open literature; or the contribution may present an alternative approach to the problem considered, which is of a similar quality as existing approaches. It is recommended that the stated improvements be proved using appropriate mathematical tools.
3. The set of conditions under which the contribution is better must be wide enough, so it corresponds to an application which is in relative demand. In other words, there must be an application for which the introduced solution represents improvement.
4. There must be a technology in which the proposed contribution can be or will be implementable.
5. In case an article presents a theoretical result, this result must be relevant enough, i.e., it must have a significant impact on further research in the field.

The previously stated requirements may have a slightly different meaning depending on the type of article being considered. For instance, for a survey article, the originality requirement means that there are no other survey articles on the chosen topic, or the survey article under review presents the topic from a different point of view. It may also be the case that the survey article under review is much newer and covers some advancement in the field not covered by previous surveys, or may include a critique of surveyed research.

For a SBC article, the originality requirement means that there is an original simulator, or a public one, with a large enough level of customization to justify originality. The covered field must be important and its importance must show a tendency to grow.

For an idea article, the originality requirement means that the idea is original, that it has potential and that is expected to grow in importance [18]. An indication (possibly a rough analysis) must be included to demonstrate that the idea has potential, and it must be said how rigorous follow-up analysis will be performed.

Finally, for a research article, the originality requirement means that the approach is significantly different from previous approaches to the same problem.

The presented work must be a significant advancement in the field and applicable widely enough to be relevant. In the case of a theoretical result, this result must have a significant impact on the research field. If an article proposes new practical solutions and algorithms for solving a problem, technology must be capable of implementing the proposed solutions (now or in the near future). In either case, the article must provide clear argumentation why the presented work is superior compared to previous approaches (this may include mathematical analysis or experimental evaluation).

4.2.2 Quality of analysis

The quality of analysis checks if the analysis given in the article is good enough to justify authors' claims about the proposed approach. Depending on the research field and the targeted journal, it may include some of the following:

1. A mathematical proof of correctness of the proposed solution or the presented result. If applicable, such proof gives the reviewer (or the reader) more confidence that the presented results are correct (for example, in the case of an algorithm, that it really does what the authors state it does).

2. A detailed analysis of computational complexity of the proposed solution. This may require some mathematical analysis which proves the stated upper bounds.

3. In some cases, the analysis of implementation complexity may be required in order to estimate the needed resources and the cost of the implementation of the proposed solution, as well as the required technology.

4. In most cases, comparison to other relevant approaches in the field is also required. This may include both theoretical consideration and experimental evaluation, whichever is applicable. The comparison must be fair and must include cases in which the proposed solution is better, as well as cases when it is not better than existing approaches. If results are theoretical, they must be related to results of other studies, and to open questions in the field (e.g., whether a result sheds a new light on certain unsolved problems).

4.2.3 Quality of presentation

Besides the originality of work and decent analysis of its substance, the work must be presented in the way which an interested reader may easily follow. For this reason, an article should have a common structure which is well established in the considered research field. This structure might be different for articles of different types. The necessary elements for the above-mentioned types of articles are shown in Fig. 4.

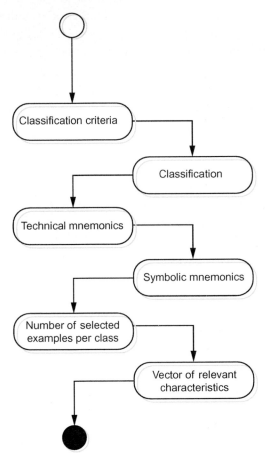

Fig. 4 Mandatory elements of survey articles, initial idea analysis articles, SBC articles, and research articles.

Since research articles are the most common category, in Figs. 5 and 6 we show an activity diagram which describes the usual structure of a research article in more detail. The first part of the article (usually called *introduction*) should provide motivation for the presented work. It should include a description of the research field, a statement of the problem being considered and justification for problem importance. Authors should then briefly describe the idea of the proposed solution and compare it to other existing approaches, clearly stating why their solution does not have the drawbacks present in other approaches.

In many cases, the introduction is followed by a *background section,* which formally introduces the notions and concepts used in the rest of the article. In the main part of the article, authors should describe their approach in detail,

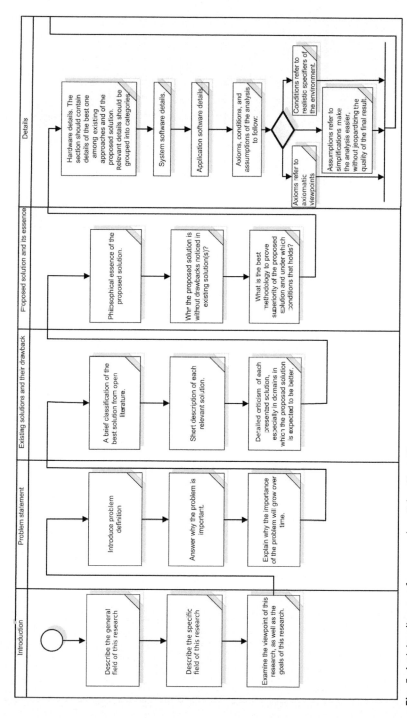

Fig. 5 Activity diagram for research articles (Part A).

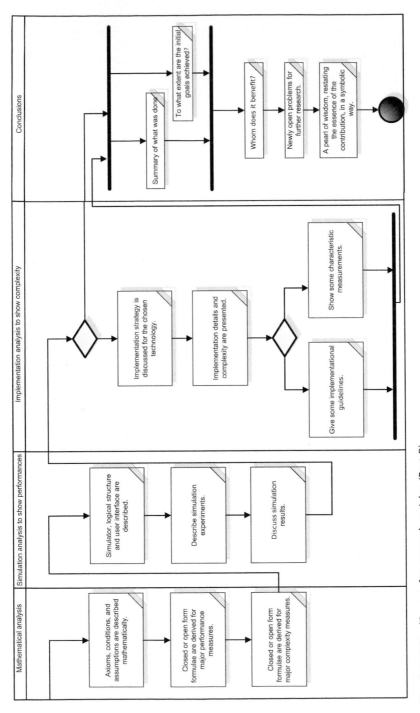

Fig. 6 Activity diagram for research articles (Part B).

state their results, and provide justification for their claims. This includes theoretical, experimental and implementation analysis of the proposed contribution. If experimental results (or other data) are required, they are usually presented in a separate section.

Comparison with related work is also commonly given in a separate section. This section may be placed after the introductory section, or somewhere toward the end of the article, after all the results of the study have been presented. The latter case is used when the reader has to know details about the proposed solution in order to adequately relate it to other approaches (e.g., when a new algorithm for problem-solving is introduced, explanation of its details might be necessary before comparing it to other existing algorithms).

An article always ends with a *conclusion section* which summarizes the results presented in the article. A part of this section usually includes directions for further work, if there are any (although this part may also be given as a separate section).

After the conclusion, zero or more appendices may be given. Appendices usually contain some details which do not need to be included into the main part of the article since they are not necessary in order to follow and understand the presented work, but they may be used to provide some additional information about the presented work which the reader might be interested in (for instance, details about proposed algorithms, additional experimental or observed data, and so on).

Many articles also include an *acknowledgment* section, in which authors express their gratitude to persons or organizations which have in some ways helped them in their work. This may include those who were willing to share their ideas with authors or give them advice, which is how they improved the presented research [19]. It is advisable to mention those who have provided infrastructure for authors' research, which may include one or more research projects in which authors have participated. Finally, gratitude should be expressed to all those who suffered by taking on everyday life responsibilities from the authors, so they could dedicate more of their time to research (this usually includes family members).

References are the last part of the article, and often are the most important one. All the referenced work must be cited correctly with enough information to make it possible to retrieve it. References may be listed alphabetically, chronologically, or they may be grouped by their topic

4.2.4 Article readability
The readability of an article is usually expressed in terms of *the readability index*. There are several methodologies for calculating such index, notably

Flesch's [20], Flesch-Kincaid's [21], Gunning's [22], etc. Flesch's index of readability (I_F) can be described with the following formula [20]:

$$I_F = 206.835 - 1.015\left(T_w/T_s\right) - 84.6\left(T_{sy}/T_w\right) \tag{1}$$

where T_w indicates the total number of words in an article, T_s denotes the total number of sentences, and T_{sy} is the sum of syllables. The scale of Flesch's index of readability is shown in Table 1.

A reviewer should include such metrics or a combination of two or more metrics in the review process. Automated analysis of text readability does not find defects in an article, but can make an article more understandable. Of course, a linguistically correctly written text is the main prerequisite for evaluation of readability.

4.3 Internal reviews

Prior to submitting an article, it is a good idea to have one internal review or more. An internal review may have a similar form as the one described earlier, but is done by a known researcher either from the same institution where the authors of the article work, or from another institution (ideally both). Internal reviews are usually written by more experienced colleagues or advisors, in case the authors are PhD students. If possible, it is advised that authors themselves do an internal review after a time distance, usually not shorter than 3 weeks. Internal reviews may help authors improve their work before they actually submit their article to a journal, and the feedback they receive from their colleagues may point to potential flaws of their approach. Furthermore, internal reviews may include discussions on the target journal where the article should be submitted. The process of internal reviewing also

Table 1 Flesh's index of article readability.

No	Score	Description
1	90–100	Easy to read
2	70–90	Easy to read by high school students
3	50–70	Easy to read by the majority of adults
4	30–50	Difficult to read by college students
5	10–30	Difficult to read, even by graduate students

Derived from J. Kincaid, R. Fishburne, R. Rogers, B. Chissom, Derivation of New Readability Formulas (Automated Readability Index, Fog Count, and Flesch Reading Ease Formula) for Navy Enlisted Personnel, Research Branch Report 8–75, Chief of Naval Technical Training: Naval Air Station Memphis, Memphis, 1975.

includes proofreading. Proofreading may include improvement of article readability as explained in the previous section. It is also a good idea to present research results to authors' research groups, before the article is submitted to the journal.

5. How to respond to reviewer's comments

After an article is submitted and the first response from anonymous reviewers is received, authors ought to prepare and submit a revised version of the article (provided, of course, that the article is not rejected). In the first response, reviewers usually require a major revision of the article, which means that significant work should be done before the results are improved in the way which is satisfactory for reviewers. This additional work may require several months (the due date is usually given by the handling editor, but in some cases it may be prolonged if needed).

After the results are improved, a revised version of the article must be written in order to include the required changes. However, it is not enough just to modify the article to meet the requirements of reviewers. A cover letter should also be written, explaining how each comment made by reviewers has been addressed.

A good reviewing practice is to itemize comments, so authors can address the review item by item (comment by comment). If that is not the case, authors must make an effort to split the reviewer's essay into a number of elementary requirements, i.e., to itemize it. After that, each item (elementary requirement) should be addressed in response to reviewers. Possible response types are:

1. The comment is accepted and it is explained what has been done or written in response to it;
2. The comment is not accepted and reasons for rejection are explained thoroughly;
3. The comment makes full sense, but is not feasible with the knowledge and resources of the author(s), and is suggested to be the subject of a follow-up article.

In all three cases, the reviewer should be given the exact text which was added to the article, or in case the added text is too long to be included in the cover letter, it should be precisely stated where the added or modified text can be found in the new version of the article (page numbers, paragraphs, lines, etc.). The added text should address all the issues raised by reviewers (the readers of the article should not ask themselves the same questions once the article is published); it should also be clearly marked/colored.

Of course, partial solutions in between the ones listed above are also possible. It is of vital importance to be honest to the reviewer, even if it is believed that full honesty may lead to rejection of the article.

6. Statistics related to the reviewing process

In order to make good decisions in the publication process, it is useful for students to be aware of certain statistics related to the process of reviewing; that is how they remember what is important. Essential guidelines are better remembered if associated with a story which contains concrete numerical values. Nowadays, most journals require authors to suggest the names of preferred reviewers. The question which most authors ask themselves is whether it is better to suggest a "friendly-oriented" person as a reviewer, or rather someone who is a top expert in the field which the article covers. Of course, the best solution is to have a friendly-oriented top expert, but this is very rarely possible. A friendly-oriented reviewer is in many cases not closely related to the topic presented in the article and such reviewer cannot help the author to improve the quality of the article. On the other hand, a review from a top expert can greatly improve article quality (regardless of the decision to accept or reject the article), but such experts often do not have time. They are typically asked to review many articles, and they often hand the articles over to their co-workers, assistants, or even students to review. Such practices can in many cases result in poorer quality of the review, denial of good articles, or acceptance of bad articles, which is not in the author's interest. Figs. 7 and 8 give some numerical values related to the probability that the suggested reviewer accepts a request for reviewing. The histograms presented in Fig. 7 also show the likelihood of chosen reviewers rejecting the article, should they agree to review it.

Fig. 7 presents the probability to accept a reviewing request (A), and the probability to be negative about a reasonably good article (B), if the potential reviewer:

(a) is referenced in the article (A = 55%, B = 0.1%),

(b) is among the most cited authors at Google Scholar for the keywords derived from the title of the article to be reviewed (A = 15%, B = 90%),

(c) is submitting an article(s) for the same conference/journal (A = 55%, B = 70%), and

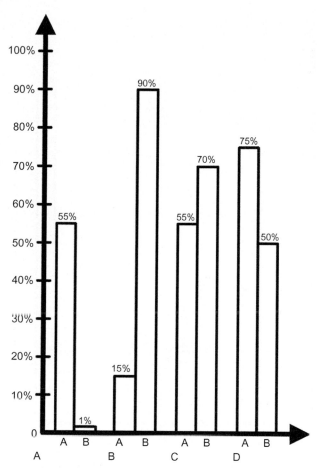

Fig. 7 Probability for a reviewer to respond to a reviewing request (A), and the probability to be negative about a reasonably good article (B), under conditions (A) through (D) described above; data generated for the year 2010. *Source: Reprinted with permission from TCCA: VIPSI series of conferences V. Milutinovic, The best method for presentation of research results, The TCCA Newsletter (1995) 1–6.*

(d) is in the carefully generated database related to the topic covered by the article to be reviewed (A = 75%, B = 50%).

Fig. 8 gives the same data for the VIPSI series of conferences as Fig. 7, but for three different time points 10 years apart [2]. There is a visible tendency that reviewers are less and less cooperative about review requests, which probably results from the fact that the number of conferences and journals has substantially increased over the past decades and the same trend is still present.

When selecting an appropriate journal, there are some more important numerical values which students have to keep in mind. These include the

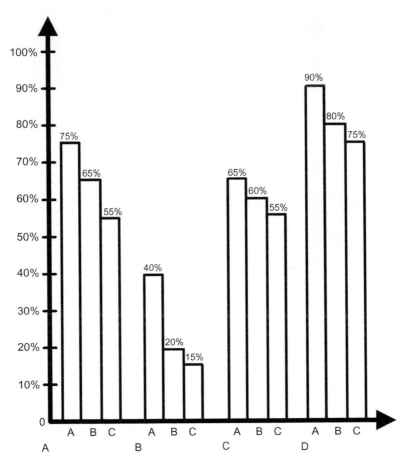

Fig. 8 Probability of a reviewer accepting a request for review for VIPSI conferences (data generated for [A] 1990, [B] 2000, and [C] 2010).

expected time of the review process, publishing delay, and expected number of citations. Review deadlines depend on journal's internal and external rules and conditions, and will likely depend on the number of submitted articles at a time, the number of reviewers per article, existence of pre-review filtering of submitted articles, willingness of original reviewers to review the revised version of an article, journal quality, etc. Some of these factors are illustrated in Ref. [16].

Although these data were obtained for articles from PubMed in 2015, there are no significant differences when they are compared to articles published later, regardless of the topic area. It is interesting that there is a correlation between the journal's impact factor and the time elapsed from submission to acceptance—the longest time is needed for publication in journals with the lowest and the highest impact factor, which is probably caused by the number of submitted articles in these journals. The amount of time which passes between submission and acceptance of an article is nearly the same as the time it takes to publish the article after it has been accepted. Suprisingly, this amount of time is expected regardless of the discipline which the article belongs to (see Ref. [15]). Variation in the accepted/published time ratio across disciplines is the result of the number of articles and the frequency of journal's publishing. The time expected to pass from acceptance to publishing must be taken into account by a student who has a publish-or-perish deadline.

When selecting a proper journal for article submission, one of the questions is whether an article should be submitted to an open access journal (OAJ) or to a subscription-based one? If an article is published in an OAJ, more researchers will read it and later use it as a reference in their articles. Some OAJs require an Article Processing Charge (APC) to fund publication of articles, which can amount to a few thousand US dollars.

Although the number of OAJs with and without APC is almost the same, there are not too many OAJs without APC with a significant scientific impact. In contrast, the majority of subscription journals do not require any publication fee, or have a reasonable level APC; however, subscription journals are read by a smaller number of researchers and hence have a smaller chance to be referenced.

The impact of OAJs and subscription-based journals differs depending on the discipline, the country where journals come from, and journal launch period [14]. Analysis showed that newly founded OAJs (in contrast to the established OAJs) have an almost equal number of citations, compared to subscription-based journals. Students are therefore not advised to base the decision on the type of access, OAJ versus not-OAJ.

7. Conclusion

This article is intended for doctoral students and scientific journal reviewers. The goal of the work presented is to prepare students for the process of writing journal articles, as well as for the process of developing critical thinking in any formal correspondence.

The article first briefly describes the process of research and publication. Such information is especially useful for a student who has never written or published any form of an article and is making the first steps in their research career. The following part of the article is related to the importance of the process of reviewing research articles and, at the same time it examines the complexity of the process. This is very important not only for those who actually write reviews, but also for those who do research and write articles. Knowing how reviewing is done and what is considered important from the reviewer's point of view is very helpful in the endeavor of writing good articles. The part of the article which comes next is on how to respond to reviewer's comments and requirements.

The article also analyzes the probability that potential reviewers will accept to do a review, which depends on the decade in which the review request occurred and the reviewer's h-index. Evaluation shows that there is a decreasing trend of reviewers' acceptance to do the review, as decades go by. We show that the response time from the moment of review acceptance conditionally depends on the journals' impact factor.

Although publishing an article is the first goal, it is also important for researchers to make sure that the article has a high chance of being referenced, meaning that it has a measurable impact on the research community. Therefore, when writing an article, authors should also make sure that the article being written has potential to be referenced.

Acknowledgments

The work presented here was supported by the Serbian Ministry of Education, Science and Technological Development (projects III44006, III47003, TR32047, 174021, 174010, 178006, COST action CA15123), and Maxeler Technologies, 3-4 Albion Place, London W6 0QT, UK.

Appendix A. Selected experiences with field variations

A.1 Experiences from the field of mathematical optimization and modeling

Researchers beginning to work in the field of optimization and mathematical modeling sometimes lack adequate mathematical background, i.e., knowledge about relevant topics: numerical methods, elements of probability theory, elements of mathematical analysis, etc.

An additional problem for these newcomers is that they often try to solve wrong problems, or too simple problems, with popular and complex methods which are inadequate. For example, combinatorial optimization has a discrete domain. There are many real life and theoretical problems with a discrete domain: sorting, finding the minimum spanning tree, routing problems, etc. An inexperienced researcher could apply wrong methodology and try to solve sorting or finding the minimum spanning tree problems by using a heuristic method (genetic algorithms [23], tabu search, electromagnetism-like metaheuristic, etc.). Such approach is not appropriate because these kinds of problems can be solved with polynomial-time algorithms.

Another very common mistake is trying to solve the most popular optimization problems at the very beginning of one's research career. Young researchers often tend to underestimate the amount of effort which other researchers have already put in, and they tend to overlook some, previously reported important results. Hence, due to the unfortunate choice of a research problem, preparations tend to be inadequate, and the effort invested does not produce satisfactory results. For example, solving the traveling salesman problem (TSP) might be very attractive due to its simple formulation and worldwide realistic application. However, making progress in terms of solving methodologies for TSP is difficult, since a large number of scientists have dealt with it previously.

Yet another common source of mistakes is the usage of new technologies based on their popularity, while the essential characteristics of the problem are neglected. For example, the map reduce paradigm is very popular and appreciated in the field of machine learning [24]. An inexperienced researcher, not fully aware of the fact that NP-hard problems have indivisible structure, might try to employ the map-reduce paradigm in solving NP-hard problems. Of course, such approach will consequently produce solutions of lower quality.

However, introducing new technologies and new problem domains into optimization have a positive impact.

Intensive progress is currently made in the field of bioinformatics, which influences other mathematical and computer science disciplines. Therefore, new optimization problems inspired by molecular biology and biochemistry arise frequently. In one such case, in a recent paper the author proposes a new problem formulation called the edge–weight k-plex partition problem, which has applications in modeling metabolic reactions and interactions in biological networks [25].

Similarly, progress in the domain of social networks raises new questions, sets new problems (as described in Ref. [26]) and creates new opportunities in the optimization domain.

A.2 Experiences from the field of combinatorial and continuous optimization

While dealing with various kinds of problems, one of the main issues that new researchers in combinatorial and continuous optimization area face consists of trying to solve the most popular problems in a particular domain.

Due to the lack of experience, one may unintentionally omit some important results from literature. Since many researchers have dealt with a particular popular problem, it might be hard for new researches in the field to list all the papers which have a huge impact on the problem itself, or the papers where the best known results are obtained using a particular set of tests. Due to the lack of comparison with results from such papers, one may face rejection of their publication. An important part of result comparison is to scale the obtained time execution values for the same test cases, since the new algorithm is tested on a machine with different performance capabilities. One way to do the scaling part is using and comparing values for different processors described in Ref. [27].

Other problems for researchers beginning to work in mathematical modeling usually arise when trying to define new problems. First, such new problems, although mathematically correct and complex enough, might not be applicable in the real world. One needs to carefully describe why a particular problem is proposed and what its main practical applications are. Note that sometimes the newly proposed problem itself has a huge impact in a particular area in the real world, which the new researcher does not consider at all. In cases where the paper is accepted with minor or major revisions, one of the main pieces of advice from reviewers is to improve the description of problem applications. Sometimes, reviewers claim that a particular problem is unimportant and lacks practical purposes, so the paper might be withdrawn. In that case, they usually list similar problems with real-world application in a particular area.

New technologies have a huge impact on defining new optimization problems, where the researcher tries to create an abstraction of a practical problem and proposes a method for its solution. Those who introduce new problems in such domains have to be careful since the problem might be equivalent to the previous one, or is not original enough to be accepted as defined. The other issue might be that, even though the practical application of the problem might seem important and concrete, it may lack complexity in a mathematical sense, which is why there is no need for its introduction. Sometimes the problem may sound complex in the way it is defined, but it may have a very well known polynomial equivalent, or there could be a non-complex polynomial time algorithm for solving the proposed problem. In that case, it means that the researcher has not invested enough time into analyzing the problem itself. It is always useful to check the complexity of the problem, which usually means that its NP-hardness or NP-completeness should be proved.

Due to the impact of social networks on our everyday lives, they have received a lot of research attention in this area. Many new optimization problems are inspired by this domain. For example, an efficient meta-heuristic approach for exploration of online social networks is introduced in Ref. [28]. The study proposes three new mathematical models for analyzing big data from social networks, based on optimization techniques for efficient exploration of information flow within the network.

Recently, researchers have frequently considered the problems where input data may vary, and they propose algorithms which consider uncertainty as an alternative to traditional optimization techniques. One of the most popular approaches in this domain is robust optimization [29]. It is typically used when exact distribution of input data is unknown. This is mostly the case in practical situations. In recent literature there are studies on robust optimization applied to various combinatorial optimization problems. For example, in Ref. [30] the famous two-stage capacitated facility location problem is extended to a robust optimization model, following the idea presented by Bertsimas and Sim [29]. The proposed model captures the uncertainty of input data and protects the network system against the worst case scenario within uncertainty sets.

A.3 Experiences from the field of formalization and automated proving of Euclidean geometry theorems

There are several approaches to formal theorem proving in geometry. We will here present some aspects of the approach which uses a combination of interactive and automated theorem proving, and we will provide an insight into formal theorem proving in general. When reviewing articles in literature, we found that researchers often lack a deep understanding of relevant issues.

Among other benefits, this area offers automated generation of formal proofs (which are verifiable with the help of a computer) and automated generation of proofs in a natural language form which can be used even by beginners in formal theorem proving (high school and university students) [31]. Automated theorem provers can even be useful in discovering certain properties of the axiomatic system used, which can result in more efficient and shorter proofs [32].

Formal theorem proving has been practiced for over 50 years. With the development of new interactive theorem provers, and tools within them, new practical applications emerge and new challenges arise. Some problems which are dealt with today were not solvable several decades ago. Powerful programs for theorem proving and large resources of (already) formalized mathematics make it one of the fastest developing areas in the past few years.

The field of geometry, especially Euclidean geometry, is very well known not only by mathematicians. It is used as a tool when introducing high school students into the notion of proof and conceptualization of formal theorem proving. Today, geometry is mostly examined through the field of interactive and automated theorem proving. Human interaction with theorem provers is essential in this area. Systems used for automated verification of high school theorem proofs can be developed [33,34].

Special challenges in the area of formal theorem proving arise from the fact that there are several different interactive theorem provers, as well as automated theorem provers, which are used by different groups of authors. Since geometry theorems and axioms used in mathematical textbooks are not formally defined (in the context of formal theorem proving), it is not uncommon that different groups of authors interpret those formulas in slightly different ways. Those differences in interpretation, despite being small, can make significant differences in the process of proving theorems and in generated formalizations of larger theories overall. Differences in individual interpretation of informal mathematics makes the process of comparison and alignment of different formalizations (of the same source) very complicated [35] and sometimes almost as demanding as the initial formulation of a mathematical textbook within the selected formal language.

The writing process starts by collecting all available information related to the problem of interest. Authors must first choose a part of the theory which they are interested in (an axiomatic system and a set of theorems). In most cases it is a mathematical textbook or a part of a textbook. In case that previous formalizations of the selected theory already exist, authors should analyze presented approaches and their results and see if they can improve previous formalizations or offer a different type of formalization. Authors must then choose a formal language for representation of theory and formulate all formulas of the proposed theory in that language. Some authors will use interactive theorem provers alone, but in the last few years a combination of interactive and automated theorem provers is used more often. A careful choice of tools is very important. Not all theorem provers use the same input and output format, and some of them offer more interaction with the user (in some cases this option is necessary for the proving process). Depending on the selected tools, the process of theorem proving will be different. Authors sometimes focus only on the statements which are being proved, while at other times authors will carefully analyze informal proofs offered by textbooks, which are being formalized and used when formulating formal proofs. In either case, authors must adapt the given information to the tools which are used.

Every step of the proving process is important, but some areas are crucial. The translation of textbook formulas into a selected language must be done very carefully. Sometimes all formulas will be easily translated and in some cases more complex translations will be necessary. Authors must address all previous formalizations on that subject thoughtfully and interpret them in the right way. Generated formalizations can be used as a basis for further development, as a starting point or for comparison. This process can also reveal errors in newly formulated or previous formalizations [35].

Progress in programs for interactive theorem proving provides the user with an increasing level of automation, even enabling an automatic verification of the axiomatic system which is used and automatic verification of certain (simpler) theorems. The development of new formats for writing theorem proofs provides new tools which make it easier to share different formalizations within several programs for interactive theorem proving [36].

A.4 Experiences from the field of interactive and automated theorem proving

Fully formally verified mathematics and software are the long-standing goals which have become practically realizable with modern computer tools. Efficient fully automated theorem provers become available for a wide set of specialized background theories (propositional and first order logic, linear arithmetic, theories used in software verification, etc.), while interactive theorem provers enable theorem proving in the very expressive framework of type theory and higher-order logic. In interactive theorem proving [37], users specify proofs in a specialized proof format, while the computer checks those proofs, automatically filling in the easier parts.

Since the very strict machinery of formal logic is used, and since all proofs are machine-checked, there is significantly less room for potential mistakes than in other areas. However, this does not mean that formalizations done within computer-supported theorem proving systems are completely error-free. Namely, a computer can verify that proofs of conjectures (lemmas, theorems) contain no errors, but it cannot check the correctness of given definitions, which can only be checked by careful manual inspection. Reviewers need to inspect if the definitions and statements of crucial theorems are aligned with traditional notions and if their formal statements are a faithful counterpart of what is claimed to be proved. A good practice which authors should follow and reviewers should encourage is to explicitly list the key definitions which the whole formalization relies upon. Notions introduced by axioms are especially dangerous, as a small error in just a single axiom could make the whole axiomatization inconsistent, from which it follows that every statement trivially holds (as false implies everything). Therefore, the definitional approach should be used and axiomatizations should be avoided whenever possible.

A frequent error of inexperienced researchers is making their formalizations monolith, focusing only on the main goals without developing abstractions. Abstractions usually make proving easier, and also make parts of the formalization useful in a wider context. Developing libraries is sometimes more important than just proving the core result.

An important paradigm shift in interactive theorem proving came with the connection of interactive provers and classic well-established fully automated provers. For example, the tool Sledgehammer [38] enables using modern SAT and SMT solvers from within the theorem prover Isabelle/ HOL. This opens up the possibility for new directions of research which young researchers are sometimes unaware of. Also, many parts of interactive proofs can be made significantly easier if integrated automated provers are employed. With this change, the central effort in the proving process shifts from finding the proofs of given conjectures to finding a proper way to encode them so that automated provers can find proofs automatically. For example, the correctness of a chess endgame strategy has been proved by encoding it in the language of linear integer arithmetic and employing an SMT solver integrated into Isabelle/HOL to prove several carefully designed lemmas [39].

This convergence of various provers is obvious in the wider field of automated reasoning (e.g., a SAT solving mechanism is being integrated into leading automated first-order theorem provers). Researchers should be aware of this and seek for methods of using newly available tools in the most efficient way.

A.5 Experiences from the field of interactive theorem proving: Some directions to young researchers

Interactive theorem provers are based on the interaction between a user and a computer, which is a topic often inadequately treated in the work of young researchers. Such systems are semi-automatic and during formal verification they help the user (usually a programmer and/or mathematician) by controlling the correctness of the proof and to some extent, when it is possible, by giving automatic evidence. Interactive provers are thus usually also called assistants for proving theorems. Today, there are many interactive provers: *Isabelle, Isabelle/HOL, Coq, Hol Light, PVS* and many more. *Isabelle/HOL* and *Coq* are widely used and years of research have generated a large collection of libraries with formally proved theorems which can be further upgraded. These systems can be used for software verification. Their application is also significant in education since they help develop and deepen mathematical knowledge.

Interactive theorem provers provide formal proofs which are considered to be correct. While interactive theorem provers are built under strict mathematical principles, there is still some room for error. Some provers enable using user-written procedures for proving. These procedures greatly shorten proofs, but also have to be formally verified and the author should offer formal justification for using these procedures. Another problem is choosing the correct mathematical setting. It has to be verified that definitions and statements correspond to what the author really wants to prove. Sometimes, statements are weakened or broadened, so the author should clearly state reasons for that and how it affects further formalizations.

In the introduction (even better in the abstract of the paper) it should be clearly stated which problem is presented and formalized and why that formalization is important. Young researchers may think that formalization is done only for the sake of formalization and that the decision to formalize a problem is self-explanatory. However, it is much better to clearly state the contribution of the formalization, i.e., what was learned from the formalization and how the formalized notions can be used.

It is said that a proof has two purposes: to explain and to justify. Although the main focus in formal proofs is on justification, a paper itself should be more oriented towards explanation. Formal proofs tend to be very long and detailed. The reviewer is interested in results and interesting steps and techniques used in the proof. A large pseudo-code could be overwhelming and lead to poor readability of the paper. It is better to use abstractions in writing definitions and statements, i.e., use the language close to the language in mathematical textbooks. Also, instead of writing all the steps of the formal proof, some key elements could be explained and highlighted. It could be interesting to emphasize which automated methods are used (for example, simplification method, smt method, Gröbner bases method). Formalization is rather widely dependent on the previous work and it should be clearly stated which results are used and which ones are newly developed. Furthermore, it is interesting to discuss whether proving could be done differently and what the advantages and drawbacks of using another approach would be (for example, in another approach proofs are longer but according to the author are easier to be conducted). However, a formal proof should always be available for inspection.

Automated theorem proving in geometry has a long history. There are many papers on this topic and many theorems have been proved using automated provers. The greatest advance was made in mid-20th century when Wu proposed his algebraic method to prove a theorem in Euclidean geometry. Modern proofs of theorems based on this method can prove hundreds of non-trivial theorems. However, the big disadvantage of these systems is that they do not produce classical evidence but only support the arguments which are not legible. In the 90s there were more attempts to get this problem solved and new methods based on the axiomatization of synthetic geometry were developed. Their main setback is that they are far less efficient than already introduced methods.

To the best of our knowledge, there are two main results that a new paper should address: (1) whether the proposed algorithm is faster than the ones already in use and (2) whether it is possible to broaden the set of theorems which can be automatically proved. Choosing non-synthetic testing datasets which enable comparison with other papers could be very important [17]. Good data will help carry out better research by preventing convergence on unrealistic solutions, while real large datasets (student homeworks, problems from competitions or exams) greatly increase the paper's chances.

Another question raised in recent years is why we should trust automated theorem provers, which are just another software which could be flawed. Thus, a reviewer should also take into account the soundness of the presented algorithm and its implementation.

A.6 Experiences from the field of software verification

Software verification is an open challenge in computer science since software bugs cost the global economy billions of dollars annually [40]. Software verification includes interactive theorem proving systems and automated bug-finding systems.

For automated approaches, there is an important theoretical limitation, which is a direct consequence of the halting problem [41]: it is not possible to create a tool which could precisely detect all the bugs in the program (i.e., without false positives or false alarms) using finite time and memory resources. This constraint means that there cannot be a perfect solution to the verification problem, which results in a trade-off between precision and efficiency. Some approaches such as abstract interpretation [42] are scalable but can introduce false alarms, while others are precise but limited in scalability, which is the case with symbolic execution [43] and model checking [44].

Model checking is an automated verification method for the analysis of software or hardware systems which can be modeled by state-transition systems [44]. It is based on research in mathematical logic, programming languages, hardware design, and theoretical computer science. At the beginning, model checking was based on systematic state space exploration and was very much limited by state space explosion (applicable only to small software and hardware systems). The first important breakthrough in model checking happened at the end of the 80s, when symbolic model checking was introduced. It enabled much larger state spaces to be searched, giving this approach the ability to efficiently handle systems with $>10^{20}$ states. However, the most important breakthrough was at the end of the last century when bounded model checking with SAT/SMT solvers pushed this edge even further [45]. The importance of efficient software verification with SAT/SMT solving introduced new challenges in the research of decision procedures, especially in theories which are necessary for modeling software properties. SAT/SMT solving has become the key enabling technology in automated verification of both computer hardware and software [45]. Model checking is nowadays widely used in industry, but is still an open research problem.

An important part of each software verification paper is comparison with related tools. There are many tools based on model checking, most of which are publicly available for academic use and some of which are open source [46]. The parameters used for running each tool should be explicitly stated in the paper (or scripts used for experimental evaluation should be publicly available). This makes all experimental results easy to reproduce and check. It is also important to make unbiased experiments on trusted benchmarks. There is an active software verification community which proposes benchmarks for comparisons of different verification techniques and approaches [47]. However, the best way to promote a new tool or technique is to use it in finding real-world bugs in an open-source industry project or in some other real-world application.

A.7 Experiences from the field of automated reasoning: SAT/SMT solving

SAT solvers are the systems which check the satisfiability of Boolean formulae. Their applications include hardware and software verification, scheduling, planning, cryptanalysis, etc.

The most important paradigm shift in the field of automated reasoning was probably triggered by the exceptional progress in SAT solving technology at the beginning of the 21st century, when the so-called CDCL (Conflict-Driven-Clause-Learning) SAT solvers were introduced [48]. Based on the well-known DPLL algorithm [49], but with significant algorithmic and implementational improvements [50], modern CDCL-based SAT solvers are now able to check the satisfiability of Boolean formulae with thousands of variables and millions of clauses. Such remarkable improvement of SAT technology made automated reasoning much more applicable in many areas, notably in hardware and software verification, but also in automated planning, scheduling, combinatorial design, and so on.

Driven by specific applications, some interesting new technologies derived from CDCL SAT solvers emerged. A notable example is the appearance of Satisfiability Modulo Theory (SMT) solvers, where an efficient SAT solving core is combined with decision procedures for some decidable first-order theories, making a significant improvement in first-order reasoning [51]. An SMT solver is a first-order satisfiability checker, and modulo is a specific theory of interest. Since SMT solvers were initially employed as an aid in software verification, theories were usually chosen to be suitable for expressing the conditions which typically appear in such applications (notably the theory of arithmetic, bit-vectors, arrays, lists and so on). However, there are also examples of incorporating decision procedures for other types of theories into SMT solvers, in order to make them applicable in other areas [52].

Another interesting example of technology derived from CDCL SAT solvers are LCG (Lazy Clause Generation) constraint solvers [53]. The emergence of LCG solvers was driven by the applications of SAT solvers in the area of constraint solving. Similarly to SMT approach, LCG combines a CDCL SAT core with the constraint propagators borrowed from classical constraint solvers, taking the best of both worlds.

An important change in the practice of SAT solving is the introduction of algorithmic portfolios—systems which rely on several SAT solvers, instead of just one, by selecting the one estimated to be the most appropriate one for a given instance, scheduling their execution, running them in parallel, etc. Many of such approaches are based on machine learning algorithms [54].

An important fact to note about automated reasoning is its multi-disciplinary character—it combines results from different areas of mathematics (e.g., mathematical logic or discrete mathematics) and computer science (programming languages, algorithms, theoretical computer science, and so on). Yet another important fact is that research in the field of automated reasoning is driven by practical applications which usually arise from industry. This imposes some specific requirements regarding research methodology, which PhD students are sometimes not quite aware of and which affects both writing the paper and writing reviews:

- The mathematical background of the research field imposes very high standards when it comes to technical accuracy. Concepts must be introduced in a consistent and well-defined manner, and notation must be very precise. It is recommended that the paper should include mathematical proofs of correctness for novel algorithms and methods which are proposed since this is usually requested by reviewers.

- Experimental evaluation of tools must be performed on real-world problems which usually arise from industrial applications. Due to the nature of problems being solved (which are usually NP-complete), the performance of a solver of a particular problem instance highly depends on some arbitrary decisions determined by the chosen configuration or a random seed. This means that the solver may be very efficient (or very slow) on some instances by chance. Therefore no reliable conclusion can be drawn from several instances. This implies that evaluation must be performed on a large and heterogeneous set of problem instances in order to get reliable information about the solver's average performance.

- Most off-the-shelf solvers come with some default configurations which are predominantly used by the researchers who apply such solvers in different fields. Although such strategy is the simplest one and often gives decent results, sometimes effort should be made to find a custom configuration which fits that particular application better.

- Due to the multidisciplinary character of automated reasoning, literature coverage must be wide enough. Authors must carefully consider all relevant approaches which could be compared or related to the proposed solution, and which arise from different areas of research. Unfortunately, students are frequently unaware of all possible connections between different subfields, which is why they may miss some important related work, which is later noticed by experienced reviewers. In portfolio design, when a researcher whose primary field is automated reasoning employs machine learning algorithms, one has to be even more on guard for common mistakes in machine learning.

A.8 Experiences from the field of machine learning

Machine learning is the area of artificial intelligence which focuses on building systems which learn from experience. The field has become popular recently, sometimes attracting researchers who are not experienced enough, which can lead to errors in the approach, and which are the issues to be discussed in the text which follows.

The input to a machine learning algorithm is most often a training set of instances (e.g., patients, images, bank transactions) described by values of some variables. The output is most often a function, a so-called model, which expresses relationships between variables, enabling the prediction of values of some variables, based on the values of other valuables. Recently, machine learning has lead to multiple breakthroughs in artificial intelligence resulting in surprising performance of computer systems, often surpassing human experts. Some of the most distinguished recent developments which heavily relied on machine learning were the development of self-driving cars and the victory of a computer system over the human world champion in the game of Go. Outstanding achievements are obtained in other areas as well, like image recognition and speech recognition, where human performance was also surpassed, modeling the semantics of natural languages, computer vision, autonomous helicopter, and quadrotor control.

The understanding and the practice of machine learning have changed several times since its origins at the end of the first half of the 20th century. There are several turning points related to the development of artificial neural networks, starting with the invention of the perceptron learning algorithm, its criticism and abandonment, the revival of ideas it is based on, the advent of feed forward neural networks and the backpropagation learning algorithm, and finally the appearance of deep neural networks in the first decade of the 21st century [55]. From a theoretical perspective, the most important turning point is the development of the statistical learning theory [56], which provided the understanding of what is sufficient for successful and efficient learning, but also yielded some very successful learning algorithms such as support vector machines. Another game changer in the field was the advent of big data [57]. Large quantities of data can alleviate some problems in machine learning such as, obviously, small sample problems, but they also give rise to other problems such as high dimensional data problems and technical problems related to storage and processing efficiency.

There are numerous mistakes inexperienced researchers make in the field of machine learning, which both researchers and reviewers should be wary of. We point to some of those mistakes, the ways in which they can be avoided and the ways in which changes in machine learning theory and practice influenced them.

Overfitting. A ubiquitous problem in machine learning is the potential of machine learning models to fit training data too well, adapting to irrelevant specifics of data and losing the capacity to generalize. This phenomenon is comparable to rote learning in humans. As explained by statistical learning theory, its cause is the high flexibility of the model employed. The problem is exacerbated by the advent of deep neural networks, which are characterized by very high flexibility. The solution consists of controlling model flexibility by means of regularization techniques [58,59]. If possible, a good alternative or complement to regularization is increasing the amount of data used for training. Prior to the advent of big data, the small quantity of available data made overfitting more likely. Afterward, as the number of variables included in the model reached hundreds of thousands or millions, the high dimensionality of data, which results in a high number of parameters of the model, became the main issue.

Off-the-shelf approach. A common mistake is made when sticking to the few most common machine learning algorithms. The No Free Lunch Theorem states that there are no machine learning algorithms which can provide the best predictive performance for all problems [60]. Moreover, in terms of performance, traditional machine learning algorithms often do not scale easily to the volume of big data. Therefore, to achieve best results, one has to design a model and an algorithm which correspond to the specifics and the volume of the problem at hand.

Evaluation. One of the most common and most persisting errors consists of improper evaluation of learned models, most notably inadequate use of cross-validation in presence of meta–parameters which control the behavior of a learning algorithm [61]. The main tenant of proper evaluation is that the data on which the evaluation is performed must not be used in model training. Practitioners are often unaware that the selection of meta–parameters is also a part of training, and they use the estimates of model performance on the evaluation data to tune meta–parameters better. In this way they use that data for training, which leads to optimistic estimates of the predictive performance of the model. The solution is to use nested cross-validation instead of the ordinary one [61]. The advent of big data and deep neural networks made the use of nested cross-validation computationally infeasible. In that context, a simpler training/validation/testing paradigm should be used to perform evaluation.

Interpretation. Machine learning models are most often used to provide predictions. However, models often express relationships between variables, which can sometimes be interpreted. One of the most common mistakes of an inexperienced practitioner is to forget that those relationships are not of causal, but of correlational nature. That means that the information about variable X can help predict the value of variable Y. However, it does not mean that one can control Y by controlling X, since X need not cause Y. Instead, both their behaviors can be caused by another variable Z. This is not to say that correlation can never help establish causation—only that it may not be enough. The advent of big data caused another twist in interpreting correlations. Namely, it was observed that in data sets with hundreds of thousands or millions of variables, some correlations appear purely by chance and are meaningless [62]. Therefore, caution and domain understanding is necessary when interpreting observed correlations. Another mistake is to interpret the models which are trained over variables measured at different scales. In such a context, variables measured at a finer scale can often appear less important than variables measured at coarser scales, as model parameters (which are subject to interpretation) adjust for scale differences. The solution is to apply standardization—to transform all variables by subtracting the mean and dividing by the standard deviation of the variable.

Social bias. Among other things, the advent of big data allowed for modeling of human behavior. Recently, it has been noticed that machine learning models exhibit racial and gender biases like connecting career success with gender or criminal behavior with race [63,64]. If such models are used in practice to help human decision making related to employment or criminal profiling, the negative effects of such biases may not only damage individuals of specific gender or race, but also reinforce such biases in humans. Therefore, tests should be performed to detect such biases. The testing and correction of such biases are still an open research topic.

A.9 Experiences from the field of natural language processing

Natural language processing (NLP) is a field of computer science, artificial intelligence and computational linguistics concerned with interactions between computers and natural (human) languages. When reviewing research papers in the NLP field, students are advised to take into account all four elements of quality analysis described in Chapter 4. In that sense, students are encouraged to do the so-called deep reading [65], meaning that they need to try to understand details of presented methods, how those methods work, and how authors implemented them. Moreover, deep reading means one should:

1. Criticize the paper and identify its gaps and limitations by
    a. examining assumptions,
    b. examining methods,
    c. examining statistics,
    d. examining reasoning and conclusions.
2. Check the quality of the benchmarking process by
    a. checking if there is a good reliable comparison with other methods,
    b. examining how this comparison is done, whether on the same datasets and under the same conditions,
    c. checking if the best-known, publicly-available datasets for the given field are used.
3. Try to propose alternative methods. Students should ask themselves how they would solve the problem if they were the authors.
4. Take notes, which includes
    a. highlighting major points,
    b. noting new terms and definitions,
    c. summarizing tables and graphs,
    d. writing a summary.

An important aspect in reviewing a paper is checking the quality of cited references. Students should examine if the authors cited good and relevant papers. In that sense, students need to consider:

1. Impact factors of the journals where papers were published; in the NLP field, it needs to be checked if authors used papers from top journals and conferences (ACL/NAACL/EMNLP/COLING); Note top 10 conferences in the NLP field: www.junglelightspeed.com/the-top-10-nlp-conferences.
2. Reputations of authors and their organizations; top NLP groups (with a very high NLP factor) come from the following universities: Cornell, Columbia, Johns Hopkins, Stanford, Edinburgh, etc.
3. How old the references are; if there is no recent literature referred to in the paper, that could be a sign that the authors do not build their research on the most recent developments.

All these suggestions can help students analyze of a research paper well, especially in the field of NLP and beyond. When doing a formal review, students are advised to apply all of the presented steps described in the article, without any changes.

Paradigm shift in NLP. Over the past years there have been a series of developments and discoveries which have resulted in major shifts in the discipline of NLP, which students must be aware of. As new and larger performance-oriented corpora became available, the use of statistical (machine learning) methods to learn transformations became the norm unlike it was the case with previous approaches where they were performed using hand-built rules. It has been shown that statistical processing could accomplish some language analysis tasks at a level comparable to human performance. At the heart of this move is the understanding that much (or most) of the work effected by language processing algorithms is too complex to be captured by rules constructed by human generalization, and it rather requires machine learning methods [66–69]. For example, early statistical part-of-speech tagging algorithms using Hidden Markov Models were shown to achieve performance comparable to humans, while a statistical parser has shown better performance than a broad-coverage rule-based parser [70].

What enabled these shifts were newly available extensive electronic resources. They first came in the form of sizable corpora, such as the Brown corpus. Later came lexical resources such as Wordnet and Treebank. Wordnet is a lexical-semantic network whose nodes are synonymous sets which first enabled the semantic level of processing [71]. In linguistics, Treebank is a parsed text corpus which annotates syntactic or semantic sentence structure. The exploitation of Treebank data has been important ever since the first large-scale Treebank, The Penn Treebank, was published. It provided gold standard syntactic resources which led to the development and testing of increasingly rich algorithmic analysis tools.

The first 30 years of NLP research was focused on closed domains (from the 60s through the 80s). The increasing availability of realistically-sized resources in conjunction with machine learning methods supported a shift from a focus on closed domains to open domains (e.g., newswire). The ensuring availability of broad-ranging textual resources on the web further enabled this broadening of domains.

In parallel with these moves toward the use of more real world data came the awareness that NLP researchers should evaluate their work on a larger scale. With this came the introduction of empirically-based, blind evaluations across systems. These efforts led to the development of metrics such as BLEU and ROUGE. BLEU, or Bilingual Evaluation Understudy, is an algorithm for evaluating the quality of a text which has been machine-translated from one natural language to another. The central idea behind BLEU is: the closer a machine translation is to a professional human translation, the better it is. ROUGE, or Recall-Oriented Understudy for Gisting Evaluation, is a set of metrics and a software package used for evaluating automatic summarization and machine translation software in NLP. The metrics compare an automatically produced summary or translation against a reference or a human-produced set of references summary or translation. These metrics are integral to today's NLP research itself, in part because they can be computed automatically and the results can be fed back into research.

At the same time with these advances in statistical capabilities came the demonstration that higher levels of human language analysis are amenable to NLP. While lower levels deal with smaller units of analysis, e.g., morphemes, words, and sentences, which are rule-governed, higher levels of language processing deal with texts and world knowledge, which are only regularity-governed. Higher levels allow for more free choice and variability in usage.

All these individual developments have resulted in the realization that NLP, by mixing statistical and symbolic methods, in conjunction with lexical, syntactic, and semantic resources, and the availability of large scale corpora for testing and evaluating approaches is gaining ground when it comes to realistic comprehension and production of human-like language understanding [70].

A.10 Experiences from the field of bioinformatics

Bioinformatics is a relatively new scientific discipline which applies various computational methods to data from life sciences. It includes several aspects of computer science, such as algorithmics, databases, data mining and machine learning, but also important fields of mathematics like probability, statistics and optimization. The "bio" part of its name is an umbrella term for different research areas from the domain of biology, biochemistry and molecular biology, to physiology, immunology, and pharmacology. A difficulty with research students here is that the two synergizing fields are relatively distant from each other, which is reflected in the fact that sometimes the issues are not presented with equal depth across articles.

Bioinformatics has experienced many paradigm shifts throughout its history, in three different dimensions:

1. The amount of data observed in the 1970s was very small, but has increased dramatically, from small sets of several sequences to modern large databases with many TB of data [72].

2. The complexity level of the observed systems has changed significantly, from individual proteins and RNA sequences to complex multi-peptide proteins and large RNA/DNA molecules, to genome and proteome of a complete organism and comparative research on different species [73].

3. The focus of modeling biological data has shifted from a simple linear structure of molecules to their 3D structure, and from simple molecular interactions to complete functional and physiological interactions in a living cell [74].

In the first decades of the area, bioinformaticians were mainly interested in creating methodological breakthroughs, while today most researchers in the field target very specific biological problems and most of them use the already developed models and tools. This is the best evidence that bioinformatics is a mature field today.

Bioinformatics is an interdisciplinary research area. Nowadays, a bioinformatics program in bachelor level education is quite rare, while master and doctoral programs are slightly more present. Consequently, the majority of current bioinformaticians are either biologists with upgraded computer science skills, or, the other way around, mathematicians or informaticians with upgraded biological skills. The width of the field is the most common source of problems in both presenting new results and understanding other people's results.

The most common mistakes specific to bioinformatics are:

1. The use of inadequate data analysis techniques. Many researchers with low or moderate mathematical or computational skills use some statistical (or modeling) methods unsuitable for the data set or the observed problem.
2. Inadequate data selection. Contrary to the previous point, good mathematicians often use good statistical and computational methods, but they work on wrong data. If some important biological attributes are not well represented in a data sample, the conclusion may be correct but not significant enough to be published.
3. The lack of presenting the "obvious." What is obvious to mathematicians and data scientists may not be so obvious to a life scientist and vice versa. It is always mandatory to check the text from the point of view of both main aspects and to provide enough details and references so that every researcher in the field is able to understand it.

A.11 Experiences from the field of database systems

The information technology (IT) industry has been led by relational databases for >30 years, but the 21st century brought a series of changes in data types, accessing data and using data. Web applications replace desktop applications, distributed servers and network storage replace database servers, and cloud computation has become reality because of its low price and high scalability [75]. Cloud databases such as Big Table are getting especially popular, and their application domains—read-intensive applications, data warehousing, data mining and business intelligence, seek for different functionality from those offered by classic databases.

Even the 1999 SQL standard announced that "Relational model is dead". In 2005, Michael Stonebraker, the designer of one of the first RDBMSs, Ingres, in his influential article [76] "One Size Fits All: An Idea Whose Time Has Come and Gone," presented database future as a broad spectrum of different database models and architectures, best suited for different domains and needs, from column storage, main memory DBMS, XML databases, Resource Description Framework (RDF), non-relational models, over different table and transaction implementations and science databases to non-relational and noSQL databases.

Data science or the data-driven scene is being developed as a new discipline, based on data, comprising all different methods and techniques for extraction of knowledge from data.

When advising students who write database papers, it is important to point out that databases are an applied discipline so each and every contribution, theoretical result, new design, query language, and accessing method has to be supported by legitimate evidence that "we will be better than others" in certain aspects—speed, cost, efficiency, width, or universality, without compromising other aspects.

Another important point is that the application domain is very important. The domain usually "sells" the obtained result and it should be carefully chosen. For example, if a new type of database, new design, or new access method has the potential to improve or speed up establishing medical diagnosis, or make it cheaper or more accurate, the result will be recognized as a very important one.

Finally, database papers are usually multidisciplinary since their "database face" often pertains to database design or implementation but it is applied to a specific domain such as business, science, social domains or humanities. It is extremely important for authors to be careful regarding the choice of the target journal and target audience—it should guide the authors in deciding what to present as a result (main contribution), and what to treat as context, background, example or appendix.

A.12 Experiences from the field of experimental (discrete) mathematics

Experimental mathematics as a separate area of study re-emerged in the 20th century, when the invention of the electronic computer vastly increased the

range of feasible calculations, with a speed and precision far greater than anything available to previous generations of mathematicians [77]. This is the
methodology of doing mathematics which includes the use of computations
for:

1. gaining insight and intuition;
2. discovering new patterns and relationships;
3. using graphical displays to suggest underlying mathematical principles;
4. testing and especially falsifying conjectures;
5. exploring a possible result to see if it is worth a formal proof;
6. suggesting approaches for formal proof;
7. replacing lengthy hand derivations with computer-based derivations;
8. confirming analytically derived results.

In previous years we used computers mainly to search for counterexamples
to some known conjectures (4), and to discover some patterns and relationships in discrete mathematics (2).

Frankl's conjecture. According to Frankl's conjecture, for each union-
closed family of subsets $\mathcal{F}$ of a finite set X, there is an element contained
in at least half members of $\mathcal{F}$. Bošnjak and Marković [78] proved by con-
sidering many separate cases that this is true if $|X| \leq 11$. The number of
union-closed families of subsets of the n-element set doubles exponentially,
and so does the number of cases to be considered. By transforming the search
through different cases and using an efficient branch-and-bound algorithm,
Vučković and Živković [79] obtained a computer assisted proof that Frankl's
conjecture is true if $|X| \leq 12$.

Factorial sums conjectures. For a positive integer n let $A_{n+1} = \sum_{i=1}^{n} (-1)^{n-i} i!$,
and $!i = \sum_{i=0}^{n-1} i!$.

There are two questions concerning these factorial sums listed in Ref. [80]:

- Problem B43: Is it true that $p \nmid A_p$ (A is not divisible by p) for all primes p?
  This question is motivated by the fact that if $p \mid A_p$ (A_p is divisible by p) for
  some prime p, then $p \mid A_n$ for all $n \geq p$, implying that there are only finitely
  many prime numbers among A_n.
- Problem B44: Is it true that $p \nmid !p$ for all primes $p > 2$? This question is
  one of the formulations of the Kurepa hypothesis [81].

By using a simple $O(n^2)$ algorithm, and by programming its critical part in an
assembly language, Živković succeeded in giving the negative answer to the
first question: if $p = 3612703$, then A_p is divisible by p [82]. Therefore, the
number of primes of the form A_n is finite: if $n \geq p$ then A_n is divisible by p.

The second question still remains unanswered. The heuristic argument
shows that, given the constant a, the probability of the existence of the prime

p such that $p \mid !n$, $a \leq n < a^2$ is about $1/2$. A computer search using an advanced algorithm shows that this prime has to be greater than 2^{34} [83].

Row space cardinalities. Denote by $\mathcal{B}_{n;m}$ the set of $n \times m$ Boolean matrices, and let $\mathcal{B}_n = \mathcal{B}_{n;n}$. Denote by $R(A)$ the row space of $A \in \mathcal{B}_{n;m}$, i.e., the subspace of $(0,1)^m$ spanned by the rows of A, with Boolean operations and, or. Denote $\mathcal{R}_n = \{a \mid a = \mid R(A), A \in \mathcal{B}_n\}$. Konieczny [84] determined the set $\mathcal{R}_n \cap [2^{n-1}, 2^n]$, and other authors determined some information about $\mathcal{R}_n \cap [2^{n-2} + 2^{n-3}, 2^{n-1}]$. Vučković and Živković [85] completely determined the set $\mathcal{R}_n \cap [2^{n-2} + 2^{n-3}, 2^{n-1}]$. The main statement which enabled the solution of this problem by exhaustive search is the following: if the matrix $A \in \mathcal{B}_n$ satisfies the condition $R(A) > 2^{n-2} + 2^{n-3}$, then there is a matrix A' obtained from A by deleting a row and a column, such that $R(A') \geq \frac{1}{2} R(A)$, hence satisfying the analogous condition $R(A') > 2^{n-3} + 2^{n-4}$.

There are some important points when reviewing papers in this area. The results obtained by computer search must be verifiable. The obtained results should be interesting enough to deserve publishing. This could be achieved by using advanced algorithms, or advanced constructions for speeding up the search. Working with young researchers in solving such problems consists of two tracks: finding efficient algorithms which lead to new results and understanding background theory, making it possible to reach new conclusions.

Appendix B. Work with students

B.1 Experiences from teaching graduate students: Development

All the above issues, including the contributions of all listed co-authors, are taught in a course titled Research and Development Methodologies with Project Management in Computer Science and Engineering. This is a one-semester course intended for senior level undergraduate students and master students of computer science and engineering, taught at the School of Mathematics and the School of Electrical Engineering of the University of Belgrade. This course has also been offered, in short or long forms, at universities in the USA (Dartmouth, Harvard, NYU, and CUNY Albany), Japan (Tokyo and Sendai), Spain (Barcelona and Valencia), Italy (Salerno and Siena), Serbia (Belgrade, at the School of Physical Chemistry, and at the School of Business Administration, University of Belgrade).

A detailed description of the course is given in Ref. [86]. For similar courses, see Refs. [87,88].

One of the homework assignments in this course is writing and reviewing an article. After writing an article, students are asked to read their colleague's homework article, analyze it, and write a formal review applying all formal review steps. At the end, they are advised about the best way to write a response to reviewers in a realistic situation as it is explained in this article.

B.2 Experiences from teaching graduate students: Research

This section presents an effort to teach computer science graduate students to focus on creativity enhancement [1] in the process of journal publication. Systematic creativity methods were taught in the above-mentioned course on Managing Sophisticated Projects in Academia and Industry (USPI). Among other issues, students were taught the methods for enhancing creativity described by Blagojevic [1] and Milutinovic [2]. We will here present top 10 representative examples of students' research project proposals, based on the creativity enhancement method from Ref. [1], described using the capability maturity model integration.

The research of SG1 (student group 1) deals with an online learning platform for blind people and their inclusion in the world of computer science and programming. Their principal aim was to provide a non-stressful learning experience for programming beginners. This research for the most part introduces originality and implantation. In that sense, the proposed methods could be classified as Implantation (I) and Mendeleyevization (M) [1]. As the group strive to be the global leader in online education and machine-to-human interactions, they need to be aware of the *quality of the original contribution*, which is described in this paper (Section 4.2.1). Also, SG1 introduced a detailed analysis of the complexity of the proposed educational model from standardization to personalization (through online platforms students have access to the best teaching software platform worldwide, 24/7, anytime, anywhere, from any device). Therefore, SG1 respected the important criterion described in this paper, which is *the quality of analysis* (Section 4.2.2).

The research of SG2 covers the field of information systems and collecting relevant information for patients with a high heart attack risk. Their aim was to develop HEALS (Heart Alarm System) and a device attached to the patient's body which periodically sends the collected data to the information system. The data is then stored, analyzed, and in case an emergency is detected, appropriate medical bodies are notified to undertake intervention. This study imports existing solutions with a varying level of granularity.

In that sense, the proposed methods could be classified under the concept of Transgranularization (T) [1]. On the other side, there is a need to apply one more existing solution, by changing the set of initial conditions. The idea is to design the architecture of an information system which will store the collected data and analyze it in order to automatically detect any anomalies which can lead to a heart attack and notify appropriate medical bodies, such as urgent care centers. Another part of the information system is the interface through which doctors can access this data in an accumulated way for regular control. This way, the study enables urgent care centers to use adaptive methodology (A) in order to act without delay. SG2 also introduced a detailed marketing plan including public relations, in order to gain public favor. Therefore, SG2 respected *the quality of the presentation* as an important criterion described in this paper (Section 4.2.3).

In Bee Electronic Sound Tracking, which is the subject of research of SG3, the mapping of inaccessible parts of the hive into a system depends on the sounds present in the hive. Experts have already determined that bees produce the same sounds in same situations. Thus, the bees' sounds can be classified as follows: the sound when they are upset, when they are ready for birth, when they do not have a queen, when they are calm, etc. But the computational model of comparing the pre-recorded sounds of bees in each specific situation and the sound collected from the hive was revitalized due to contemporary technology. The proposed comparative analysis is developed using the method (R) [1]. After the analysis, if a sound is matched, the information is sent to the beekeeper's mobile phone. This enables the beekeeper to receive pieces of information about his hives at any location. Having in mind our goal to bring more efficiency to publishing research results, SG3 have started a quality review on their own, particularly in terms of the readability index. Their work is still in progress.

The research of SG4 focuses on hardware methods which should enable children with dyslexia to attend school normally and do homework without any problems resulting from their impairment. Their system can overcome difficulties caused by this disease by reading the text to the user. There is also a panel with all math symbols to help the user write a solution to a math problem as additional support for solving math problems. The proposed communication protocol and its evaluation analysis follow methods M + H + G [1]. Various learning and teaching styles in engineering education were explored to meet the requirements stated in this paper defined as *the quality of original contribution* (Section 4.2.1).

The research of SG5 proposes a new way for resource-efficient eco-innovative water production. The key functionality of the study is that users increase their water intake by drinking water instead of juices. The SG5 system purifies water and adds flavor to it, which is all done at home, without the need to go to a store to buy ready-made flavored waters. The research is motivated by existing solutions and introduces an original approach which generalizes the essence of these solutions—method G in Ref. [1]. The existing solutions were analyzed by biologists from Institute of Biochemistry, regarding the materials which are not harmful and which can easily enter the body, in order to help the participants in the project select the materials from which flavors will be made. It is a new paradigm which tackles identified problems—method M in Ref. [1]. Finally, testing the product itself is one of the activities in the methodology, from which it can be seen whether the product behaves in accordance with the expected results, whether water is sufficiently purified, and whether the taste is satisfactory. It calculates the similarity between two materials by including, among other parameters, the quality, taste, and harmfulness—method E in Ref. [1].

Student group SG6 explores the Internet of Things and ported the mechanism and automate control into their idea—a prototype of a smart robot house cleaner. The initial solution was highly motivated by smart system experiments and analytical protocol design, statistical analysis and interpretation. The research comprised multiple fields as in method C [1]. Also, SG6 managed to apply the methodologies for research innovation and journal publications in science and engineering described in Section 4 of this paper.

The research of SG7 involves developing a system for monitoring self-driving cars in Serbian mountains. Their approach connects theory and practice in the domain of the Internet of Things. Their monitoring system will be developed as a multi-layer complex system, where each layer is responsible for a particular type of processing and communication, which is done in accordance with methods E and M [1]. Since their system is built on the top of representative production activities described in this paper, they were highly motivated to carefully read all the existing methods of innovation at a very early stage of their research.

Finally, the most popular homework among students involves developing microkernel architecture. One group developed microkernel architecture for constraint programming, while three other groups were inspired by open source mathematical software. They tried (using methods M and H [1]) to develop mathematical software which can use a mobile device to control actions on a central screen. All those research groups explored user experience in relation to social networks and digital media. Following the obligatory elements described in Section 4 of this paper, they developed flexible systems with an adaptive user interface in order to meet the requirements of future critical reviews.

References

[1] V. Blagojevic, et al., A systematic approach to generation of new ideas for PhD research in computing, in: Advances in Computers, vol. 104, Elsevier, 2016, pp. 1–19.

[2] V. Milutinovic, The Best Method for Presentation of Research Results, The TCCA Newsletter, 1995, pp. 1–6.

[3] J. Schimel, Writing Science: How to Write Papers That Get Cited and Proposals That Get Funded, Oxford University Press, USA, 2012.

[4] Elsevier, How to Conduct a Review, https://www.elsevier.com/reviewers/how-to-conduct-a-review, 2017, Elsevier.

[5] Elsevier, Tools and Resources for Reviewers, https://www.elsevier.com/reviewers/tools-and-resources-for-reviewers, 2017, Elsevier.

[6] I. Hames, Peer Review and Manuscript Management in Scientific Journals: Guidelines for Good Practice, Blackwell Publishing Ltd, 2007. ISBN: 978-1-4051-3159-9.

[7] N. Kimberly, G. Wendy, A quick guide to writing a solid peer review, Eos 92 (2011) 233–234.

[8] J.M. Provenzale, R.J. Stanley, A systematic guide to reviewing a manuscript, J. Nucl. Med. Technol. 34 (2) (2006) 92–99.

[9] Wiley, How to Perform a Peer Review, https://authorservices.wiley.com/Reviewers/journal-reviewers/how-to-perform-a-peer-review/index.html, 2017, Wiley.

[10] W.S. Noble, Ten simple rules for writing a response to reviewers, PLoS Comput. Biol. 13 (10) (2017) e1005730. https://doi.org/10.1371/journal.pcbi.1005730.

[11] W.C. Hywel, How to reply to referees' comments when submitting manuscripts for publication, J. Am. Acad. Dermatol. 51 (1) (2004) 79–83. ISSN 0190-9622. https://doi.org/10.1016/j.jaad.2004.01.049.

[12] B.W. Taylor, Writing an effective response to a manuscript review, Freshw. Sci. 35 (4) (2016) 1082–1087.

[13] Thinkscience, Writing Effective Response Letters to Reviewers: Tips and a Template, https://thinkscience.co.jp/en/articles/2016-06-WritingResponseLetters.html, 2017, Thinkscience.

[14] B.-C. Björk, S. David, Open access versus subscription journals: a comparison of scientific impact, BMC Med. 10 (2012) 73.

[15] B.C. Björk, S. David, The publishing delay in scholarly peer-reviewed journals, J. Informetr. 7 (4) (2013) 914–923. ISSN: 1751-1577. https://doi.org/10.1016/j.joi.2013.09.001.

[16] K. Powell, Does it take too long to publish research? Nature 530 (7589) (2016) 149–151.

[17] D. Fanelli, How many scientists fabricate and falsify research? A systematic review and meta-analysis of survey data, PLoS One 4 (5) (2009) e5738.

[18] M. Pearl, Getting good ideas in science and engineering, IPSI Trans. Adv. Res. 3 (2) (2007) 3–7.

[19] J. Friedman, Why we need basic research, Trans. Adv. Res. 4 (1) (2008) 2–5.

[20] R. Flesch, A new readability yardstick, J. Appl. Psychol. 32 (1948) 221–233.

[21] J. Kincaid, R. Fishburne, R. Rogers, B. Chissom, Derivation of New Readability Formulas (Automated Readability Index, Fog Count, and Flesch Reading Ease Formula) for Navy Enlisted Personnel, Research branch report 8–75, Chief of naval technical training, Naval Air Station Memphis, Memphis, 1975.

[22] R. Gunning, The Technique of Clear Writing, McGraw-Hill International Book Co., NY, New York, 1952.

[23] V. Filipović, Fine-grained tournament selection operator in genetic algorithms, Comput. Inform. 22 (2) (2003) 143–161.

[24] C.-T. Chu, S.K. Kim, Y.-A. Lin, Y.Y. Yu, G. Bradski, K. Olukotun, A.Y. Ng, Map-reduce for machine learning on multicore, in: Proceeding NIPS'06 Proceedings of the 19th International Conference on Neural Information Processing Systems, Canada, December 04–07, 2006, pp. 281–288.

[25] P. Martins, Modeling the Maximum Edge-Weight K-Plex Partitioning Problem, arXiv, 2016. preprint arXiv:1612.06243.

[26] Kartelj, A., Filipović V., and Milutinović, V., "Novel approaches to automated personality classification: ideas and their potentials," in: Proceedings of the Conference: 35th International Convention MIPRO, Opatija, Croatia, 2012, Vol.: pp. 1017–1022.

[27] J.J. Dongarra, Performance of Various Computers Using Standard Linear Equations Software, University of Tennessee Computer Science, 2004. Technical report, 89–85.

[28] Z. Stanimirović, S. Mišković, Efficient metaheuristic approaches for exploration of online social networks, in: Big Data Management, Technologies, and Applications, IGI Global, 2014, pp. 222–269.

[29] D. Bertsimas, M. Sim, Robust discrete optimization and network flows, Math. Program. 98 (1) (2003) 49–71.

[30] S. Mišković, Z. Stanimirović, I. Grujičić, Solving the robust two-stage capacitated facility location problem with uncertain transportation costs, Opt. Lett. 11 (6) (2017) 1169–1184.

[31] S. Stojanović, V. Pavlović, P. Janičić, A coherent logic based geometry theorem prover capable of producing formal and readable proofs, in: P. Schreck, J. Narboux, J. Richter-Gebert (Eds.), Automated Deduction in Geometry, Lecture Notes in Computer Science, vol. 6877, Springer, Berlin, Heidelberg, 2011, pp. 201–220. ADG 2010.

[32] S. Stojanović, Preprocessing of the axiomatic system for more efficient automated proving and shorter proofs, in: T. Ida, J. Fleuriot (Eds.), Automated Deduction in Geometry, Lecture Notes in Computer Science, vol. 7993, Springer, Berlin, Heidelberg, 2013, pp. 181–192. ADG 2012.

[33] S. Stojanović Đurđević, Automatsko proveravanje neformalnih dokaza teorema srednjoškolske geometrije, Info M 15 (58) (2016) 11–19.

[34] S. Stojanović Đurđević, From informal to formal proofs in Euclidean geometry, Ann. Math. Artif. Intell. 85 (2019) 89–117. https://doi.org/10.1007/s10472-018-9597-7.

[35] S. Stojanovic Đurđević, J. Narboux, P. Janicic, Automated generation of machine verifiable and readable proofs: a case study of Tarski's geometry, Ann. Math. Artif. Intell. 74 (2015) 249.

[36] S. Stojanovic, J. Narboux, M. Bezem, P. Janicic, A vernacular for coherent logic, in: S. Watt et al., (Ed.), Intelligent Computer Mathematics—CICM 2014, Lecture Notes in Computer Science, vol. 8543, Springer, 2014, pp. 388–403.

[37] F. Marić, A survey of interactive theorem proving, Zbornik Radova 18 (26) (2015) 173–223. http://elib.mi.sanu.ac.rs/files/journals/zr/26/zrn26p173-223.pdf.

[38] J.C. Blanchette, S. Böhme, L.C. Paulson, Extending sledgehammer with SMT solvers, in: N. Bjørner, V. Sofronie-Stokkermans (Eds.), 23rd International Conference on Automated Deduction: CADE-23, Springer, 2011, pp. 116–130. LNAI 6803.

[39] F. Marić, P. Janičić, M. Maliković, Proving correctness of a KRK chess endgame strategy by using Isabelle/HOL and Z3, in: 25th International Conference on Automated Deduction—CADE, Lecture Notes in Computer Science, vol. 9195, 2015, pp. 256–271.

[40] G. Tassey, The Economic Impacts of Inadequate Infrastructure for Software Testing, Technical report, National Institute of Standards and Technology, 2002.

[41] A. Turing, On computable numbers, with an application to the entscheidungs problem, Proc. Lond. Math. Soc. 2 (42) (1936) 230–265.

[42] P. Cousot, R. Cousot, Abstract interpretation: a unified lattice model for static analysis of programs by construction or approximation of fixpoints, in: Symposium on Principles of Programming Languages (POPL), ACM Press, 1977, pp. 238–252.

[43] J.C. King, Symbolic execution and program testing, Commun. ACM 19 (1976) 385–394.

[44] E.M. Clarke, T.A. Henzinger, H. Veith, R. Bloem (Eds.), Handbook of Model Checking, Springer, 2017. ISBN: 978-3-319-10574-1.

[45] A. Biere, A. Biere, M. Heule, H. van Maaren, T. Walsh, Handbook of satisfiability, in: Frontiers in Artificial Intelligence and Applications, vol. 185, IOS Press, Amsterdam, The Netherlands, 2009.

[46] M. Vujošević-Janičić, V. Kuncak, Development and evaluation of LAV: an SMT-based error finding platform, in: R. Joshi, P. Müller, A. Podelski (Eds.), Proceedings of the 4th International Conference on Verified Software: Theories, Tools, Experiments (VSTTE'12), 2012, Springer-Verlag, Berlin, Heidelberg, 2012, pp. 98–113.

[47] D. Beyer, Competition on software verification, in: C. Flanagan, B. König (Eds.), Proceedings of the 18th International Conference on Tools and Algorithms for the Construction and Analysis of Systems (TACAS'12), Springer-Verlag, Berlin, Heidelberg, 2012, pp. 504–524. https://doi.org/10.1007/978-3-642-28756-5_38.

[48] M.W. Moskewicz, et al., Chaff: engineering an efficient SAT solver, in: Proceeding DAC '01 Proceedings of the 38th Annual Design Automation Conference, Las Vegas, USA, 2001, pp. 530–535.

[49] M. Davis, et al., A machine program for theorem-proving, Commun. ACM 5 (7) (1962) 394–397.

[50] J. Marques-Silva, et al., Conflict-driven clause learning SAT solvers, in: A. Biere, M. Heule, H. van Maaren, T. Walsh (Eds.), Handbook of Satisfiability, IOS Press, 2009, pp. 131–153.

[51] R. Nieuwenhuis, et al., Solving SAT and SAT modulo theories: from an abstract Davis–Putnam–Logemann–Loveland procedure to DPLL (T), J. ACM 53 (6) (2006) 937–977.

[52] M. Bankovic, Extending SMT solvers with support for finite domain alldifferent constraint, Constraints 21 (4) (2016) 463–494.

[53] O. Ohrimenko, et al., Propagation via lazy clause generation, Constraints 14 (3) (2009) 357–391.

[54] M. Nikolic, F. Maric, P. Janicic, Simple algorithm portfolio for SAT, Artif. Intell. Rev. 40 (4) (2013) 457–465.

[55] I. Goodfellow, Z. Bengio, A. Courville, Deep Learning, MIT Press, 2016.

[56] V. Vapnik, Statistical Learning Theory, Wiley, 1998.

[57] P. Buhlmann, P. Drineas, M. Kane, M. van der Laan, Handbook of Big Data, CRC Press, 2016.

[58] N. Srivastava, G. Hinton, A. Krizhevsky, I. Sutskever, R. Salakhutdinov, Dropout: a simple way to prevent neural networks from overfitting, J. Mach. Learn. Res. 5 (June) (2014) 1929–1958.

[59] R. Tibshirani, Regression shrinkage and selection via the lasso, J. R. Stat. Soc. Ser. B 58 (1996) 267–288.

[60] D. Wolpert, The lack of a priori distinctions between learning algorithms, Neural Comput. 8 (7) (1996) 1341–1390.

[61] D. Krstajic, L.J. Buturovic, D. Leahy, S. Thomas, Cross-validation pitfalls when selecting and assessing regression and classification models, J. Chem. 6 (2014) 10.

[62] C. Calude, G. Longo, The deluge of spurious correlations in big data, Found. Sci. 22 (2017) 595–612.

[63] A. Caliskan, J. Bryson, A. Narayanan, Semantics derived automatically from language corpora contain human-like biases, Science 356 (6334) (2017) 183–186.

[64] A. Romei, S. Ruggieri, A multidisciplinary survey on discrimination analysis, Knowl. Eng. Rev. 29 (5) (2014) 582.

[65] P.Q.N. Minh, Research Methods in Natural Language Processing, FPT Technology Research Institute, FPT University, 2017.

[66] J. Graovac, A variant of N-gram based language-independent text categorization, Intell. Data Anal. 18 (4) (2014) 677–695.

[67] J. Graovac, J. Kovačević, G. Pavlović-Lažetić, Language independent n-gram-based text categorization with weighting factors: a case study, J. Inf. Data Manag. 6 (1) (2015) 4–17.

[68] J. Graovac, J. Kovačević, G. Pavlović-Lažetić, Hierarchical vs. flat N-gram-based text categorization: can we do better? Comput. Sci. Inf. Syst. 14 (1) (2017) 103–121.

[69] J. Graovac, M. Mladenović, I. Tanasijević, NgramSPD: exploring optimal N-gram model for sentiment polarity detection in different languages, Intell. Data Anal. 23 (2) (2019) 279–296.

[70] E.D. Liddy, et al., Natural language processing, in: Report Based on the MINDS Workshops, 2007. http://www.itl.nist.gov/iaui/894.02/MINDS/FINAL/NLP.web.pdf.

[71] G. Pavlović-Lažetić, J. Tomašević, Ontology-driven conceptual document classification, in: Proceedings of the International Conference on Knowledge Discovery and Information Retrieval, Valencia, Spain, 2010, pp. 383–386.

[72] D. Zou, L. Ma, J. Yu, Z. Zhang, Biological databases for human research. Genomics Proteomics Bioinformatics 13 (1) (2015) 55–63, https://doi.org/10.1016/j.gpb.2015.01.006.

[73] A. Glass, T. Karopka, Genomic data explosion—the challenge for bioinformatics? in: P. Perner (Ed.), Advances in Data Mining, Lecture Notes in Computer Science, vol. 2394, Springer-Verag, Berlin Heidelberg, Heidelberg, 2002, pp. 80–98.

[74] G. Pavlovic-Lazetic, et al., Bioinformatics analysis of disordered proteins in prokaryotes, BMC Bioinf. 12 (2011) 66.

[75] G. Feuerlich, Database trends and directions: current challenges and opportunities, in: Proceedings of the Dateso 2010 Annual International Workshop on Databases, Texts, Specifications and Objects Stedronin-Plazy, Czech Republic, April 21, 2010, vol. 567, 2010, pp. 163–174. CEUR Workshop Proceedings.

[76] M. Stonebraker, U. Cetintemel, One size fits all: an idea whose time has come and gone, in: ICDE '05 Proceedings of the 21st International Conference on Data Engineering, 2005, pp. 2–11. April 05 − 08.

[77] J. Borwein, D. Bailey, Mathematics by Experiment: Plausible Reasoning in the 21st Century, second ed., CRC Press, 2008.

[78] I. Bošnjak, P. Marković, The 11-element case of Frankl's conjecture, Electron. J. Comb. 15 (2008) R88.

[79] B. Vučković, M. Živković, The 12-element case of Frankl's conjecture, IPSI BgD Trans. Internet Res. 13 (1) (2017) 65–71.

[80] R. Guy, Unsolved Problems in Number Theory, third ed., Springer-Verlag, New York, 2004.

[81] D.J. Kurepa, On the left factorial function, Math. Balkanica (N.S.) 1 (1971) 147–153.

[82] M. Živković, The number of primes $\sum_{i=1}^{n}(-1)^{n-i}i!$ is finite, Math. Comput. 68 (225) (1999) 403–409.

[83] V. Andrejić, M. Tatarević, Searching for a counterexample to Kurepa's conjecture, Math. Comput. 85 (2016) 3061–3068.

[84] J. Konieczny, On cardinalities of row spaces of Boolean matrices, Semigroup Forum 44 (1992) 393–402.

[85] B. Vučković, M. Živković, Row space cardinalities above $2^{n-2} + 2^{n-3}$, IPSI BgD Trans. Internet Res. 13 (1) (2017) 72–84.
[86] V. Milutinovic, S. Vujicic Stankovic, A. Jovic, D. Draskovic, M. Misic, D. Furundzic, A new course on R&D Project Management in Computer Science and Engineering: subjects taught, rationales behind, and lessons learned, in: A.R. Hurson, V. Milutinović (Eds.), Advances in Computers, vol. 106, Academic Press, UK, 2017, pp. 1–19. ADCOM.
[87] B. Burnett, D. Evans, Designing Your Life, http://www.stanford.edu/class/me104b/cgi-bin/, 2011.
[88] D. Patterson, How to Have a Bad Career in Research/Academia, https://people.eecs.berkeley.edu/~pattrsn/talks/badcareer.pdf, 2001, University of California; Berkeley.

About the authors

Milan Banković was born in Petrovac na Mlavi, Serbia, on May 3, 1982. He received his Bachelor degree in computer science from the Faculty of Mathematics, University of Belgrade, Serbia, in 2006. He defended his PhD thesis in the field of computer science at the same faculty in 2016. He has worked as a teaching assistant at the Department of Computer Science, Faculty of Mathematics, University of Belgrade, Serbia, from 2007 to 2017. Since September 2017 he works as an assistant professor at the same department. His research interests are in the field of automated reasoning, with emphasis on SAT and SMT solvers and their applications.

Vladimir Filipović is Associate Professor at the Faculty of Mathematics, University of Belgrade, Serbia. He received his PhD degree at the same faculty in 2006. His research interests include combinatorial optimization, operational research, computational intelligence, graph algorithms, and big data.

Jelena Graovac is an Assistant Professor in the Department of Computer Science, Faculty of Mathematics, University of Belgrade. She received MSc (2008, Computer Science) and PhD (2014, Computer Science) degrees from the Faculty of Mathematics, University of Belgrade. Her theses titles were "XML database management of lexical resources" and "Contribution to text categorization methods: mathematical models and applications" for the MSc and PhD degrees, respectively. Her research interests include natural language processing using machine learning and knowledge-based approaches. She co-authored many scientific papers as book chapters and articles in journals and conference proceedings. She is a co-founder of the Language Resources and Technologies Society.

Jelena Hadži-Purić is a Researcher in the Computing Lab at the Faculty of Mathematics, University of Belgrade. Also, she teaches informatics at Mathematical Grammar School in Belgrade, Serbia. She is chairman of the Serbian National Committee for Olympiads in Informatics and team leader of the Serbian teams for many international competitions. A lot of her students are in the final rounds of the national programming competitions, some on BOI and IOI. Her current research interest includes algorithms and data structure, complex network analysis, and project management.

Ali R. Hurson is a Professor of Computer Science at the Missouri University of Science and Technology (S&T), after having served as department chair from 2008 to 2012. Before joining Missouri S&T, he was a Professor of Computer Science and Engineering at the Pennsylvania State University. His research for the past 30 years has been on the design and analysis of general, as well as special-purpose computer architectures. His research has been supported by NSF, DARPA, the Department of

Education, the Air Force, the Office of Naval Research, Oak Ridge National Laboratory, NCR Corp., General Electric, IBM, Lockheed Martin, Pennsylvania State University, and Missouri S&T. He has published over 300 technical papers in areas including multidatabases, global information sharing and processing, applications of mobile agent technology, object-oriented databases, mobile and pervasive computing environments, sensor and ad-hoc networks, computer architecture and cache memory, parallel and distributed processing, dataflow architectures, and VLSI algorithms. He has served as a member of the IEEE Computer Society Press Editorial Board, an IEEE Distinguished speaker, editor of the IEEE Transactions on Computers, editor of the Journal of Pervasive and Mobile Computing, and as a member of the IEEE/ACM Computer Sciences Accreditation Board. He is currently serving as an ACM distinguished speaker, area editor of the CSI Journal of Computer Science and Engineering, and editor-in-chief of Advances in Computers.

Aleksandar Kartelj completed his PhD in Computer Science at Faculty of mathematics in year 2014. His research interests cover areas of optimization, mathematical programming, and data mining. His publications are mostly related to metaheuristic optimization methods, data classification, and dimensionality reduction.

Jovana Kovačević finished her undergraduate and PhD studies at the University of Belgrade, Faculty of Mathematics, Department for Computer Science where she has the position of an assistant professor at the moment. She is a member of bioinformatics research group at the same institution. Her research interests include bioinformatics, text processing and machine learning.

Nenad Korolija is with the School of Electrical Engineering, University of Belgrade, Serbia. He received a PhD degree in electrical engineering and computer science in 2017. His interests and experiences include developing software for high performance computer architectures and DataFlow architectures. During 2008, he worked on the HIPEAC FP7 project at the University of Siena, Italy. In 2013, he was an intern at the Google Inc., Mountain View, CA, USA. In 2017, he was working for Maxeler Ltd., London.

Miloš Kotlar is a PhD student at School of Electrical Engineering, University of Belgrade. His general research field includes implementation of machine learning algorithms using the DataFlow paradigm (FPGA accelerators). The most recent research is comparing ControlFlow and DataFlow for machine learning and tensor calculus through different aspects such as speed, power, complexity, and MTBF.

Nenad B. Krdžavac is a research scientist with extensive experience conducting ground-breaking research, contributing to knowledge of Semantic Web technologies. He has strong research interests spanning ontology development, semantic web services development, description logics, automated reasoning, and model-driven engineering. He previously held roles at National Institute of Standards and Technology (NIST) in United States, Cork University in The Republic of Ireland, Michigan State University in United States, and the University of Belgrade, Serbia. He is a co-inventor of systems and methods for establishing semantic equivalence between concepts (US patent pending number US20160224893A1).

Filip Marić is currently an Associate Professor at Faculty of Mathematics, University of Belgrade. His main research interests are in interactive theorem proving and its applications in formalization of mathematics and software verification. He is also interested in SAT and SMT solving and their applications and in teaching programming at introductory level.

Saša Malkov is an Associate Professor at the Department of Computer Science, Faculty of Mathematics, University of Belgrade. He received his PhD from the University of Belgrade, Faculty of Mathematics. His research interests include bioinformatics, functional programming, and software development.

Veljko Milutinović (1951) received his PhD from the University of Belgrade in Serbia, spent about a decade on various faculty positions in the United States (mostly at Purdue University and more recently at the University of Indiana in Bloomington), and was a co-designer of the DARPAs pioneering GaAs RISC microprocessor on 200 MHz (about a decade before the first commercial effort on that same speed) and was a co-designer also of the related GaAs Systolic Array (with 4096 GaAs microprocessors). Later, for almost 3 decades, he taught and conducted research at the University of Belgrade in Serbia, for departments of EE, MATH,

BA, and PHYS/CHEM. His research is mostly in datamining algorithms and dataflow computing, with the emphasis on mapping of data analytics algorithms onto fast energy efficient architectures. Most of his research was done in cooperation with industry (Intel, Fairchild, Honeywell, Maxeler, HP, IBM, NCR, RCA, etc.). For 10 of his books, forewords were written by 10 different Nobel Laureates with whom he cooperated on his past industry sponsored projects. He published 40 books (mostly in the United States), he has over 100 papers in SCI journals (mostly in IEEE and ACM journals), and he presented invited talks at over 400 destinations worldwide. He has well over 1000 Thomson-Reuters WoS citations, well over 1000 Elsevier SCOPUS citations, and about 4000 Google Scholar citations. His Google Scholar H index is equal to 30. He is a Life Fellow of the IEEE since 2003 and a Member of The Academy of Europe since 2011. He is a member of the Serbian National Academy of Engineering and a Foreign Member of the Montenegro National Academy of Sciences and Arts.

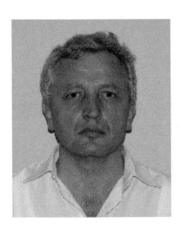

Nenad Mitić is a full Professor at the Department of Computer Science, Faculty of Mathematics, University of Belgrade. He received his BSc, MSc and PhD from the University of Belgrade, Faculty of Mathematics. From 1983 to 1991 he was system analyst on IBM mainframe in Statistical Office of the Republic of Serbia, Belgrade. From 1991 till now he is with Faculty of Mathematics, Belgrade. His research interest cover areas of bioinformatics, data mining, big data, and functional programming.

Stefan Mišković is a Teaching Assistant at the Department for Informatics and Computer Science, Faculty of mathematics, University of Belgrade. At the same faculty, he graduated as a Bachelor of Mathematics and Computer Science in 2010, and defended his master thesis "Genetic algorithm for the Two-stage capacitated facility location problem" in 2011. In 2016, he obtained his PhD thesis "Solving the min-max robust discrete optimization problems with applications." His research is mainly focused on mathematical modeling, discrete optimization, metaheuristic methods, and computer vision.

Mladen Nikolić got his PhD from the Faculty of Mathematics, University of Belgrade in year 2013. in the field of artificial intelligence. Since 2014. he works as an Assistant Professor at the Faculty of mathematics. Some of the courses he taught are machine learning, scientific computing, data mining, artificial intelligence, and construction and analysis of algorithms. Areas of his research interests are machine learning and automated reasoning. He published 16 research papers at journals and conferences. He participated in several international research projects funded by Defense Advanced Research Projects Agency (DARPA), Swiss National Science Foundation (SNF), European Research Council (ERC), and Ministry of Science of the Republic of Serbia.

Gordana Pavlović-Lažetić is a full Professor at the Department of Computer Science, Faculty of Mathematics, University of Belgrade. She received her BSc, MSc, and PhD from the University of Belgrade, Faculty of Mathematics. She spent 2 years at the University of California, Berkeley, as a PhD student, visiting scholar and a consultant at Relational Technology, Inc., working on extending the Relational Database Management System INGRES to support textual data. Her research interests include databases, bioinformatics, and natural language processing.

Danijela Simić (born Petrovic) was born on September 26, 1986. in Valjevo. In 2005 she enrolled at the Faculty of Mathematics, University of Belgrade, Department of Computer Science and Informatics. The study ended in 2009 with an average grade of 9.86 (out of 10). In October 2009, she enrolled PhD studies in computer science at the Faculty of Mathematics, University of Belgrade. The doctoral dissertation titled "Formalization of Various Geometry Models and Applications in Verification of Automated Theorem Provers"

defended in August 2017 with the mentor prof. Filip Maric. The basic field of interest is automatic reasoning, with an emphasis on automatic and interactive proving in geometry. She had published several papers in international journals and had exhibited several times at international conferences. Since October 2009, she has been employed at the Department of Computing and Informatics, Faculty of Mathematics, University of Belgrade.

Sana Stojanović Djurdjević is a Teaching Assistant at the Department for Computer Science, Faculty of Mathematics, University of Belgrade. She defended her PhD thesis in the field of automated reasoning: "Formalization and automated proving of Euclidean geometry theorems" in 2016. Her field of interests are automated and interactive theorem proving, coherent logic and formalization of axiomatic systems for geometry. Courses that she taught are: Interactive theorem proving, Algorithms and Data Structures, Programming language C, Programming language C++, Programming Language C#, Cryptography, Computer Graphics, Teaching Methods in Computer Science (several courses on undergraduate and master studies).

Staša Vujičić Stanković, PhD, is a Teaching Assistant at the University of Belgrade, School of Mathematics, Serbia. She holds PhD (2016) in Computer Science and Informatics from the same University. She published several scientific and conference papers, and gave numerous talks at conferences. Her research fields include natural language processing, and in particular information extraction, information retrieval and web search, formal languages and automata theory, databases, data mining, semantic web, as well as computer science education and educational tools.

Milena Vujošević Janičić is an assistant professor at the Department of Computer Science, Faculty of Mathematics, University of Belgrade, Serbia. She received her PhD from the University of Belgrade in 2013, in the area of software verification with the thesis "Automated generation and checking of verification conditions." Her main research interests are in automated bug finding, model checking, and application of software verification techniques in different fields. She authored or coauthored one book and more than fifteen research papers in scientific journals and at conferences. She has participated in several national and international research projects.

Miodrag Živković was born in Kovin, Serbia, in 1956. He received BE degree in electrical engineering from the Faculty of Electrical Engineering, Belgrade University in 1979 and and MSc and PhD degrees in mathematics from the Faculty of Mathematics, Belgrade University in 1985 and 1990, respectively. From 1980 to 1995 he was with the Institute for Applied Mathematics and Electronics, Belgrade. From 1995. till now he is with Faculty of Mathematics, Belgrade. His current research interests include discrete mathematics.

CHAPTER TWO

ALGATOR—An automatic algorithm evaluation system

Tomaž Dobravec
Faculty of Computer and Information Science, University of Ljubljana, Slovenia

Contents

Abstract

The design of algorithms for solving various kinds of computer-related problems can generally be divided into two parts: the theoretical evolution of the algorithms and the practical assessment of their behavior. While the aim of the first part is to derive algorithms and to prove their theoretical characteristics, the second part struggles to discover the usefulness of the presented solutions when it comes to solving existing problems on real-world computers. Due to the necessary simplifications used in the

Advances in Computers, Volume 116
ISSN 0065-2458
https://doi.org/10.1016/bs.adcom.2019.07.002

models of the theoretical analysis, the results of both parts of the process are not always consistent. From a theoretical point of view, it is very important to develop good algorithms, theoretically prove their efficiency and thus sometimes show the theoretical boundaries of a particular problem. However, if these algorithms do not perform well in practice, all the theoretical work is not much more than "art for art's sake." Therefore, an unbiased and holistic algorithm evaluation is a very important element of the algorithm-design process. To create a fair and reproducible environment in which the algorithm's implementations can be compared is a big challenge; therefore, researchers normally use more or less reasonable simplifications, which leads to results that are not comparable and therefore sometimes misleading. To overcome this problem we introduce a system called ALGATOR that was developed to facilitate and automate the algorithm-evaluation process. ALGATOR creates an environment in which the time complexity and other project-specific indicators can be measured and provides tools for the automatic and semi-automatic analyses of the results of the algorithms' execution. In this chapter we present the implementation and the functionalities of the ALGATOR system and give several practical examples of its use in the evaluation of real algorithms.

1. Introduction

This chapter addresses the problem of assessing the quality of algorithms and presents a system called ALGATOR that facilitates and automates the algorithm–evaluation process. We believe that this system provides a unique solution, since to the best of our knowledge nobody else has yet proposed or implemented such a holistic evaluation system. In addition, ALGATOR is easy to use and it offers many out–of–the–box mechanisms that simplify the assessment process. In order to demonstrate the usability of the system, we implemented several algorithms for different problem domains and present them here.

The development of a given problem-solving algorithm consists of several phases, which include the design of the solution, the analysis of the correctness and efficiency, the implementation in a selected programming language and the practical experiment of the implementation using the representative test cases. These phases usually appear in a circular, rather than a linear, order since some (or even all) of them have to be repeated several times. Ideally, the design is followed by an analysis (possibly simultaneously), the implementation and the testing. When the tests show that all the results are correct, that the quality of the algorithm is optimal and that the theoretical time complexity is consistent with the measurements, the developed algorithm is declared correct and optimal, and the development process is considered complete. Unfortunately, however, we are not living in an ideal world, and during the development of the algorithm, several complications might arise that make the above process much longer and more demanding. Often, only careful testing reveals the weaknesses of the algorithm and then the next iteration of the development has to be initiated. This process then has to be repeated until the test results are satisfactory, which means the quality of the developed algorithm is heavily influenced by the quality of the testing.

Another barrier to the algorithm-design process lies in the complexity of the theoretical analysis of the algorithm's correctness and efficiency. This analysis is frequently very difficult to perform for the most general case. To be able to complete the analysis and to provide (at least some) theoretical results, additional assumptions (for example, relating to the distribution of the input data) have to be made. Therefore, the results of the analysis are not usually applicable to all the test cases that occur in practice. Besides, the analysis is always performed with the assumptions of the selected computational model that (more or less successfully) imitates the real environment in which the algorithm will be implemented and executed. Again, for the sake of simplification, not all the parameters of the model are usually taken into account. Thus, for example, theoretical models do not usually include assumptions about the concrete implementation of the memory management (and, more specifically, the influence of the cache), which in practice greatly affect the speed of the implementation. The results of the analysis are only valid within the selected model and, as such, they only represent an approximation of the real-life behavior of the algorithm.

Besides the gap between the model and the real-life environment, there are also other factors that condition the effectiveness and correctness of the analysis. Some of them are listed in the following. Due to the complexity of the computational methods used during the analysis, a fully precise formula is rarely derived. Instead, the result of the analysis is often expressed with asymptotic symbols, which (at least) hide the undefined constant factors. To classify an algorithm into a complexity class, such an assessment is satisfactory; however, in practice the constants can make a real difference. For example, the average time complexities of the Quicksort and Heapsort algorithms (for sorting arrays of data) are both in the $O(n \log n)$ [1], but the constants that are hidden in the big-O notation are very different, which makes the Quicksort algorithm much faster. In the analysis we often take into account only the average and the worst cases of the algorithm's behavior and we (intentionally or unintentionally) forget to consider the special cases. It is true that analyzing these cases is usually the hardest job, but in practice they might have a significant impact on the average execution time. If, when executing the algorithms, the "bad special cases" do not occur often (or they do not occur at all), the theoretical average complexity is too pessimistic; and vice versa, if they occur very often (or if the input is almost always one of the bad special cases) the theoretical average complexity is too optimistic. A good example of this phenomenon is the Simplex algorithm for solving linear programming problems, which theoretically has an exponential worst-case time complexity, but in practice it turns out that the bad test cases do not appear at all, which makes the Simplex algorithm surprisingly fast in practice. Another partial (and therefore not totally accurate) analysis takes place when some of the parameters that affect the implementation are not considered. An example of such a situation is a detailed analysis of the Quicksort algorithm, which takes into account the number of comparisons (C), the number of swaps (S) and the number of recursive calls (R), performed during the sorting process [2]. This very interesting research estimates the time complexity of the Quicksort algorithm as a function of C, S and R, which accurately predicts the algorithm's behavior, but only when the sequence being sorted consists of approximately n different numbers. In practice, however, it turns out that the time complexity decreases with the decreasing diversity of the sequence and it can be up to three times lower for sequences containing only a few distinct elements (see Section 4.1 for details). The above-mentioned examples are just some of the reasons why the results produced by the theoretical analysis might be questionable, and sometimes useless in practice, since the algorithm can behave (at least for some test cases) completely differently than predicted by the analysis. From a theoretical point of view these results are interesting and completely legitimate, as they enable the (theoretical) comparison and classification of algorithms. However, applying these results in practice (where the actual environment usually differs from the theoretical assumptions) might raise problems. In the case of an incomplete theoretical analysis, both can occur: the results of the analysis ensure that the algorithm is theoretically fast, while it turns out to be practically slow (for example, an algorithm that frequently uses pairwise far-distant memory locations, while the model used by the theoretical analysis does not consider a slow forward and backward memory-flow rewinding) and vice versa, according to the results of the analysis, it seems that the algorithm is slow and at the same time it is practically fast (like the above-mentioned Simplex algorithm). Note that not only classic computer models are used in practice. In recent times some new models (like data-flow computing [3,4], for example) are upcoming and they deserve theoretical as well as practical attention when considering an algorithm to solve a particular problem. A good example of the analysis for the Simplex algorithm in the data-flow environment is given in Ref. [5].

Before using an algorithm in a real-world application, concrete information about its behavior in practice should be provided to support and upgrade the theoretical results. Basically, this means that the algorithm should be implemented and executed on real-world test cases, while the results of the execution should be examined using algorithm-engineering methods to perform the experimental and empirical analyses [6]. The experimental analysis uses experimental algorithmics to control the parameters of the components in the model structure and to perform the statistical analysis [7,8]. Its results are more accurate than those produced by the theoretical analysis, and at the same time more general than those induced by the empirical approach, which, by measuring the use of computer resources, assesses the complexity of the algorithm in the real-world environment. Even though the selection of a testing environment (the computer with all of its characteristics) determines the algorithm's behavior, the results of the empirical analysis show (a part of) the realistic nature of the algorithm. The empirical analysis is used to measure the quality and to judge the correctness of the algorithm's results. Although the correctness of the algorithm can usually only be proved theoretically (to prove it practically, the algorithm should be run on all the test cases, which is generally impossible), running the algorithm on the selected test cases (and comparing the results with the correct ones) can prove its (partial) correctness or its incorrectness (if the comparison yields false matching). If the algorithm is solving a problem for which more than one solution is possible, and if there exists a criterion function to assess a given solution, then the empirical analysis can measure the quality of the result produced by this algorithm. In addition, it can also measure the quality of the algorithm's implementation, which is reflected in the amount of resources used. Usually, the fewer resources that are used, the better algorithm. The most commonly measured resources are the time, the space (memory), the number of involved processors, the consumed electricity, the generated noise, etc.

To choose the best algorithm for solving a given problem, the results of a theoretical analysis might help in the first round of selection, where the algorithms with better asymptotical boundaries are selected. Nevertheless, a real distinction between the selected algorithms with the same theoretical boundaries can only be made by comparing their behavior in the real environment, which should be prepared with great care to ensure unimpeachable results.

Some of the many reasons why the preparation of such an environment is not an easy job are listed in the following:

- **Large overhead of coding.** To prepare a good testing environment it is possible to spend more time than is needed to implement the testing algorithm.
- **Programming language barriers**. Even with good programming skills, the implementation of a proper and reliable testing environment (that can, for example, accurately measure the algorithm's execution time) requires the use of (sometimes undocumented) hacks and tricks.
- **Diversity of algorithms.** To compare an algorithm with other algorithms, it is necessary to implement not only our own algorithm, but also all the algorithms included in the test.
- **Unexplored nature of the problem.** Preparing a comprehensive set of test cases is a difficult job. Even with a huge effort we cannot ensure that the selected test cases cover all the possible classes of real cases and that the test cases are not somehow tailored to favor our implementation.
- **Arduous repeatability.** Tests conducted in a "garage" environment cannot be repeated easily to verify the reliability of the results.
- **Incomparable results.** The result of tests cannot be assuredly compared with the results of similar tests made by other researchers. Even though the same algorithms and the same test sets are used (which in practice is very difficult to achieve), the testing environment (i.e., the hardware and software used to perform the tests) can never be exactly the same.

To overcome all of the above-stated problems (and many more) we designed a system called ALGATOR that facilitates the algorithm's evaluation and assessment process. The ALGATOR system can be used as a single-user or as a multi-user tool. A single-user version is intended to be used by researchers in the last phase of the algorithm-development process to check the correctness of the results, to acquire different indicators of the quality and to estimate the algorithm's practical complexity. Since the installation of this version and the preparation of a testing project are relatively simple tasks, the system's usage-time overhead is minimal. Besides the implementation of the algorithm and the preparation of the sets of test cases, only a few manageable tasks are to be performed and ALGATOR is ready to start the evaluation.

On the other hand, the multi-user version of ALGATOR provides an environment in which several users (i.e., researchers) are able to compare their solutions (i.e., the implementations of their algorithms) for solving the instances of predefined problems. ALGATOR was designed to support the evaluation process for several problems at the same time. For each problem domain an administrator (i.e., a person who is interested in solving this problem and who possesses enough knowledge about the problem domain) creates a project and specifies the problem's characteristics, including the representative sets of test cases to be used in the testing process. All the other users of ALGATOR can only contribute their algorithm's implementations and thus they cannot influence the testing process. This makes ALGATOR **efficient** (due to the low coding overhead), **fair** (the tests are made independently of the algorithms' authors), **unbiased** (the tests are not tailored to any concrete solution), **reliable** (the tests are made using a carefully designed system), **transparent** (the testing environment, the tests and the algorithms' implementations are publicly available) and **repeatable** (the tests can always be repeated) **algorithm–evaluation system** that can be used as a reference point for assessing the algorithms developed by different researchers all around the world.

According to the classification given in Ref. [9], the ALGATOR system was developed using the Implantation (I) technique. In the field of algorithms' assessment research there are some tools, but none of them cover this research area completely. ALGATOR should be seen as an implant that extends and supplements the missing functionality of the existing tools.

In this chapter we present ALGATOR from the three different perspectives: we present the idea that defines a problem as a set of abstract entities, we present some details about how this idea was implemented into a working system and finally we give some examples of the uses of the system when it comes to solving real problems. More precisely, in Section 2 we present a model used by ALGATOR to define a problem on as abstract a level as possible. We reveal the meaning of terms like project, user role, test case, test set, test-set iterator, algorithm class and measurement. We show their roles in the system and present the ways that they interact with each other in order to support the implementation of the algorithms for a selected problem. In Section 3 we leave the abstract and enter the real world—we present some details of ALGATOR's implementation and use, which include information about the organization and configuration of the system's and projects' data, about the mechanisms used to execute the algorithms and to measure the desired indicators, about the different versions of the system and about the extraction of the results using ALGATOR's query engine. To be even more concrete, in Section 4 we present three examples of the use of the ALGATOR system for solving real–life problems. Each of the three ALGATOR measurement types are presented on a different problem domain. Namely, to present the em measurements (i.e., the time consumption and the algorithm's quality indicators) we use the classic Sorting Problem in Section 4.1; to present the use of the cnt measurements (i.e., the counters of program-code use) we analyze the algorithms implementing the Fourier transform in Section 4.2; while the jvm measurements (i.e., the counters of the Java bytecode use) are presented in Section 4.3 using the matrix-multiplication problem domain. We conclude the article with some final remarks and open questions in the Conclusions section.

2. Problem-presentation model

To make the system useful for as many problems as possible, ALGATOR treats the problem and all the subordinate entities (the algorithm, the test case, the test set and the test-set iterator) as abstract objects. To make a project "come alive," all of these entities must be implemented as concrete (non-abstract) classes or configuration files. In this section we reveal the abstract model that drives the ALGATOR system by presenting the problem and all of its subordinate abstract entities, and we show how to concretize these entities in order to implement a selected problem.

2.1 Use of the system

ALGATOR can be used by many different users, each of which has different roles and access rights. An unauthenticated user (i.e., a guest) can only view and browse the publicly available data, and can only use those actions that do not change the characteristics of the system. For all the other activities the user must be logged into the system and must have one of the following roles:

System administrator. The system administrator installs and manages the whole system (the software and hardware parts), and has access to all the resources of the system.

Project administrator. The project administrator sets a project by defining the interfaces to describe the problem and the structure of the algorithm, implementing the project-specific methods to read the data and execute the algorithms and by defining sets of test cases on which algorithms will be executed. In addition, the project administrator characterizes the format of the input and the output of the algorithms (i.e., defines the parameters of the input and the indicators of the output). The project administrator has access to all the project's resources. If the project is made public, all the users can see the project's data, while only the project and system administrators can see private projects.

Researcher. The researcher defines an algorithm within the selected project, triggers the predefined tests and analyzes the algorithm by comparing its results with the results of other algorithms. The information about the public algorithms is available to every user of the system, while only the owner (i.e., the researcher) and the project administrator can browse the private algorithms.

The ALGATOR system can be used by one person on a single computer or by several people with several computers involved. In the case of a single-computer installation, the user of the system plays all the roles described above, while on a multi-user and multi-computer installation the typical usage of the system is as follows:

- The system administrator prepares the system by providing the hardware, installing the software packages (e.g., the executing machine, the web server, the task server/client system, etc.), preparing the data-storage system, and by publishing the Internet address of the installed system.

- A project administrator adds a new project and defines all the project's properties. When the project is completely defined and declared as public, ALGATOR automatically generates an Internet subpage with the project presentation and with usage-guide sections.

- The project administrator adds some state-of-the-art algorithms for solving the problem of the project, which will be used as a reference for the evaluation process (i.e., the results of the algorithms added by the researchers will be compared with the results of these reference algorithms).

- In accordance with the rules, which are available on the project's website, a researcher adds a new algorithm. ALGATOR automatically runs this algorithm on predefined tests. The researcher then checks for the correctness and compares the results of his/her algorithm with the results of other algorithms defined in the project. The researcher can also decide to make the algorithm public (by default, the new algorithms are private).

- All unauthorized users (i.e., guests) browse the results and print the graphs and other data produced by ALGATOR. They can also perform some actions (like customization of the presentation) that do not alter the project's configuration. At any time a guest can register with the system and contribute as a project administrator or as a researcher.

2.2 Project description

The basic data unit of the ALGATOR system is the so-called *project*, which is created by a project administrator and used by other users (to add algorithms and/or to view the results of the execution, for example). The purpose of the project is to combine the information about a selected problem domain, which includes: the problem, the algorithm and the test-case description, the test sets with the input data, the algorithm implementations and the results (see Fig. 1). The information describing the project is given in the configuration files (in the JSON or CSV formats) and with the Java code. For example, for the system to be able to verify the correctness of the execution results, the administrator of the project should provide a Java code (i.e., a method) that performs a project-dependent result verification. At the end of the algorithm's execution, ALGATOR calls this method and according to its result generates an appropriate indicator of the algorithm's correctness.

The core of the project's configuration is the so-called atp file, in which the following information is written:
- the name and a short description of the project,
- the name of the author and the date of creation,
- a list of algorithms and test sets,
- a list of libraries (JAR files) files used to execute the algorithms,
- the names of the computer families that can perform the execution.

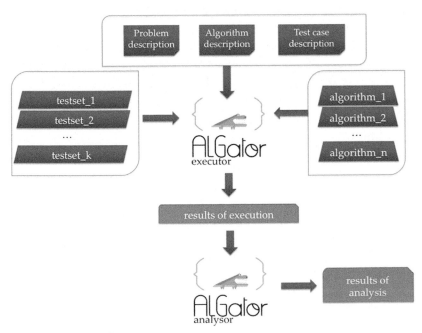

Fig. 1 Components of an ALGator project.

All the project's properties are given in a textual format (JSON file) and are programming-language independent, as presented in the example in Fig. 2.

Note that in this and in the following examples we will frequently use a simple project called the Sorting Project. The problem in this project is to sort the given array of integers into ascending order. The input to the project's algorithms is an array and it is expected that, at the end of the algorithm's execution, this array will be sorted. There are several well-known algorithms to solve this problem, for example, the Wirth and Hoare algorithms (they both use one pivot), and the Yaroslavskiy algorithm (which uses two pivots in the partitioning phase), and many more. For this problem it is easy to verify the correctness of the solution with a simple scan through the output array comparing the subsequent elements. The test sets of the input data consist of arrays of integers of different sizes and different data distributions (e.g., arrays of integers mixed uniformly at random, sorted arrays, arrays sorted in reverse order and the like). The problem of sorting data was chosen since it is reasonable to assume that it is a well-known problem with well-defined concepts and that everybody (or at least every reader of this chapter) knows what a sorting algorithm should do and how its input and output should look.

Besides the textual information (like the description of the project) and the list of algorithms and test sets, there are two important pieces of data in the configuration of the project. First, there are the java libraries (jar files) that can be attached to the project using the ProjectJARs and AlgorithmJARs variables. The jars listed in the ProjectJARs are used by the project's code in the phases of the test-case creation and the results verification to facilitate the project's administration. For example, if the result-verification process requires copying files, the administrator of the project can add an appropriate jar (e.g., commons-io.jar) to the ProjectJARs and use methods that support

```
"Project": {
    "Name":             "Sorting",
    "Description":      "Sorting the arrays of integers",
    "Author":           "td",
    "Date":             "October 2017",
    "Algorithms":       ["Wirth", "Hoare","Yaroslavskiy"],
    "TestSets":         ["TestSet1","TestSet2","TestSet3"],
    "ProjectJARs":      ["commons-io-2.5.jar"],
    "AlgorithmJARs":    ["sorting-helpers.jar"],
    "EMExecFamily":     "F1"
}
```

Fig. 2 Configuration of a project—the content of the Sorting.atp file.

copying the files (e.g., `FileUtils.copyFile()`). The `jars` listed in the `ProjectJARs` are used in the administrator's code, while in the algorithm's code only the `jars` listed in the `AlgorithmJARs` could be used. By adding a `jar` to this variable, the project administrator can provide common data structures and methods that are available to all the algorithms of the project.

The second important data in the configuration file are the so-called *execution families* that are listed in the `EMExecFamily`, `CNTExecFamily` and `JVMExecFamily` variables. With these variables the administrator defines the family of computers that could be used to execute the algorithms of the project (see Section 3.9).

The other components making up the project are described in the following sections.

2.3 Presenting test cases and test sets

The type of input and output of the algorithms can differ greatly from project to project. For example, in the sorting project, the input and output of the problem-solving algorithms are arrays of integers, while in the shortest-path project (the aim of the problem-solving algorithm in this case is to find the length of the shortest paths between two given vertices of a given graph) the input would be the graph and the two vertices, and the output would be an integer. Therefore, an important task of the project administrator is to define the input and output data structures and to introduce them to the system. For these purposes ALGATOR provides two generic placeholders, the `AbstractInput` and the `AbstractOutput` classes, which have to be extended to the `[Project]Input` and `[Project]Output` classes with the appropriate member fields to hold the input and output data. An example of a project-specific input class for the sorting project (i.e., the `SortingInput` class) is presented in Fig. 3. Since the input of the sorting algorithms is represented with an array of integers, the only data field used in the `SortingInput` class is an `arrayToSort` array. If the data structure required in the project is too complex to be presented in a single class (for example, the graph data structure), the project administrator is encouraged to

```
class SortingInput extends AbstractInput {
    /**
     * An array of integers to be sorted.
     */
    int [] arrayToSort;
}
```

Fig. 3 `SortingTestCase` class implementation.

implement the data structure in separate classes, pack them into a `jar` and use this `jar` in the project. In this way the `[Project]Input` class remains reasonably small and readable, while at the same time the researcher gains the full capability of the data structure and (if available) its helping methods.

Both the `[Project]Input` and the `[Project]Output` classes are joined in the `[Project]TestCase` class, in which the former holds information about the input instance and the latter about the expected output (when applicable). The information stored in the instance of the `[Project]TestCase` class facilitates the process of checking the correctness and/or the quality of the algorithm's results. It is worth mentioning that the instance of the `[Project]Output` is never passed to the execution phase of the algorithm (otherwise the developers of the algorithms could possibly cheat by returning the hold data as a result). In the preparation phase (before executing the algorithm) the required input data has to be extracted from the test-case object and packed into a form that is suitable for the algorithm's invocation (see Section 2.4 for details).

The test cases of the project are written in files and grouped together into the so-called test sets, which are the minimum execution units of the system. The test sets should be prepared in such a way as to expose a particular behavior of the algorithm on different types of input. For example, some algorithms are very fast (or very slow) on a subset of all possible inputs (like the problems described in Ref. [10]). To adequately present the differences between the algorithms, at least one test set should be created to cover each special subset of the input.

When ALGATOR is executed, it runs the given algorithm on all the test cases collected in a given test set, one by one. A standard way to describe a test set is by using a text file in which each line represents one test case. The content of these lines is strictly project-specific and it has to be interpreted using a project-specific programming code. An example of a text file describing a set of eight test cases for the Sorting Problem is given in Fig. 4. In this file each line contains three or more values separated by a

```
test1:1000:RND
test2:1000:SORTED
test3:1000:INVERSED
test4:1000:FILE:numbers.rnd:412317
test5:2000:RND
test6:2000:SORTED
test7:2000:INVERSED
test8:2000:FILE:numbers.rnd:23423
```

Fig. 4 Content of a text file describing a set of eight test cases for the Sorting Problem.

semicolon. The first and second values describe the identification (i.e., the name) and the size (i.e., the number of elements in the array to be sorted) of a test case. The third value defines the type of an input array, with the following possible values and meanings: RND (random array), SORTED (sorted array), INVERSED (array sorted in inverse order) and FILE (array of numbers that should be read from the given file starting at the given offset). The structure of the file presented in Fig. 4 was defined by the administrator of the Sorting Problem.

To iterate through the text file associated with a given test set, ALGATOR uses the methods of the DefaultTestSetIterator class. For each line read from the text file the getTestCase() method of the [Project]TestCase class is called. This method parses the input line and creates a set of parameters that describe the test case. Using these parameters it calls the generateTestCase() method, which creates the instance of the test case.

In the AbstractTestCase class the getTestCase() method is declared as an abstract method; therefore, the project administrator (who defined the format of the text file and therefore the only person who knows how to interpret its lines) has to provide the implementation of the project-specific getTestCase() method in the [Project]TestCase class. Similarly, it is the task of the project administrator to implement the generateTestCase() method of this class.

Examples of concrete implementations of the getTestCase() and the generateTestCase() methods of the SortingTestCase class are presented in Figs. 5 and 6. Due to the different complexities of the representation of the input and output data, the implementations of these methods in different projects might differ a lot. It is only important that using these methods

```
class SortingTestCase extends AbstractTestCase {

    @Override
    public SortingTestCase getTestCase(String testCaseDescriptionLine, String path) {
        // create a set of variables ...
        Variables inputParameters = new Variables();
        inputParameters.setVariable("Path", path);

        // ... add the values of the parameters to the set ...
        String [] fields = testCaseDescriptionLine.split(":");
        inputParameters.setVariable("Test", fields[0]);
        inputParameters.setVariable("N",     Integer.parseInt(fields[1]));
        inputParameters.setVariable("Group",fields[2]);
        //... add the rest of the parameters ....

        // ... and finally, create a test case determined by these parameters
        return generateTestCase(inputParameters);
    }
```

Fig. 5 Part of the SortingTestCase class with a simplified getTestCase() method implementation.

```
@Override
public SortingTestCase generateTestCase(Variables inputParameters) {
    int     probSize = inputParameters.getVariable("N"    ).getIntValue();
    String group    = inputParameters.getVariable("Group").getStringValue();

    // prepare an array of integers of a given size ...
    int [] array = new int[probSize];
    // ... and fill the array according to the group value
    switch (group) {
      case "RND":
        Random rnd = new Random();
        for (int i = 0; i < probSize; i++) array[i] = Math.abs(rnd.nextInt());
        break;
      case "SORTED":
        for (int i = 0; i < probSize; i++) array[i] = i;
        break;
      // ...
    }

    // create a test case ...
    SortingTestCase sortingTestCase = new SortingTestCase();
    sortingTestCase.setInput(new SortingInput(array));
    int [] expectedResultArray = getSortedArray(array);
    sortingTestCase.setExpectedOutput(new SortingOutput(expectedResultArray));
    // ... and return
    return sortingTestCase;
}
```

Fig. 6 Part of the implementation of the `generateTestCase()` method.

```
"TestSet" : {
    "Name"              : "TestSet1",
    "ShortName"         : "TS1",
    "N"                 : "95",
    "TestRepeat"        : "10",
    "TimeLimit"         : "1",
    "DescriptionFile"   : "TestSet1.txt"
}
```

Fig. 7 Configuration of a test set—the content of the `TestSet1.atts` file.

ALGATOR can achieve its main goal, i.e., to create an instance of a test case according to the data written in a single line of a test–set text file.

Each test set is associated with an `atts` configuration file in which the test set's parameters are given. An example of the `TestSet1.atts` file presented in Fig. 7 describes a test set with `N`=95 test cases; each test in this test set will be executed `TestRepeat`=10 times; the time limit for each test is `TimeLimit`=1 s; and the name of the text file for this test set is `TestSet1.txt`.

2.4 Algorithm description

The algorithm in the ALGATOR system is implemented as an `AbstractAlgorithm` class with the three important methods, `init()`, `run()` and `done()`, as depicted in Fig. 8.

```
abstract class AbstractAlgorithm implements Serializable {
  AbstractTestCase currentTestCase;
  AbstractOutput   algorithmOutput;

  public void init(AbstractTestCase testCase) {
    currentTestCase = testCase;
  }

  public abstract void run();

  public Variables done() {
    AbstractOutput expected = currentTestCase.getExpectedOutput();
    Variables result = new Variables(expected.getIndicators());

    for (EVariable eVariable : result) {
      Object value = expected.getIndicatorValue(
          currentTestCase, algorithmOutput, eVariable.getName());

      if (value != null)
        eVariable.setValue(value);
    }
    return result;
  }
}
```

Fig. 8 ALGATOR's implementation of the abstract algorithm.

```
[Project]TestCase testCase = new_instance_of_test_case();
[Project]AbsAlgorithm  alg = new_instance_of_algorithm();

alg.init(testCase);
timer.start();
alg.run();
timer.stop();
indicators = alg.done();
```

Fig. 9 Simplified program flow for the algorithm's execution.

The three methods (among which at least the run() method is supposed to be overridden in the project-specific [Project]AbsAlgorithm class) are used during the ALGATOR's execution process (i.e., when ALGATOR runs an algorithm to solve a given test case), as shown in Fig. 9.

The init() method is used to prepare the test case to be instantly applicable to the algorithm. It can be considered as the preparation phase (that is performed just before the execution phase) used to extract the required input data from the test case and to pack it into a form suitable for the algorithm's invocation. To protect possibly incorrect data uses it is important that the execution part of the algorithm does not access the test-case instance directly. Instead, the init() method should prepare a copy of the data needed for the execution and store this data to the local fields. An example of this data extraction for the Sorting Problem is depicted in Fig. 10.

```
abstract class SortingAbsAlgorithm extends AbstractAlgorithm {

  protected abstract SortingOutput execute(SortingInput sortingInput);

  SortingInput currentInput;

  @Override
  public void init(AbstractTestCase testCase) {
    super.init(testCase);
    currentInput = (SortingInput) testCase.getInput();
  }

  @Override
  public void run() {
    algorithmOutput = execute(currentInput);
  }
}
```

Fig. 10 Implementation of the `SortingAbsAlgorithm` class.

Here, the overriding `init()` method prepares a `currentInput` field to be used in the algorithm's invocation phase.

The central role in the ALGATOR's execution process is played by the `run()` method, in which the problem for the given input is actually solved. ALGATOR treats the execution time of this method as the execution time of the algorithm. To ensure the correct timing results all the data-preparation tasks should be performed in the `init()` method and all the post-processing tasks (like checking for the correctness of the result) in the `done()` method. The `run()` method is reserved to execute only the pure algorithm's code.

Supposing that the algorithm consists of n phases to perform the task, the project administrator should define n abstract methods, each used to execute one phase of the algorithm. The signatures of these methods (i.e., the number and the type of the parameters) are strictly project-specific and have to be defined by the project administrator. Similarly, the calling sequence of the phases' methods has to be defined by the project administrator in the `run()` method (since only the administrator knows how these methods are supposed to be used). Before each method call, the phase's switch has to be signaled to the system (by calling the `timer.Next()` method) to ensure that the timing of each of the phases will be measured separately. By default, the timer is activated just before the `run()` method is invoked and it is stopped when the method returns. As a consequence, a manual timer maintenance is not required if the algorithm consists of only one phase. In this case the `run()` method consists of one statement, i.e., the call of the single execution method, as depicted in the example in Fig. 10. An example of the `run()` method for a slightly more complicated problem, i.e., the File Compression Problem, is given in Fig. 11. Here, the algorithm consists of the two phases: the compression phase (in which an input file is compressed) and the decompression phase (in which the compressed file is decompressed). Both phases are required to ensure the correctness of the algorithm (i.e., the algorithm is correct if and only if the decompressed file equals the original file). In this implementation of the `run()` method ALGATOR uses two timers: the first (default) one for the compression phase and the second one for the decompression phase. The switch between the timers is made by calling the `timer.next()` method.

For each phase of the algorithm the administrator has to define a signature for the corresponding method. For the algorithms that contain only one phase, this signature usually follows a common pattern and is defined with

```
protected abstract [Project]Output execute([Project]Input input),
```
but it can also be changed by the administrator. The signatures of all the phases are introduced to the end user (i.e., the researchers who will implement their versions of the algorithm) in a simplified version of the `[Project]AbsAlgorithm` class. An example of these classes for the sorting and the file-compression projects is presented in Fig. 12.

Each sub-class of the simplified version of the `[Project]AbsAlgorithm` class, in which all the abstract methods are implemented, is considered as a legal algorithm for solving the given project's problem. A simple (but fully functional) example of the algorithm for the Sorting Problem is given in Fig. 13.

```
public void run() {
    compress(inputFilename, compressedFilename);
    timer.next();
    decompress(compressedFilename, referenceFilename);
}
```

Fig. 11 Implementation of the `run()` method for the file-compression problem.

```
abstract class SortingAbsAlgorithm extends AbstractAlgorithm {
    protected abstract SortingOutput execute(SortingInput sortingInput);
}

abstract class CompressionAbsAlgorithm extends AbstractAlgorithm {
    protected abstract void compress  (String input, String output);
    protected abstract void decompress(String input, String output);
}
```

Fig. 12 Simplified versions of the `[Project]AbsAlgorithm` classes for the sorting and file-compression problems.

```
class JavaDefaultSortingAlgorithm extends SortingAbsAlgorithm {
  @Override
  protected SortingOutput execute(SortingInput input) {
    // sort the array using java default method ...
    Arrays.sort(input.arrayToSort);

    // ... and return the result
    SortingOutput result = new SortingOutput();
    result.sortedArray = input.arrayToSort;
    return result;
  }
}
```

Fig. 13 Implementation of the algorithm for the Sorting Problem.

The last method used in the algorithm's execution process is the done() method. This method is used to verify the correctness of the algorithm's results and to prepare the indicators of the execution. For each of the required indicators (which are defined in the corresponding atrd file; for details see Section 2.5) an instance of the EVariable has to be generated, initialized and appended to the set of indicators (which will be returned as a result of the done() method). To find the values of the indicators the done() method (which is defined in the AbstractAlgorithm class and can be overridden in the [Project]AbsAlgorithm class, if needed) uses the getIndicatorValue() method. Since the done() method is not included in the timing mechanism (i.e., the timers are stopped before the invocation of this method), its actions do not influence the measured execution time of the algorithm. Nevertheless, to prevent delays and to provide a rapid testing environment, the administrator should avoid greedy actions in the project-specific getIndicatorValue() method. Instead, the administrator should pre-compute the correct results, pack them into the corresponding [Project]TestCase instance and use them as the reference values to be compared with the algorithm's results.

An example of the getIndicatorValue() method is presented in Fig. 14. In the Sorting Problem we have only one indicator, namely the "Check" indicator, which has the value "OK" if the array returned by the algorithm is sorted and "NOK" otherwise. To determine this value, the getIndicatorValue() method uses both arrays: the one returned by the algorithm and the one prepared in the test–case instance (in the expectedOutput variable) and the Arrays. equals() method.

2.5 Results of the tests

During the evaluation process ALGATOR generates a plethora of indicators for the time and the use of other resources, and for the quality of the

```
@Override
protected Object getIndicatorValue(
                    AbstractTestCase testCase,
                    AbstractOutput algorithmOutput,
                    String indicatorName)
{
  SortingTestCase sortingTestCase = (SortingTestCase) testCase;
  SortingOutput   sortingOutput   = (SortingOutput)   algorithmOutput;

  int [] output         = sortingOutput.sortedArray;
  int [] expectedOutput = sortingTestCase.getExpectedOutput().sortedArray;

  switch (indicatorName) {
    case "Check" :
      return Arrays.equals(output, expectedOutput) ? "OK" : "NOK";
  }

  return null;
}
```

Fig. 14 Implementation of the `getIndicatorValue()` method.

algorithm's results. To generate the indicators ALGATOR performs three independent measurements: the em measurement to generate the timing and project-specific indicators, the cnt measurement to generate the counters and the jvm measurement to generate the indicators of the Java bytecode instructions' usages. Since the test set is a minimum execution unit, one measurement produces the indicators for all the test cases of a given test set. The measurements are run independently to prevent their interactions and corruption of the results (for example, measuring the time and the counters of execution in one test would produce totally useless timing indicators, since even a small change in the algorithm's code (like incrementing a counter in a loop) can change the algorithm's behavior dramatically). At the end of the measurement the results are written to the output files—one file for each tuple (algorithm, testset, measurement)—whose content is defined by the administrator of the project in the [Project]-[mtype].atrd file (here [mtype] stands for one of the em, cnt or jvm). The output file [Algoritem]-[TestSet].[mtype] contains one line of a text data for each test case of the TestSet; the line contains the indicators in the order prescribed in the corresponding atrd file.

2.5.1 Measuring the time and the project-specific indicators

ALGATOR's main measurement (i.e., the em measurement) is used to measure the time and (optionally) to produce the project-specific indicators.

The time, as the most interesting resource being observed, is measured automatically. Even though by default only one timer is engaged in the measurements, the system can use as many timers as required. If the algorithm in question consists of more than one phase, the project administrator could decide to use more timers (for example, to measure each phase separately). To include a new timer in the measurement, the administrator of the project simply calls the `timer.next()` method inside the `run()` method.

To produce as accurate time measurements as possible, ALGATOR runs each test case several times (the number of times is controlled by the `TestRepeat` parameter in the corresponding test-set configuration file) and collects the execution times for each run. At the end of all the executions ALGATOR returns the first, the best and the worst execution times. It can also preform some basic statistics and return an average, a variance and a sum of the times of all the executions. By defining the parameters of the output configuration `atrd` file, the administrator of the project selects which timers are to be included in the output file. In most cases the minimum time is the most informative, since it indicates the execution time for which the influences of the environment were the smallest.

An example of an `atrd` is presented in Fig. 15, where one parameter of the test (i.e., `N`—the size of the input array) and two indicators

```
"Result": {
    "ParameterOrder"    : ["N"],
    "IndicatorOrder"    : ["FOmin", "F1avg"],

    "Parameters":    [
      {
        "Name":          "N",
        "Description": "The size of the test",
        "Type":          "int"
      }
    ],
    "Indicators":    [
      {
        "Name"          : "FOmin",
        "Description" : "The minimal time of the phase 0",
        "Type"          : "timer",
        "Subtype"      : "0 MIN"
      },
      {
        "Name"          : "F1avg",
        "Description" : "An average time of the phase 1",
        "Type"          : "timer",
        "Subtype"      : "1 AVG"
      }
    ]
}
```

Fig. 15 Example of a `Sorting-em.atrd` configuration file.

```
JWirth;TestSet1;Test-1;DONE;100;79;12
JWirth;TestSet1;Test-2;DONE;200;142;23
JWirth;TestSet1;Test-3;DONE;300;202;34
JWirth;TestSet1;Test-4;DONE;400;265;37
JWirth;TestSet1;Test-5;DONE;500;336;42
JWirth;TestSet1;Test-6;DONE;600;458;51
```

Fig. 16 Example of the output file for the em measurement.

(i.e., FOmin and F1avg) are defined. The first indicator measures the minimum execution time for the phase 0 and the second the average execution time for the phase 1. For each phase i and for each statistical function f being one of (FIRST, MIN, MAX, SUM, AVG, VAR) an indicator of the subtype "i f" can be listed in the atrd file. An example of the output file produced by ALGATOR using the atrd file from Fig. 15 is depicted in Fig. 16.

The first four values in each line are automatically added to the output file by the system and they stand for the name of the algorithm (JWirth), the name of the test set (TestSet1), the test-case identification (Test-i) and the execution status (can be one of DONE, KILLED, FAILED). The following three values in each line are the values of the parameter N and the indicators FOmin and F1avg.

Besides the time indicators, the em measurements include the project-specific indicators that can measure any measurable resource or value (like the memory and the disk use or the correctness and the quality of the result). The administrator of the project defines the name and type of the project-specific indicator in the atrd file and determines its value with the done() method. For example, in order to include an indicator named Check with the possible values "OK" or "NOK" (indicating the correctness of the result) for the Sorting Problem, the project administrator has to define the indicator by adding the indicator's definition to the list of the indicators in the atrd file and by defining the indicator's value as presented in the done() method in Fig. 10 (here the correctness of the result is checked by calling the isArraySorted() method; depending on the value returned by this method, the indicator is given the value "OK" or "NOK").

2.5.2 Measuring the counters

ALGATOR's counters' mechanism makes it possible to measure the uses of the marked lines of the algorithm's code (i.e., how many times the flow of the algorithm comes to these lines during the execution on a given test case). If, for example, all the lines in which a method a() is called, are marked, then ALGATOR will count the number of times the method a() was called. Using this mechanism, ALGATOR can measure the distribution of the uses of the (logical) operations of an algorithm. For example, in the Sorting Problem, every algorithm performs the following operations: read an array element, write an array element, compare two elements, swap two elements, make a recursive call, etc. Using the counters mechanism, ALGATOR can measure the number of times each of these operations was performed.

To mark a line of a code, an additional line of the format

```
//@COUNT{counter_name, value}
```

has to be added just before the line. Each time the flow of the algorithm comes to the so-marked line, the value of the counter counter_name will be increased by the value. ALGATOR can use any number of counters, but they have to be defined by the project administrator in the corresponding [Project]-cnt.atrd file. After the execution of the algorithm, the values of all the counters, listed in the IndicatorOrder variable of this file, are written to the output file called [Algorithm]-[TestSet].cnt. The measurement of the counters is run once for each test case (the value of the TestRepeat parameter is ignored).

Note that the cnt marking line (i.e., the line used to mark a selected line of code) is a Java comment. The reason for this is that ALGATOR aims to keep the em and the cnt measurements strictly independent. In the em measurement, when ALGATOR measures time, the cnt marks are left as Java comments and therefore ignored during the execution. This ensures that the measurement of the counters does not influence the measured time of the execution. On the other hand, when ALGATOR measures the counters (and the time of execution is not observed), the original Java source file is replaced by a new one, in which all the marks (java comments) are replaced by the corresponding Java instructions, as follows:

(1) all the marking lines

```
//@COUNT{counter_name, value}
```

are replaced with

```
Counters.addToCounter("counter_name", value);
```

(2) all the lines containing //@REMOVE_LINE, are removed.

An example of the use of the counter CMP to count the number of comparisons performed by the algorithm is presented in Fig. 17.

In this code the numbers are compared in the if clause and in the condition of the while loop. The if clause contains two comparisons, but in some cases (if the first comparison returns false) only the first one is performed.

```
/*//@REMOVE_LINE                          Counters.addToCounter("CMP",  1);
//@COUNT{CMP, 1}                          if (a[e1] != a[e2]) {
if (a[e1] != a[e2]) {                       Counters.addToCounter("CMP",  1);
  //@COUNT{CMP, 1}                         }
}
*///@REMOVE_LINE
if (a[e1] != a[e2] && a[e2] != a[e3])     if (a[e1] != a[e2] && a[e2] != a[e3])
{                                         {
  //@COUNT{CMP, 1}                          Counters.addToCounter("CMP",  1);
  while (a[++less] < pivot) {               while (a[++less] < pivot) {
    //@COUNT{CMP, 1}                          Counters.addToCounter("CMP",  1);
    ...                                        ...
  }                                         }
}                                         }
```

Fig. 17 Example of the counters' use—the original code (left) used in the em measurement and the altered code (right) used in the cnt measurement.

To count the number of comparisons accurately, an additional compare was added in the marking lines. The number of comparisons performed by the while loop equals the number of performed iterations plus one; therefore, the //@COUNT{CMP, 1} mark has to be added twice (before and after the while condition).

Note again that in the case of the em measurements, the code depicted on the left-hand side of Fig. 17 will be executed, while in the case of the cnt measurements, the code from the right-hand side will be generated and used.

2.5.3 Measuring the Java bytecode instruction uses

The measurements described in the previous sections are used in the case of both the Java and the C/C++ algorithms: in both we can measure time, project-specific indicators and counters. On the other hand, the Java bytecode instruction use is (probably not surprisingly) limited to the Java algorithms. Before the execution of these algorithms, they are compiled into bytecode that is later used by the Java virtual machine. Using a dedicated virtual machine, which was developed especially for this purpose, ALGATOR is able to inspect the execution of the algorithm and count the uses of the bytecode instruction. At the end of the execution (which is performed separately and independently of the em and cnt measurements) the statistics for each of the 200+ bytecode instructions is written to the corresponding output file. Analyzing the statistics obtained by this measurement and combining them with knowledge of the behavior of the Java virtual machine (some information about this is presented in Ref. [11]) complementary knowledge of the observed algorithm might be derived. The configuration of the project for the jvm measurement is not required; everything is performed automatically by ALGATOR. The details about the implementation of the Java virtual machine used in these measurements are given in Section 3.6, while an example of the use of the generated statistics is presented in Section 4.3.

3. Implementation of the ALGATOR system

3.1 Organization of the system

The ALGATOR system consists of several independent applications that use the same input data (information about the projects) and that communicate through different channels. In this section we introduce all the components, their roles in the system and the ways in which they interact with each other.

Fig. 18 presents the organization of the ALGATOR system. In order to make the system as scalable as possible, its parts were designed as independent components (each box in Fig. 18 represents one such component) that can co-exist on the same computer or they can be installed on different computers that are connected via network connections. In the most usual arrangement, one computer hosts the data (data_root storage), the server (the TaskServer and the QueryEngine), the web server (WebPage) and the terminal, and one or more computers host the client executors (computers on which the algorithms are executed). In the case of the standalone single-user installation all the necessary components (the data_root and the algator.Execute program) are installed on a single computer.

The backbone of the complete ALGATOR system is the components depicted in the middle of Fig. 18: the data_root storage (the data from all the projects are stored here) and the server, which is used as a communications bridge between the components.

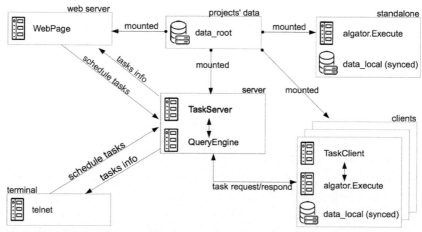

Fig. 18 Organization of the ALGATOR system.

On the left-hand side of Fig. 18 there are two components used to manipulate and present the data (the web server and the terminal), while on the right-hand side the executors (i.e., components used to execute the tasks) are depicted.

The heart of the system is the so-called data_root storage, used to store all the information about the projects, including the problem definitions, the algorithms' implementations, the test sets and the execution results. The data_root is currently implemented as a folder structure with a precisely defined meaning for the folders and files, as described in Section 3.2. The implementation of the data_root as a relational database could have some benefits (like rapid search using the predefined indices) and it might be considered in the future as a possible solution. The current implementation as a folder structure was selected because of its simplicity and transparency, since it enables an easy manual manipulation without special tools.

All the applications depicted in Fig. 18 use the data_root storage to read the existing and produce new information about the projects. The precondition for most of them to work properly is the data_root folder to be mounted and mapped into the $ALGATOR_DATA_ROOT environment variable. Even though such an approach enables unlimited access to all the data required, it is relatively slow (especially if the data_root and the application are installed on remote computers) and it might cause congestions with the data access. To prevent any distortion of the results resulting from this issue, each computer uses a data_local copy of the data_root storage. Before the tasks (i.e., the executions of the algorithms) are executed, the storages are synchronized so that only the data stored on the local computer is used during the execution.

The central role in the ALGATOR system is played by the so-called TaskServer, which provides a bridge between the data and its presentation and production. TaskServer provides the data and schedules the task execution for most of the system's applications (except for the standalone installation, where the algator.Execute program directly accesses the data and controls the tasks to be executed). The TaskServer is tightly connected with the QueryEngine; therefore, it can also provide query results quickly. The TaskServer holds the information about the tasks to be executed and schedules them to the appropriate computers. The WebServer and the terminal position the task in the TaskServer's queue, from where they are consumed by the TaskClients. The TaskServer does not search for the clients to assign them tasks; instead, it simply waits for the requests. While a TaskClient is idle it repeatedly sends task requests to the

TaskServer, and if an appropriate task exists, it is delivered to the client. ALGATOR uses this mechanism of "requesting a task" (instead of "searching for clients") to ensure that the clients are not disturbed while they execute the tasks. For more details about the communication between the TaskServer and the TaskClients see Section 3.4.

The web server with the WebPage is used to present all the data about the projects in a user-friendly format (i.e., using charts, graphs and diagrams) and to enable the data's manipulation (i.e., creating a new project, adding a new algorithm, executing tasks, etc.). All this can also be done using the terminal and the telnet program, but such an approach requires a deeper understanding of the whole system and it is thus reserved only for experts (not for general use).

3.2 System-configuration folder

Each computer in the ALGATOR system stores (retrieves) settings and other configuration data to (from) the so called ⟨algator_root⟩ folder. To facilitate the use of this folder by ALGATOR's applications, the folder has to be mapped into the $ALGATOR_ROOT environment variable. The ⟨algator_root⟩ contains the following subfolders: local_config, data_root, data_local and app, as depicted in Fig. 19.

The most important folder in this structure is the data_root, which contains all the information about the projects. Since this folder is normally used by several computers, it is common to place it on one computer and to mount it to the others. This enables the use of the same synchronized data at different locations. The mounted volumes are usually slow and might slow down the execution; therefore, there is always a local copy of the project's

```
<algator_root>
├ local_config          // the computer's configuration
│ └ config.atlc
├ data_root
│ ├ global_config       // the system's configuration
│ │ └config.atgc
│ └ projects            // information about the projects
│   ├ log               // logging folder
│   │ └ tasks           // communication folder
│   └ schema            // folder for schema
├ data_local            // local copy of the data_root
└ app                   // programs and services
```

Fig. 19 Structure of the ⟨algator_root⟩ folder.

data in the `<data_local>` folder. The content of this folder is automatically synchronized with the (mounted) `<data_root>` at the beginning of each execution; during the execution the data is always retrieved from this local copy. The `local_config` folder contains information about the current computer, like the name and the family of the computer, the execution capabilities, etc. To stress the difference: the `data_root` contains the global projects' data (common to all the computers), and the `local_config` contains the data that is strictly computer dependent. The `app` folder is used to store ALGATOR's applications and services (like `algator.Execute`, `algator.TaskServer`, etc.).

The `projects` folder in the `data_root` contains the configuration information for all the projects that are defined in the system. For each project `[Project]` it contains a folder named `PROJ-[Project]` with the structure presented in Fig. 20. The main project's configuration is in the `[Project].atp` file (besides the basic information like the name of the project and the name of the project's author, here all the included test sets and algorithms are listed). For each algorithm `[Algorithm]` of the project, the `algs` folder contains the subfolder `ALG-[Algorithm]` with the configuration file `[Algorithm].atal` and the algorithm source `[Algorithm]Algorithm.java`. For each test set the `tests` folder contains an `atts` and a `txt` file with the configuration and listing of the test cases, respectively.

```
PROJ-[Project]
├ proj
|   └ [Project].atp
|       ├──── [Project]-em.atrd
|       ├──── [Project]-cnt.atrd
|       └──── src
|               ├──── [Project]AbsAlgorithm.java
|               ├──── [Project]TestCase.java
|               └──── [Project]Input.java
|               └──── [Project]Output.java
├ algs
|   └ ALG-[Algorithm]
|       ├ [Algorithm].atal
|       └ src
|           └ [Algorithm]Algorithm.java
├ tests
|   └ [TestSet].atts, [TestSet].txt
├ results
├ queries
└ reports
```

Fig. 20 Structure of the `data_root/projects` folder.

3.3 Standalone application use

For a single user who wants to test his/her own solutions for a selected problem on a personal computer, a standalone version of ALGATOR was developed. This application is a bridge to ALGATOR's execution engine and facilitates all the phases of the algorithm's development and testing process.

To install ALGATOR as a standalone application a user should perform the following simple steps:

- Create three folders: the root folder for the ALGator files (we will refer to this folder as `<algator_root>`), a folder for the application (`<algator_root>/app`) and a folder for project's data (`<algator_root>/ data_root`).
- Download the `ALGator.jar` package from Ref. [12] and place it in the `<algator_root>/app` folder.
- Modify the two environment variables: set the value of the `ALGATOR_ROOT` to `<algaator _root>` and add `<algator_root>/app/ALGator.jar` to the `CLASSPATH`.

After the installation, ALGATOR can be launched from a console using the following commands:

- `java algator.Version`
  Checks the correctness of the installation and prints the version of ALGATOR.
- `java algator.Admin`
  Creates a new project, a new algorithm or a new test set.
- `java algator.Execute`
  Lists the status or executes the algorithms.
- `java algator.Analyse`
  Runs the GUI-based tool for analyzing the results of the execution.

By default, all the commands use the `$ALGATOR_ROOT/data_root` folder as a project-configuration folder. A user can change this setting by using the `-dr` switch.

When the system is successfully installed, a project has to be created. Here, ALGATOR can help only partially—the `java algator.Admin` command creates the folder structure (in the `data_root` folder) with the corresponding configuration and source files. To finish the project's creation and configuration a user should edit these files manually. Basically, this means that after running the commands

    java algator.Admin -cp ProjectName

(which creates the project named `ProjectName`) and

    java algator.Admin -ca ProjectName AlgName

(which creates the algorithm named `AlgName`) the user should edit

- the project configuration file `proj/ProjectName.atp`,
- the project source files
    - ○ `src/ProjectNameAbsAlgorithm.java`,
    - ○ `src/ProjectNameTestCase.java`,
    - ○ `src/ProjectNameTestSetIterator.java`,
- the algorithm configuration file `algs/ALG-AlgName/AlgName.atal`, and the algorithm source file `AlgNameAlgorithm.java` in the `algs/ALG-AlgName/src` folder.

In addition, for each test set that will be used, the user should run

> `java` ALGATOR.`Admin -ct ProjectName TestsetName`,

and edit the files `tests/TestsetName.atts` and `tests/TestsetName.txt`. For more details about the format and the content of these files see Section 2.

When the project, algorithm(s) and test set(s) are created and configured, the execution of the algorithms is triggered by the `algator.Execute` command, in which the user specifies which algorithm should be run and which test set should be used. Running `algator.Execute` without additional parameters (note that the name of the project is always mandatory) will force ALGATOR to run all the outdated tests (i.e., tests that have never been run and tests for which some configuration has changed).

An example of running the algorithm `JHoare` (Java version of the Hoare algorithm) on a test set with five test cases is presented in Fig. 21.

To analyze the results of the execution, ALGATOR offers a tool called `algator.Analyse` in which the user can create queries to get, filter, group and sort the data produced by the `algator.Execute` and to present the data as graphs. An example of using this tool is presented in Fig. 22.

Running the ALGATOR (`project=Sorting, algorithm=JHoare, testset=TestSet3`)

```
$ java algator.Execute Sorting -a JHoare -t TestSet3
ALGator Execute, version 0.85 (Oktober 2017)

[JHoare, TestSet3, em]: test  1 / 5  - DONE
[JHoare, TestSet3, em]: test  2 / 5  - DONE
[JHoare, TestSet3, em]: test  3 / 5  - DONE
[JHoare, TestSet3, em]: test  4 / 5  - DONE
[JHoare, TestSet3, em]: test  5 / 5  - DONE
```

The result produced – the content of the `JHoare-TestSet3.em` file

```
JHoare;TestSet3;Test-1;DONE;test0;RND;100;273;24;125;673;1256;OK
JHoare;TestSet3;Test-2;DONE;test1;RND;200;377;44;261;771;1616;OK
JHoare;TestSet3;Test-3;DONE;test2;RND;300;449;97;305;849;2051;OK
JHoare;TestSet3;Test-4;DONE;test3;RND;400;775;130;519;1075;3191;OK
JHoare;TestSet3;Test-5;DONE;test4;RND;500;1235;192;784;1235;4842;OK
```

Fig. 21 Example of running the algorithm named `JHoare` on a test set `TestSet3` containing five test cases and the result of the execution.

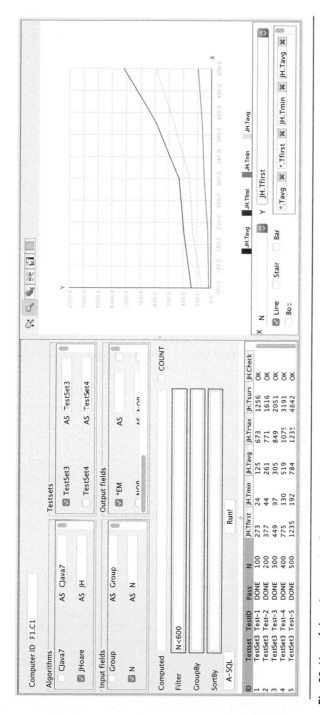

Fig. 22 Use of the algator.Analyse program.

The standalone installation of the ALGATOR system has several advantages. Primarily, the fact that all the files (the program and the project's data) are located on a single computer simplifies the maintenance of the system. On the other hand, using only one computer for all the tasks (configuration, execution and analyses) can cause several problems, for example, poor computer responsiveness during the execution of the tests and bad influences on the execution results if the computer is used simultaneously for any other job. Therefore, the standalone installation should be used only for small preliminary tests or on a computer that is used only for ALGATOR tests (which means that during the execution of the tests no other jobs of any kind are allowed). The second important drawback of the standalone installation is that the projects are closed to the public and that other researchers cannot browse or contribute results. To overcome the problems of the standalone installation, a client/server version of ALGATOR should be used.

3.4 ALGATOR as a client/server system

To enable the use of several computers in the ALGATOR system, each dedicated to its own job (i.e., data storage, web server, algorithm execution, etc.), the concept of client/server was implemented. In a client/server ALGATOR system there is one server computer and as many client computers as needed. The server computer runs the `algator.TaskServer` program and listens for the requests sent by the clients. The task of the server is to control and dispatch the data of the registered projects. For each project the server maintains a queue of tasks to be performed (i.e., which algorithms are to be run and which test sets are to be used; a task is given with a triple `(project, algorithm, testset)`). In addition, the server is capable of running the queries and forwarding their results. The following list contains some of the possible requests that might be sent to the `TaskServer`:

`ADDTASK project algorithm testset mtype`; Adds the task to the task queue.

`LIST`; Lists the statuses of the tasks in the queue.

`REMOVETASK taskid`; Removes a task from the task queue.

`GETNEXTTASK`; Returns the description of a task to be executed or null if the task is not available.

`GETQUERYRESULT project query`; Returns the result of a given query.

`PRINTLOG`; Returns the content of a log file.

A simple way to communicate with the server is by using the `algator.Request` program. If the program is run from the console, the arguments are sent to the server as a request and the server's response is written to the screen.

```
do forever
  task = ask_server(GETNEXTTASK)
  if (task != null) {
    algator.run(task)
  }
}
```

Fig. 23 The logic "asking for a task" performed by the `TaskClient`.

On each computer that will be used to execute algorithms (i.e., on each execution machine), the `algator.TaskClient` program should be run. This program constantly communicates with the task server, asking it for the task to be executed, as shown in the listing in Fig. 23. ALGATOR uses the "asking for a task" logic since the opposite logic (i.e., "searching for a computer to execute a task") causes too much network traffic and disturbs the execution machines during the execution of the algorithms (which would make the timing results unreliable). Using the "asking for a task" logic the `TaskClients` ask for a task only when the hosting machine is ready to run. If a task is available in the `TaskServer`'s queue, it will be executed immediately after the server's response.

The `TaskServer` is also used by the `WebServer` that hosts ALGATOR's web page, which offers the interfaces for both the task administration (adding, removing and showing the status of the tasks) and the presentation of the results (presenting the query results as tables and graphs). The `WebServer` and the `TaskServer` do not interfere with each other and could be installed on the same computer.

The installation of the client/server programs is simple: like for the standalone installation, on each computer the <`algator_root`> folder structure (with the `ALGator.jar` in the `app` folder and the `data_root` subfolder) has to be established. The only difference is that the `data_root` folder should be physically present only on one of the computers (usually on the server), everywhere else only a link (mount) to that folder should be made. The package `ALGator.jar` contains all the programs described in this section. The ALGATOR task server is established by running.

```
java algator.TaskServer &
```
while the client starts with
```
java algator.TaskClient.
```
Note that with the & character we indicate that `TaskServer` should be run as a server program (in the background), since this computer can also be used for other programs (like `WebServer`). On the other hand, to ensure quality measurements, the `TaskClient` should be run as the only program on the hosting computer.

3.5 Executing Java algorithms

To execute the algorithms and to accurately measure the indicators of quality and efficiency, certain conditions have to be satisfied. In order to compare the results of the different executions of the same algorithm, the system
must ensure an equivalent testing environment for all the executions. In the
time-measuring process, the system must eliminate all the unwanted side
effects, so that the measured time indicates (as much as possible) the real time
needed to solve a given task. In this section we present the mechanisms used
by ALGATOR to produce accurate, reliable and repeatable results.

In the algorithm-execution process two Java virtual machines are involved. The
first one (JVM1) is run at the beginning of the execution and it serves as a controller of
the process. It iterates the given test set, generates test cases and for each test case runs
a fresh copy of the Java virtual machine (JVM2) to execute the algorithm.

This ensures that each test is run in a clean environment where all the
possible influences of other executions are eliminated: at the beginning of
the execution of each test case, the memory is free (so that the garbage
collector will have to collect only the garbage produced by the testing algorithm) and the JIT compiler has no previous information for optimization.
In addition, the fact that JVM2 is a sub-process of JVM1 makes it possible to
limit the execution time. If JVM2 does not finish in the expected time,
JVM1 kills it and proceeds to the next test case, as depicted in Fig. 24.

To exchange the data, the two JVMs use a common communication folder,
which is created (and removed at the end of the execution) by JVM1 and given to
JVM2 as an input parameter. This folder contains the following three files:

(a) `alg.ser.test`

This file contains an instance of the algorithm in a serialized form; to
pass the required parameters of the execution to JVM2, the algorithm

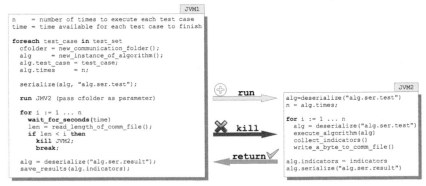

Fig. 24 Execution of the algorithm using two Java virtual machines.

encompasses data structures to store the test case and the input parameters (e.g., the number of times to execute the algorithm with a given test case). The file is produced by JVM1 (just before JVM2 is invoked) and it is read by JVM2 several times.

(b) `alg.ser.result`

In this file the serialized algorithm contains data structures to hold information about the indicators of the executions. The file is produced by JVM2 at the end of all the executions and read by JVM1 when it regains control.

(c) `comm.data`

This communication file is used by both JVMs to control the execution time. At the end of each execution of the algorithm, JVM2 writes a byte to this file to signal its progress. While waiting for JVM2 to finish all the executions, JVM1 repeatedly checks the length of this file; if the length remains unchanged for too long, JVM1 kills JVM2 to finish the blocked process.

The procedure described above is used in the case of measuring the em and the cnt indicators. The only difference is in the source file being used to instantiate the algorithm. Namely, in the case of the em measuring, the original algorithm's source file (containing possible counters as the Java commented lines) is used, while in the case of the cnt measuring, a new file is generated from the original algorithm's source file by replacing each commented line

    //@COUNT{name,value}

with the line

    Counters.addToCounter("name", value);

This newly created file is then compiled and used to instantiate the algorithm's objects. During the execution of the algorithm, the `addToCounter()` method increases the value of the counter in a local dictionary, from where the values are read and transformed to the indicators' values at the end of the execution.

3.6 External executor for the bytecode measurements

To measure the jvm indicators ALGATOR uses a Java virtual machine called vmep [13] that was developed especially for the needs of these measurements. The vmep is based on an open-source Java virtual machine jamvm [14] in which some instructions that enable Java bytecode use counting were added. The source code of the vmep implementation of jamvm (including the usage instruction) is available at [15].

jamvm is a program written in the C programming language that compiles into a Java virtual machine with support for OpenJDK. This machine was selected since it supports several important features, including the support of POSIX threads, the class loading, the garbage collector with mark/sweep and unloading, soft references, a hash table for the runtime constant pool, dynamic JNI loading from a standard library, etc. In addition, the structure of the jamvm sources enables the efficient integration of the bytecode counting extension. This extension was inserted into an existing jamvm C code where the Java interpreter inlines the machine instructions and gives a label to each of them. During the jamvm's execution of a Java program, this code is used just before each Java bytecode instruction is invoked, which enables the inserted code to increase an appropriate counter of the bytecode instruction use.

In the normal use of vmep we do not inspect the behavior of the whole Java program, because we are only interested in its part (e.g., in the behavior of a single method). Therefore, vmep enables the counting mechanism to be switched on and off during the execution of a program. To focus on a selected method's execution (like in the case where ALGATOR focuses on the method that executes the algorithm), the counting has to be switched on just before the call of this method and switched off immediately after the program flow is returned. The user of vmep should be aware that in this case the resulting bytecode statistics includes some additional operation codes that do not belong to the called method (e.g., one invokespecial that is used to call the method, one or more load instructions (like aload_0) that are used to load the reference of the current object or parameters to the stack and the like), but the number of these instructions is usually negligible in comparison with the number of the method's instructions. The communication between the Java program and the interpreter involves using the native methods that were incorporated into the vmep and introduced to Java's vmep library (which is packed into a classes.jar file) through the JNI. A typical use of vmep is presented with the code in Fig. 25.

Assuming that the vmep library is in a CLASSPATH any Java compiler can be used to compile the VmepTest.java. On the other hand, to execute the resulting VmepTest.class the vmep virtual machine has to be used.

The use of vmep is presented with the following example. Let the algorithm.run() method contain a simple code to add the first 100 numbers, as depicted in Fig. 26A. The resulting Java bytecode uses only 12 different bytecode instructions (see Fig. 26B).

```
1  import jamvm.vmep.InstructionMonitor;
2  import jamvm.vmep.Opcode;
3
4  public class VmepTest {
5    public static void main(String[] args) {
6        InstructionMonitor monitor =new InstructionMonitor();
7
8        monitor.start();   // start counting
9          algorithm.run(); // run inspected algorithm
10       monitor.stop();    // stop counting
11
12       // print the statistics for the run() method
13       int[] freq = monitor.getCounts();
14       for(int i = 0; i < freq.length; i++){
15           println(Opcode.getNameFor(i) + " : " + freq[i]);
16       }
17   }
18 }
```

```
1  import jamvm.vmep.InstructionMonitor;
2  import jamvm.vmep.Opcode;
3
4  public class VmepTest {
5    public static void main(String[] args) {
6        InstructionMonitor monitor =new InstructionMonitor();
7
8        monitor.start();   // start counting
9          algorithm.run(); // run inspected algorithm
10       monitor.stop();    // stop counting
11
12       // print the statistics for the run() method
13       int[] freq = monitor.getCounts();
14       for(int i = 0; i < freq.length; i++){
15           println(Opcode.getNameFor(i) + " : " + freq[i]);
16       }
17   }
18 }
```

Fig. 25 Typical use of the vmep's instruction counters.

A

```
1  void run() {
2    int result = 0;
3    for(int i = 1; i <= 100; i++) {
4      result += i;
5    }
6  }
```

Java code

B

```
void run();
 0: iconst_0
 1: istore_0
 2: iconst_1
 3: istore_1
 4: iload_1
 5: bipush      100
 7: if_icmpgt   20
10: iload_0
11: iload_1
12: iadd
13: istore_0
14: iinc        1, 1
17: goto        4
20: return
```

bytecode

Fig. 26 (A) Simple Java algorithm and (B) its bytecode representation.

```
iconst_0  : 1        iconst_1  : 1        iload_0    : 100
iload_1   : 201      istore_0  : 101      istore_1   : 1
iadd      : 100      iinc      : 100      if_icmpgt  : 101
goto      : 100      bipush    : 101      return     : 1
```

Fig. 27 Output of the VmepTest program.

The output of the VmpeTest program using the run() method from Fig. 26 is presented in Fig. 27. During the execution of the run() method, most of the instructions are used inside a loop (which spans from line 4 to line 17; lines 4–7 are used to check the loop condition, lines 10–13 represent the loop body, lines 14–17 are used to maintain the loop variable and return the flow to the next iteration). The body of the loop is repeated exactly 100 times, which reflects, for example, in the use of the iadd instruction—it was used exactly 100 times. Checking for the loop condition is performed 101 times (the last check fails and the flow is transferred to line 20); therefore, the instructions used in this block of code (like if_icmpgt) were used 101 times.

The program vmep is incorporated into the ALGATOR system so that users can simply measure the Java bytecode uses of their algorithms. When ALGATOR is used in the jvm mode, the algorithm to measure is run similarly to that described in Section 3.5 for the em and cnt measurements. The only difference is that in the jvm measurements the vmep virtual machine is used (instead of the system's default java) and that the invocation of the algorithm is enclosed with the monitor.start() and monitor.stop() commands. When the algorithm finishes, the results returned by the monitor.getCounts() method are transformed into ALGATOR's indicators' values (for each bytecode instruction an indicator is generated; the name of the indicator equals the instruction's mnemonic and its value represents the use of this instruction during the execution of the algorithm). The indicators are written in the corresponding output file and the user can employ them to analyze the algorithm's behavior.

3.7 Data-storage organization

The data associated with ALGATOR's projects is used by different applications: the executors (standalone application or the TaskClient) use the projects' configuration data to execute the algorithms and to generate the results; the analysis tools (like algator.Analyse) use the data to evaluate the results and to produce the reports; the WebPage uses the data to present it in different forms. All these actors should use and produce the data in a way that does not interfere and influence the process flow and efficiency of the other actors. In the case of the standalone installation all the data is installed on a hosting computer and there is no need for sharing. On the other hand, if ALGATOR is installed and used as a client/server application, there are many computers involved and they all need to access the same data of the registered projects. To minimize the network traffic and to prevent delays in accessing the data, the client/server version of ALGATOR uses the following logic. All the projects' data is stored in the data_root folder on a single computer (this computer might be the computer on which the TaskServer and the WebServer are installed or it might be a special computer dedicated to data storage). On all the other computers the data_root folder is mounted with the read–write access. This enables simultaneous access to the files, but it is quite slow and therefore only useful for small files. The speed of accessing the mounted drive could affect the measuring results; therefore, each client makes a copy of the data to a local folder before it executes the algorithm. To copy the data it uses the synchronizing mechanisms (like rsync program) so that only fresh (new or changed) data is transferred. The results of the execution are written directly to the data_root (not to a local copy, but to the mounted drive) so that they are immediately visible to all the parties involved. Using this mechanism everyone receives their data quickly and at the same time every piece of (public) data is synchronized and constantly available to everyone.

3.8 Query language description

The results of the executions are stored in text files as comma-separated values. Prior to the analysis these files have to be reorganized and the data have to be collected in structures that enable a fast search and comparison. This process has to be flexible since the organization of the data required for an ongoing analysis depends on the properties of the project. An important part of ALGATOR, the so-called query engine, which facilitates data manipulation and the preparation of the quality input for the analysis, is presented in this section.

To demonstrate ALGATOR's query-engine capabilities, let us present the following example. Suppose that we have a project P with three algorithms, A1, A2 and A3, and two test sets, T1 and T2. In each test set we have 100 test cases with 20 different input sizes (i.e., 5 test cases for each input size). For this project ALGATOR measures the following indicators: Tmin (minimum execution time of a test case), Tavg (average execution time of a test case) and Check (correctness of the result with the possible values OK and NOK). In addition, the result files also contain the value of the parameter N (the size of the input). After running ALGATOR on the project, the output folder will contain 6 files (one for each pair (algorithm, testset)), each of which will contain 100 lines (one line for each test case) with the following content.

```
algorithm:testset:testid:status:N:Tmin:Tavg:Check
```

where the algorithm and the testset are the corresponding names of the algorithm and the test set, testid is a unique identifier of the test case and the status is one of DONE, FAILED or KILLED. The values of N, Tmin, Tavg and Check correspond to the measured values of the test case's parameters and indicators. Some lines of the output files A1-T1.em and A1-T2.em are presented in Fig. 28.

While analyzing the algorithms the output files in their pure form are not useful since they contain only a limited amount of information. For a comprehensive view of the behavior of an algorithm (in comparison with other algorithms) we need to combine several output files in a structured common file—and this is where ALGATOR's query engine takes action. For example, in order to produce an array of minimal execution times for the test cases of the test set T2 for the algorithms A1 and A2, the following query should be used:

```
{"Query": {"Algorithms": ["A1","A2"],
    "TestSets": ["T2"],
    "Indicators": ["Tmin"]
}}
```

This query combines the results from files A1-T2.em and A2-T2.em and generates an output text file containing 100 lines with the following format:

```
resultid:testset:testid:A1.Tmin:A2.Tmin
```

A
```
A1:T1:Test0:Done:10:25:28:OK
A1:T1:Test1:Done:10:29:30:OK
A1:T1:Test2:Done:10:31:30:OK
A1:T1:Test3:Done:10:26:29:NOK
A1:T1:Test4:Done:10:20:27:OK
A1:T1:Test5:Done:20:51:56:OK
A1:T1:Test6:Done:20:58:60:OK
A1:T1:Test7:Done:20:61:62:OK
...
...
```

B
```
...
...
A1:T2:Test92:Done:190:59:57:OK
A1:T2:Test93:Done:190:49:55:NOK
A1:T2:Test94:Done:190:38:51:OK
A1:T2:Test95:Done:200:102:120:OK
A1:T2:Test96:Done:200:116:124:NOK
A1:T2:Test97:Done:200:102:114:OK
A1:T2:Test98:Done:200:108:122:OK
A1:T2:Test99:Done:200:89:113:OK
```

The first lines of the
output file A1-T1.em

The last lines of the
output file A1-T2.em

Fig. 28 Part of the content of the output files: (A) first lines of A1-T1.em and (B) last lines of A1.T2.em.

The last two columns in this file are the minimum execution times for each algorithm. Since the test sets in this example contain several test cases for each size of the input N, the output generated by this query will also contain several lines for each N, which is a little inconvenient, if we want to use the output, for example, to plot a graph. To overcome this problem, the ALGATOR query offers a GroupBy attribute, with which we can group several output lines into a single line. The grouping criteria is given in the following format:

```
group_field {; [field:]field_stat}
```

The group_field specifies the field by which the lines are going to be grouped: all the lines with the same value of the group_field field will be grouped into a single line. The value of the other fields will be calculated using the field_stat statistical formula (one of the MIN, MAX, AVG, SUM, FIRST, STDDEV); if no field_stat is given, the default (MIN for numbers and FIRST for strings) will be used. The field_stat function that is given without a field value replaces the default statistical formula for all the numerical values.

In the example above the setting "GroupBy": ["N; Tmin:MIN; Tavg:AVG; MAX"] will group lines with identical values of the field N and produce one line for each N; the value of the Tmin (Tavg) field in this line will contain the minimum Tmin (the average Tavg, resp.) value for all the grouped lines; all the other fields will be grouped using the MAX function.

Using this GroupBy setting in the example above the first line of the output would be: 5;T1;Test0;Done;10;20;28;OK; the value 20 in the Tmin field represents the minimum value of 25, 29, 31, 26, 20; the value 28 in the Tavg field represents an average of 28, 30, 30, 29, 27; the value 5 in the resulted field is the maximum value of the values 1, 2, 3, 4, 5.

Besides the GroupBy, ALGATOR's query language also uses the following attributes to manipulate the generated output:

- SortBy: the string value of this attribute lists the names (separated by the semicolon) of the fields to be used as the sorting keys to sort the query's output; to sort the output (using the given order of keys: first the output will be sorted by the first key, then by the second and so on) ALGATOR uses a stable sorting algorithm.
- Filter: this attribute is used to filter out the query's output lines; it contains several conditions like

```
field_name operator value
```

joined together with the logical operators & and |. The operator in each condition is one of <, >, >=, <=, ==.

The filter is used before the output is grouped (i.e., the lines are first filtered out and then grouped by the GroupBy criteria).

In our example the setting "Filter": "Tmin>25 & Tmin<100" will filter out all the results with Tmin <=25 and Tmin>=100. The filter could also be used, for example, to filter out all the failed executions, by setting "Filter": "Check==OK".

— Count: if the value of this boolean attribute is true, ALGator will count the number of lines generated by the query; when using this attribute, the output will contain one line with the number of lines generated for each algorithm included in the query. The Count attribute is usually combined with the Filter attribute to count the number of lines satisfying the filter criteria.

For example, if the filter is set to be "Check==OK" and Count to be true, the query will produce the following output:

```
#;A1.COUNT;A2.COUNT
   1;20;42
```

(i.e., 20 tests of A1 and 42 tests of A2 returned the correct result).

The Count attribute is especially interesting if used with different filters—for each filter the counting produces one line of the output. By combining these lines we get an array of values for each algorithm. To enable efficient counting with different filters, ALGATOR uses a notion to define a parameterized value of the Filter. In this notion the range of values for the parameter used in the filter is defined.

For example, the setting "Filter":"Tmin<$1 @(10,55,5)" represents 10 different values for the Filter: "Tmin<10," "Tmin<15," ..., "Tmin<55." Using this filter in combination with the Count=true will produce 10 lines of output, each corresponding to one of the filters.

An example of the use of this feature is in the Graph-Isomorphism Problem [16], where the quality of the algorithm is measured by the number of successfully solved instances in a given time frame. By setting "Count":"true" and "Filter":"Check==OK& N<$1 @(10,100,10)" the query returns the information about how many successfully solved instances there were when N (the size of the input graph) was 10, 20, 30, ..., 100.

3.9 Scalability of the ALGATOR system

The results of the algorithm executions are only comparable if the executions were performed on the same or on comparable computers. To increase the efficiency of the system, ALGATOR allows the use of more than one execution computer. The computers of the system are classified into so-called computer families, where all the computers of one family use the same essential hardware and their performance is thus comparable.

```
FamilyID     : "F1",
Platform     : "Ubuntu 16.04",
Hardware     : "2,66GHz Intel Core 2Duo, 32GB DDR3",
SystemType   : "64",
Computers    : [
    { ComputerID   : "C1",
      IP           : "193.2.168.120",
      Capabilities : ["EM", "CNT"]
    },
    { ComputerID   : "C2",
      IP           : "193.2.168.110",
      Capabilities : ["EM", "CNT", "JVM"]
    }
]
```

Fig. 29 Configuration of a computer family.

The configuration of the ALGATOR system includes information about the computer families used and a list of computers belonging to each of them. An example of the configuration for one family containing two computers is given in Fig. 29. In this family both computers are able to execute the EM and CNT measurements, while only the second computer can perform the JVM measurement.

The administrator of the project decides which family is suitable to execute the project's algorithms and sets the EMExecFamily, CNTExecFamily and JVMExecFamily attributes in the project's configuration file. This setting is used by the TaskServer, which only assigns the tasks to suitable computers.

The results of the execution are placed in an appropriate folder, named after the computer family that performed the execution. To enable additional comparisons of the algorithms' performance (for example, to compare the impact of the type (i.e., 32 or 64 bit) of computer system) the computer family to run an algorithm might be forced to override the default settings. Furthermore, the ALGATOR query engine makes it possible to specify the computer family whose results will be used in the query. Using these mechanisms ALGATOR enables both a comparison of the algorithms run on equally configured systems and cross-platform comparisons.

To extend the ALGATOR system with a new execution computer the system administrator decides to which family the computer belongs (if none of the existing families is appropriate, a new family must be defined), changes the system configuration files and installs TaskClient software on the new computer (see Section 3.4). This makes the ALGATOR system very flexible and prepared to be used for a large number of different projects and algorithms.

4. Using ALGₐₜₒʀ in real applications

In this section we present some of possible uses of the ALGATOR system. For each measurement type (em, cnt and jvm) we introduce a use case with a relatively simple problem domain. The purpose of this section is to give some clues about how the results of ALGATOR's measurements could be used in practice.

4.1 Using EM indicators in the sorting problem

Measuring the time is one of the most frequently used features OF ALGATOR. To illustrate this powerful and yet simple feature, we will use the well-known Sorting Problem. The aim of the sorting algorithm is to change the order of the elements in the input array A so that for each $0 \leq i < j < A.\ length$ the element $A[i]$ will be "less or equal to" the element $A[j]$. There are several approaches to solving this problem, but in this section we will focus on the comparison-based algorithms (note that in certain circumstances the algorithms that perform an element classification instead of a cross-element comparison (like Radix sort, for example) could run faster). It is not hard to show that the lower bound for any comparison-based sorting algorithm is $O(n\log n)$ and there is a plethora of algorithms achieving this bound, at least on average. In this section we will compare three (two fast and one slow) sorting algorithms. Since the theoretical results for all three algorithms are well known, we present this use case only as a proof of concept.

For the first algorithm to be used in this comparison we chose a Bubblesort algorithm, which compares all the consecutive elements and changes their position if they appear in the wrong order. To sort the array, this procedure has to be repeated n times (where n is the number of elements in the array). This algorithm is trivial to implement, but it has a quadratic time complexity. The second algorithm used in the comparison is the classic Hoare's one-pivot quick-sort algorithm [17], which has an average time complexity of $O(n\log n)$. For the third algorithm we chose the algorithm that is used in the standard Oracle's Java 7 implementation as a sorting algorithm called by the Arrays. sort(int[] a) method. This algorithm uses several different approaches to optimize the time consumed for sorting small arrays and it uses a variation of the Quicksort algorithm (i.e., the Yaroslavskiy's two-pivot Quicksort algorithm [18]) for large arrays. Since the implementation of the sorting algorithm used in Java 7 was optimized for many different cases [19], it is reasonable to expect that it will be faster than Hoare's basic Quicksort algorithm. The tests performed by ALGATOR confirm this hypothesis and give the answer to the question of "how much faster" the optimized Java 7 algorithm is for the selected test cases.

The implementation of the Sorting Problem in the ALGATOR system is quite simple. The test cases are arrays of integers and their implementation in Java (as in many other languages, including the C language) is straightforward (using the one-dimensional primitive array). To enable repeatable tests, all the test cases are encoded in the input files using two different formats; the small test cases (i.e., the test cases with <10 million elements) are defined by a file (containing several million random numbers) and an offset (the starting number in the file); the larger tests are defined by a pseudo-random number generator and by a seed used to start a sequence. Using these two encodings, every test case can be quickly reproduced when needed.

We named the method to perform the sorting the execute() method; this method requires only one parameter—the array to be sorted—and it sorts the given array in-place. Every Java class extending the SortingAbsAlgorithm class (and overriding the execute(int [] data) method) is called an algorithm for solving the Sorting Problem. In our test we implemented three sorting classes: the BubbleSortAlgorithm, the HoareAlgorithm and the Java7Algorithm. The first two are presented in Fig. 30. The implementation of the third one is trivial, since its execute() method contains only the call to the Java's sort() method like this: public void execute(int[] data) {Arrays. sort(data);}.

The test of the correctness of the algorithms' results in this project is trivial—we only need to scan the output array and compare every two consecutive elements. The array is sorted if the first element in the comparison is never bigger than the second one.

To compare these algorithms we created three test sets. The first one contains small test cases (with sizes of the input array from 10,000 to 100,000) and it was used to show the huge difference between the quadratic and linearithmic time complexities. As is clear from Fig. 31, the BubbleSort algorithm consumes a huge amount of time, even for these relatively small arrays (around 13s to sort an array of 100,000 numbers). As such, the BubbleSort algorithm is practically useless. It is also clear from Fig. 31 that the Hoare and Java7 algorithms sort the arrays of small sizes so quickly that the time cannot be adequately presented on this scale (the blue and red lines do not even stray from the abscissa).

To find the difference between the Hoare and Java7 algorithms we used another test set with arrays having from 10^6 to 10^7 elements. This test set is suitable for comparing our linearithmic algorithms since the execution time

```java
public class BubbleSortAlgorithm extends SortingAbsAlgorithm {

    @Override
    public void execute(int[] data) {
        for (int i = 0; i < data.length; i++) {
            for (int j = 0; j < data.length - 1; j++) {
                if (data[j] > data[j + 1]) {
                    swap(data,i,j);
                }
            }
        }
    }
}
```

```java
public class HoareAlgorithm extends SortingAbsAlgorithm {

    void hoare(int[] arr, int left, int right) {
        int i = left, j = right, tmp;
        int pivot = arr[(left + right) / 2];

        while (i <= j) {
            while (arr[i] < pivot) i++;
            while (arr[j] > pivot) j--;

            if (i <= j) {
                swap(arr,i,j);
                i++; j--;
            }
        };
        if (left < j) hoare(arr, left, j);
        if (i < right) hoare(arr, i, right);
    }

    @Override
    public void execute(int[] data) {
        hoare(data, 0, data.length - 1);
    }
}
```

Fig. 30 Implementation of the Hoare and the BubbleSort algorithms.

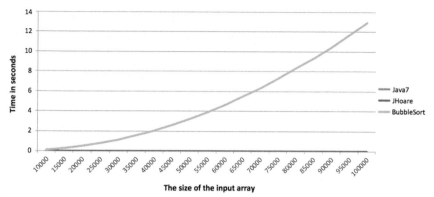

Fig. 31 Execution time for the three sorting algorithms.

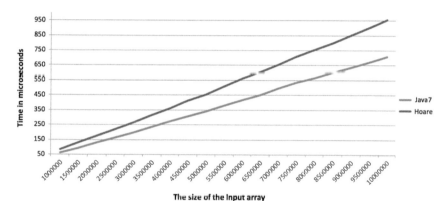

Fig. 32 Execution time for the Hoare and Java7 algorithms.

when running its test cases varies from approximately 0.1 to 1 s in the worst case (note that the `BubbleSort` algorithm would need around 100 days to sort the array of 10^7 elements).

By executing the `Hoare` and `Java7` algorithms on this test set we confirmed the hypothesis that the `Java7` algorithm is much faster than the `Hoare` algorithm (see Fig. 32). The average ratio between the execution time for the test cases of this test set was 1.37 in favor of `Java7`. It is also not surprising that this ratio is decreasing with the increasing size of the input array, since the main optimization of the `Java7` algorithm (compared to the `Hoare`) is in the sorting of small arrays (it stops the recursion and sorts the remaining array iteratively). With increasing size of the input array the effect of this optimization decreases. In some of our tests this ratio was even bigger (it increased up to 3.2 in the case of sorting the arrays of a million (almost) equal elements), but it never got below 1.3 (even for the test cases with 10^8 elements). Therefore, we can conclude that the Java7 algorithm is much better than the (theoretically not so bad) Hoare algorithm.

By observing the results of the execution we noticed that the behavior of the algorithms might depend on the number of different elements in the input array. Therefore, we created a test set containing test cases of equal sizes (i.e., all the arrays contained 10^6 elements), differing only in the so-called range, i.e., the size of the random number pool. The test set contained test cases with ranges from 2^3 (i.e., test cases could possibly contain only eight different numbers) to 2^{30} (test cases could contain the whole range of random numbers). Running ALGATOR on this test set yielded very interesting results, for which we still cannot find suitable explanation. But the fact is that both algorithms, Hoare and Java7, perform much better if the range of numbers in the input array is smaller (see Fig. 33).

To sort an array of a million elements with only eight different values the Hoare algorithm needs 28.6 µs and to sort a million different elements it needs 84.8 µs. The ratio between the best and the worst performance is 2.96, which means that the diversity of the elements in the input array slows down the Hoare algorithm by almost three times. For the Java7 algorithm this slow-down ratio is even greater—the ratio between the best (9.3 µs for range 8) and the worst performance (61.9 µs for range 2^{30}) is 6.6.

It is not hard to explain the change in the shape of both graphs in Fig. 33 approximately at the range 2^{18} (from linear growth to an almost constant value), since this change is because all the input arrays contain 1 million elements and therefore they cannot contain more than 1 million different

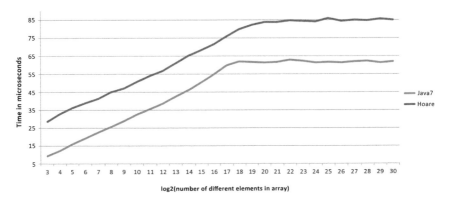

Fig. 33 Time complexities of the Hoare and Java7 algorithms on test cases with different ranges (from 2^3 to 2^{30}).

elements. Regarding the diversity of the elements, all the input arrays in this test set with ranges from 2^{19} to 2^{30} are similar, which implies a similar execution time. We confirmed this observation with another test set containing arrays of 10^7 elements. For this test set the shape of the graphs changed slightly later, at the range of approximately 2^{21}—which was not a surprise, for a 10-times larger array we need 10 times more elements for an equal diversity.

Even though these results are interesting and consistent over the several measurements that we performed, we still do not have a reasonable explanation as to why the number of different elements influences the behavior of both algorithms so greatly. We post this as an open question left for further research.

4.2 Using CNT indicators in the DTF problem

In this section we will show how to use ALGATOR's counters mechanism to find a relation between the measured counters and the behavior of the algorithm. In particular, we will focus on the selected implementations of the discrete Fourier-transform (DFT) algorithms and search for the relation between the number of recursive calls and the execution time. Using the measured counters, it is not hard to find the relation between the size of the input (n) and the number of recursive calls and, consequently, between n and the execution time.

DFT is a function, DFT: $\mathbb{C}^n \to \mathbb{C}^n$, which transforms the complex vector $X = (x_0, x_1, \ldots, x_{n-1})$ to the complex vector $\overline{X} = (\overline{x}_0, \overline{x}_1, \ldots, \overline{x}_{n-1})$, where

$$\overline{x}_k = \sum_{r=0}^{n-1} x_k \omega^{kr}, \text{ for every } k = 0, 1, \ldots, n-1.$$

Here, ω denotes the primitive n-th root of unity (that is, $\omega^n = 1$ and $\omega^r \neq 1$ for all $r = 1, 2, \ldots, n-1$). Even though this definition seems very simple (and, to tell the truth, is kind of boring), the DFT is a very important and powerful function. As a consequence of the interesting properties owned by the primitive n-th root of unity, the DFT is able to map numbers and polynomials from a space where their multiplication is very slow (i.e., it can be performed in quadratic time) to a space where they can be multiplied much faster (i.e., in linear time). To use this property of the DFT for fast multiplication, the calculation of the DFT must also be fast enough. In other words, if the time complexity for calculating the DFT were to be quadratic, then the advantage gained with fast multiplication would be canceled by the cost of the DFT transformation. The DFT can be calculated by definition (see formula above), but this calculation takes $O(n^2)$ for vectors of size n, which is unfortunately too slow to be useful. However, there are several other ways to calculate the DFT more quickly, for example, a well-known Fast Fourier Transform (FFT) algorithm takes $O(n \log n)$ time to calculate the DFT. Using the FFT, the speed-up factor of multiplication compared with the usual multiplication is $O\left(\frac{n}{\log n}\right)$.

The FFT algorithm (also known as the Cooley-Tukey algorithm [20–22]) divides the main problem into several sub-problems, which can be solved independently (in the same way), and then generates the solution to the main problem by combining the solutions of the sub-problems. This algorithm can be implemented recursively (as a typical divide-and-conquer problem) or iteratively. Since the iterative solution is faster, it is commonly used in FFT implementations. In our tests, however, we use the recursive implementation in order to study the relations between the number of recursive calls and the practical time complexity of the algorithm. The algorithms presented in the following are not the fastest solution for the DFT Problem, but they serve well as an example of recursive algorithms to be used and analyzed in the ALGATOR system.

There are several variations of the FFT algorithm differing in the number of sub-problems (called the *basis of the algorithm*) generated in the first stage of the algorithm. In this section we focus on the FFT algorithms with basis 2, 3, and 4. The algorithm with the basis 2 (4) divides the given problem into 2 sub-problems of sizes $n/2$ (4 sub-problems of sizes $n/4$, resp.), while the algorithm with basis 3 divides it into 2 sub-problems with sizes $n/4$ and one with size $n/2$, where n is the size of the original problem. Besides the difference in the number and the sizes of the sub-problems (which reflects in the number of recursive calls), the different-based FFT algorithms also differ in the way they combine the solution of the sub-problems and consequently in the number of complex multiplications, which makes them more or less efficient.

The three versions of the FFT algorithms were implemented in the ALGATOR system by Zorman [23]. In the implementation of these algorithms (Java methods) ALGATOR's notions for the two counters were added: to count the number of recursive calls the `//@COUNT{CALL, 1}` command was added as the first line of the recursive method and to count the number of complex multiplications the `//@COUNT{MULTIPLY, 1}` command was added before each call to the complex multiplication method `multiply()`. In the following we will refer to the three algorithms as FFT2, FFT3 and FFT4 (the number in the name denotes the basis used in the algorithm). For each algorithm FFTi ($i=2, 3, 4$) the measured counters will be denoted by $C_i(n)$ (for calls) and $M_i(n)$ (for multiplications). The algorithms were run on a test set containing vectors of random complex numbers, each vector representing a particular test case. For FFT2 and FFT3 the dimensions of these vectors were 2^k ($k \in \{1, 2, ..., 19\}$), and 4^k for FFT4 ($k \in \{1, 2, ..., 9\}$).

In the implementations of FFT2 and FFT4 the recursion continues until the dimension of the problem is 1. In this trivial case the algorithms return the input value (i.e., if $dim(x) = 1$ then FFTi$(x) = x$ for $i=2, 4$). This is possible since the dimensions of the input vectors are always powers of 2 and 4, respectively. On the other hand, for FFT3 the recursion stops when the dimension of the input equals 1 or 2. In the trivial case the algorithm returns the identity, while in the case of $n=2$ the FFT2 algorithm is used to stop the recursion. The recurrence tree for FFT3 in the case of $n=2$ and $n=4$ is depicted in Fig. 34.

Due to the fact that the three algorithms are totally deterministic, with the trace of their execution being dependent only on the dimension of the input vectors (and not on the values of the vector's components), the number of recursive calls used to calculate the DFT can be found in two ways: using ALGATOR's counters mechanism and by solving the appropriate recurrence equations. This gives us a good opportunity to prove the correctness of ALGATOR's counters mechanism—if the results obtained using the above-mentioned methods are the same, we can reasonably conclude that ALGATOR counts correctly. This result is very important and can be used in cases where the theoretical (analytical) approach is not possible.

In FFT2 each call to the recursive function divides into two calls to the same function with the problem size $n/2$. Therefore, the number of all the calls to the recursive function FFT2 equals $C_2(n) = 2C_2(n/2) + 1$ with

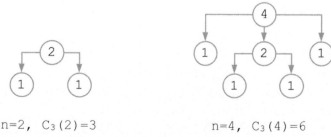

$$n=2, \quad C_3(2) = 3 \qquad\qquad n=4, \quad C_3(4) = 6$$

Fig. 34 Recurrence tree for the FFT3 in the case of $n=2$ and $n=4$.

the initial condition $C_2(1) = 1$. This recurrence resolves into $C_2(n) = 2n - 1$. Similarly, for FFT4 we have $C_4(n) = 4C_4(n/4) + 1$ with $C_4(1) = 1$, which resolves into $C_4(n) = \frac{4n-1}{3}$.

For FFT3 we have a slightly more complicated recurrence relation with two initial conditions: $C_3(n) = 2C_3(n/4) + C_3(n/2) + 1$, $C_3(2) = 3$ and $C_3(4) = 6$ (see Fig. 34). Solving this recurrence (using the Acra-Bazi theorem [24], for example) gives $C_3(n) = \frac{5}{3}n - \frac{(-1)^k}{6} - 1/2$, where $k = \log_2 n$.

To summarize, the number of all the calls to the recursive functions equals

$$
C_i(n) = \begin{cases}
2n - 1 & \approx \frac{6}{3}n; \ i = 2 \\
\frac{5}{3}n - \frac{(-1)^k}{6} - \frac{1}{2} & \approx \frac{5}{3}n; \ i = 3 \\
\frac{4n - 1}{3} & \approx \frac{4}{3}n; \ i = 4
\end{cases}
\tag{1}
$$

To verify these results we run ALGATOR, which returned the counters as presented in the table in Fig. 35 and in the graphs in Fig. 36. It is easy to check that the measured counters for the number of recursive calls C_i are identical to those calculated by (1).

N	FFT2		FFT3		FFT4	
	C2	M2	C3	M3	C4	M4
2	3	1	3	1	/	/
4	7	4	6	3	5	3
8	15	12	13	9	/	/
16	31	32	26	23	21	24
32	63	80	53	57	/	/
64	127	192	106	135	85	144
128	255	448	213	313	/	/
256	511	1024	426	711	341	768
512	1023	2304	853	1593	/	/
1024	2047	5120	1706	3527	1365	3840
2048	4095	11264	3413	7737	/	/
4096	8191	24576	6826	16839	5461	18432
8192	16383	53248	13653	36409	/	/
16384	32767	114688	27306	78279	21845	86016
32768	65535	245760	54613	167481	/	/
65536	131071	524288	109226	356807	87381	393216
131072	262143	1114112	218453	757305	/	/
262144	524287	2359296	436906	1601991	349525	1769472

Fig. 35 Measured counters for the three FFT algorithms.

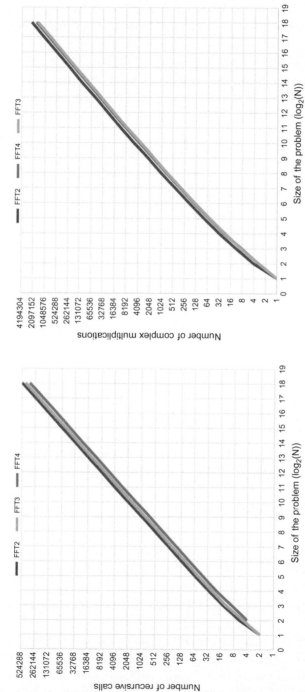

Fig. 36 Number of recursive calls (C_i) and number of complex multiplications (M_i) performed by the three FFT algorithms.

The other counter measured by ALGATOR is the one that counts the number of complex multiplications made by the algorithm. The values of this counter are also presented in the table in Fig. 35.

By observing the measured results a simple relation between both counters can be derived. First calculations show that the quotient between the C_i and M_i (for $i=2$, 3, 4) is "almost" constant and it is relatively easy to figure out that the "almost" factor in this case is the logarithm, namely

$$M_i = \log(n) \; C_i^* \begin{cases} \dfrac{1}{5}; & i=2 \\[2mm] \dfrac{1}{4}; & i=3 \\[2mm] \dfrac{9}{32}; & i=4 \end{cases} \tag{2}$$

This result represents an asymptotic behavior of the counters' values (the larger is n, the smaller is the error), but even for small instances the error is relatively small. Combining (1) and (2) we get

$$M_i = \frac{1}{24} n \log(n)^* \begin{cases} 12 & i=2 \\ 8 & i=3 \\ 9 & i=4 \end{cases} \tag{3}$$

Thus we can present the number of the recursive call and the number of the complex multiplication as follows:

	$C_i(n)$	$M_i\ (n)$
FFT2	$6/3\,n$	$12/24\,n\ \log(n)$
FFT3	$5/3\,n$	$8/24\,n\ \log(n)$
FFT4	$4/3\,n$	$9/24\,n\ \log(n)$

The algorithm FFT2 performs the most recursive calls and the most complex multiplications; therefore, it is not surprising that this algorithm is the slowest one. This result can also be seen in Fig. 37, where the measured time complexities of the three algorithms are presented. In this figure we also see that FFT3 and FFT4 are changing the leading position (from this graph it is hard to say which of them is faster). This is due to the fact that one of them performs more recursive calls and the other more complex multiplications than the other. Obviously, the influence of the benefit nullifies the influence of the drawback of each algorithm, with a certain (limited) error, of course.

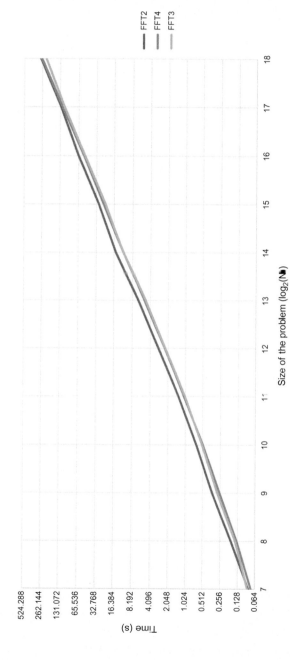

Fig. 37 Time complexities of the FFT2, FFT3 and FFT4 algorithms.

To make the knowledge of the counters as useful as possible, we tried to find the relation between the number of recursive calls and the time complexities of the algorithms. Since it is known that the time complexity of the FFTi algorithm (i =2, 3, 4) is linearithmic to n, we fitted the time $T_i(n)$ (i = 2, 3, 4) with

$$T_i(n) = a_i + b_i * C_i(n) \log(C_i(n)),$$

where a_i and b_i are constants and $C_i(n)$ is the number of recursive calls per-formed. We used the first half of the measurements (the second half was used after fitting to check the efficiency of the method) and the method of least squares to calculate the constants a_i and b_i. The result of the fitting for FFT3 is presented in Fig. 38 (for the other algorithms the results are very similar). From this graph we can see that the fitting was very successful. An average relative error of the fitting curve for this case was <4%.

In the example in this section we showed how ALGATOR's ability to count the user-defined counters can help to analyze the algorithm and to predict its behavior. Using the counters we fitted the time complexity of the FFTi algorithms (i=2, 3, 4) relatively accurately. Since we also know the relation between the size of the input (i.e., the dimension of the input vector) and the number of recursive calls (in this case we derive this relation analytically, but in other, more complicated cases, the relation (or at least its approximation) could be derived from the measured data), we can predict the execution time using only the size of the input. This result is important, especially for the cases where the execution time is long and the approxima-tion of the execution time can help to decide if it is worth waiting for an accurate result or another method should be used.

4.3 Using JVM indicators in the matrix-multiplication problem

In this section we will introduce the standard algorithm for solving the prob-lem of matrix multiplication and some of its variations. Using these algo-rithms and the ALGATOR's ability to count the Java bytecode instructions used during the algorithm's execution we will present results showing the relation between the used instructions and the execution time of the algorithm. We will also present the estimation for the execution time of one simple bytecode instruction and for the imul and iadd instructions. Using these estimations we will give the predicted overall execution time of the algorithm and the error of the prediction.

To explore the jvm measuring capabilities of ALGATOR we chose a simple Matrix-Multiplication Problem: given two square matrices A and B, each

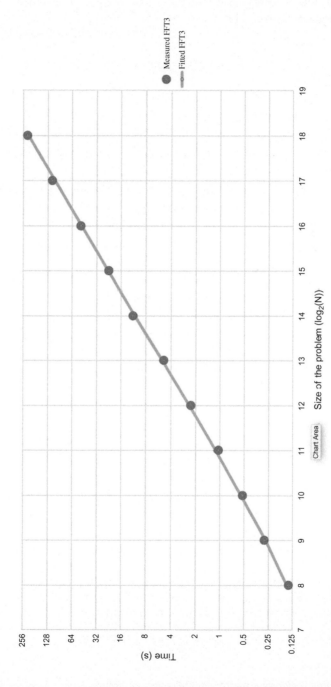

Fig. 38 Measured and fitted time complexity of `FFT3`.

containing n^2 elements (a_{ij} and b_{ij} for $i, j = 0, 1, \ldots, n-1$), calculate the elements of a square matrix C by

$$c_{ij} = \sum_{k=0}^{n-1} a_{ik} b_{kj}.$$

Since the number of operations in this formula is cubic to n, we can reasonably expect that the time complexity of any algorithm implementing this formula would have the time complexity $\Theta(n^3)$. The simple implementation of this formula is presented in the listings in Fig. 39.

We named this implementation the MUL algorithm since its main (and the most resource-consuming) operation is the multiplication. Using ALGATOR's time-complexity indicators we measured the time needed to execute this algorithm on a set of test cases with the dimensions n ranging from 200 to 500. We run all the tests described in this section on a personal computer with an Intel(R) Core(TM) i7-6700 CPU running at 3.40 GHz with 32 Gb of memory. The execution of the matrix multiplication for smaller inputs ($n=200$) was made in 6000 μs and for larger matrices ($n=500$) in 140,000 μs. To eliminate the impact of the real environment we executed all the tests (i.e., we calculated each product) 500 times and we took the minimum time of all the executions (obviously, this is the time in which the execution can be performed if the environmental influences are as small as possible). The time of the execution for the Matrix-Multiplication Problem is depicted in Fig. 40 with a blue line.

In the same graph a simple prediction for the execution time is also depicted. It was calculated using a simple method called Calc1. In this method we calculated a multiplication factor $c = avg(time_i/n^3)$. The red dots in Fig. 40 represent the graph of a function cn^3. Obviously, the red dots are of the same shape as the blue line (which is because our algorithm has an $\Theta(n^3)$ time complexity) but it is not very accurate. An average error (i.e., the difference between the measured (blue) and the calculated (red) time

```
void MUL(int[][] A, int[][] B, int[][] C) {
  for (int i = 0; i < A.length; i++)
    for (int j = 0; j < A.length; j++)
      for (int k = 0; k < A.length; k++) {
        C[i][j] += A[i][k] * B[k][j];
      }
}
```

Fig. 39 Implementation of a matrix-multiplication algorithm.

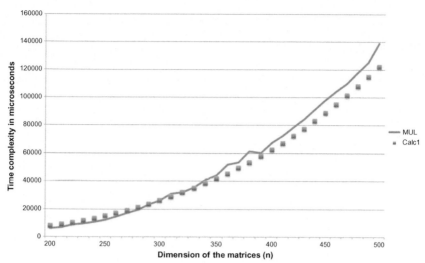

Fig. 40 Time complexity of the MUL algorithm (blue line) and the performance prediction calculated by a simple Calc1 method (red dots).

divided by the measured time) is 11.25%. This error is slightly smaller (i.e., 7.1%) if we take only the larger dimensions of the input matrices (from 300 to 500), but it is still relatively large. Therefore, the method Calc1 cannot be considered to be a very successful method.

In order to find a better performance prediction for this algorithm, we used ALGATOR's capability for measuring uses of the Java bytecode instructions (the bytecode for the algorithm MUL is listed in Fig. 41).

The results show that only 16 (out of 202) Java bytecode instructions are used during the execution of this algorithm: 10 instructions for the stack manipulation (ICONST_0, ILOAD, ALOAD_1, ALOAD_2, ALOAD_3, IALOAD, AALOAD, ISTORE, IASTORE, DUP2), 2 instructions to control the flow of the program (IF_ICMPGE, GOTO), the ARRAYLENGTH instruction used to determine the size of an array, and three arithmetic instructions (IADD, IMUL, IINC).

The use frequencies of these instructions for the matrices of sizes from 10 to 50 are presented in the table in Fig. 42.

It is clear from the data in the table that for most of the instructions their use is of the order $\Theta(n^3)$. The only exceptions are the instructions ICONST_0 and ISTORE with the order $\Theta(n^2)$. From the data presented in Fig. 42 we calculated the overall number of the instructions $INST(n)$ used in the MUL algorithm:

$$INST(n) = 25n^3 + 12n^2 + 12n + 6.$$

```
void MUL(int[][], int[][], int[][]);
 0: iconst_0          |   36: dup2
 1: istore 4          |   37: iaload
 3: iload 4           |   38: aload_1
 5: aload_1           |   39: iload 4
 6: arraylength       |   41: aaload
 7: if_icmpge 73      |   42: iload 6
10: iconst_0          |   44: iaload
11: istore 5          |   45: aload_2
13: iload 5           |   46: iload 6
15: aload_1           |   48: aaload
16: arraylength       |   49: iload 5
17: if_icmpge 67      |   51: iaload
20: iconst_0          |   52: imul
21: istore 6          |   53: iadd
23: iload 6           |   54: iastore
25: aload_1           |   55: iinc 6, 1
26: arraylength       |   58: goto 23
27: if_icmpge 61      |   61: iinc 5, 1
30: aload_3           |   64: goto 13
31: iload 4           |   67: iinc 4, 1
33: aaload            |   70: goto 3
34: iload 5           |   73: return
```

Fig. 41 Java bytecode for the MUL algorithm.

n	ICONST_0	ILOAD	ALOAD_1	ALOAD_2	ALOAD_3	IALOAD	AALOAD	ISTORE
10	111	7221	2221	1000	1000	3000	3000	111
20	421	56841	16841	8000	8000	24000	24000	421
30	931	190861	55861	27000	27000	81000	81000	931
40	1641	451281	131281	64000	64000	192000	192000	1641
50	2551	880101	255101	125000	125000	375000	375000	2551

n	IASTORE	DUP2	IADD	IMUL	IINC	IF_ICMPGE	GOTO	ARRAYLENGTH
10	1000	1000	1000	1000	1110	1221	1110	1221
20	8000	8000	8000	8000	8420	8841	8420	8841
30	27000	27000	27000	27000	27930	28861	27930	28861
40	64000	64000	64000	64000	65640	67281	65640	67281
50	125000	125000	125000	125000	127550	130101	127550	130101

Fig. 42 Java bytecode instructions use in the matrix-multiplication algorithm.

This means that in the case of $n=500$, for example, the JVM performs $25 \times 500^3 + 12 \times 500^2 + 12 \times 500 + 6 = 3{,}128{,}006{,}006$ bytecode instructions to execute the MUL algorithm. Since this execution requires approximately $140{,}000$ μs, the average time to execute one Java bytecode instruction is 0.044 ns.

Analyzing the results presented in the table in Fig. 42 (extended with measurements for $n=60, \ldots, 500$) a natural question arises: can we calculate an average time (over all the measurements) used to execute one bytecode instruction and use this average to predict the behavior (i.e., the time consumption) of the MUL algorithm for a given n? To find an answer to this question, we propose the following method Calc2: calculate the average time I_n

used for one bytecode instruction while performing MUL on the matrix of size n (e.g., $I_{500} = 0.044$) and calculate I as an average of I_n. Then use I to estimate the execution time of MUL by $T(n) = I*INST(n)$ Using this method we calculated $I = 0.039$ ns (note that we used only measurements for $n = 300, \ldots, 500$ since we assume that the measured times are much more accurate for larger inputs). Surprisingly, the Calc2 method gives very similar results to the method Calc1: the average difference between those two methods is 0.03% for $n = 300, \ldots, 500$. In other words, calculating the uniform average time per bytecode instruction yields another useless method for estimating the time consumption.

The main reason for bad results is that some bytecode instructions are much more expensive than the others. For example, we can reasonably assume that the IMUL instruction executes for much longer than the ILOAD instruction (the first instruction multiplies two integers, while the second one loads an integer onto a stack). The question is, how many different types of instructions (instructions of the same type take approximately the same time to execute) are included in the MUL algorithm. To answer this question we implemented two algorithms, both of them very similar to MUL. The first one, the ADD algorithm, is an exact copy of MUL, with the only difference being in line 5, where it uses an addition instead of a multiplication (C[i][j] += A[i][k] + B[k][j];). The execution of this algorithm results in the uses of exactly the same Java bytecode instructions. The only difference is that instead of both IMUL and IADD, only the IADD instruction is used. As a consequence, in the MUL we have n^3 IADDs and n^3 IMULs, while in the ADD algorithm we have $2 \times n^3$ IADDs and zero IMULs. The number of all the other instructions is equal in both algorithms. In the second algorithm (called the SET algorithm), we deleted the line 5 of MUL and replaced it with 4 lines, as shown in the listings in Fig. 43. The resulting SET algorithm

```
void SET(int[][] A, int[][] B, int[][] C) {
  int x = 0, y;
  for (int i = 0; i < A.length; i++)
    for (int j = 0; j < A.length; j++)
      for (int k = 0; k < A.length; k++) {
        y = A[i][j];
        B[i][j] = x;
        C[i][j] = y;
        x++;
      }
}
```

Fig. 43 Java code for the SET algorithm.

compiles into a Java bytecode program with exactly the same number of instructions as the `MUL`, which means that for executing `SET` on matrices of size n, JVM also performs $INST(n)$ bytecode instructions. The only difference is that the algorithm `SET` does not use the `IADD` nor the `IMUL` instructions; it uses only the following 12 instructions: `ICONST_0`, `ILOAD`, `ALOAD_1`, `ALOAD_2`, `ALOAD_3`, `IALOAD`, `AALOAD`, `ISTORE`, `IINC`, `IF_ICMPGE`, `GOTO`, `ARRAYLENGTH`. We made an assumption that these instructions are all equally resource-consuming, we named them "simple instructions," and we used the method `Calc2` to calculate their average execution time I. Using the resulting $I = 0.0248$ ns (subsequently denoted as I_s) and formula $T_{SET}(n) = I_S * INST(n)$ we found that the `Calc2` method in this case yields an almost perfect estimation. Fig. 44 shows the measured time of the `SET` method (blue line) and its estimation provided by the `Calc2` method. The average error ($n = 200, \ldots, 500$) of this method is 0.4%. This means that the calculated $I_s = 0.0248$ nanoseconds is a reasonably good estimation for the execution time of a simple Java bytecode instruction on this computer.

To make a good estimation for the `MUL` algorithm we now only have to determine the estimation for the time complexities of the `IMUL` and `IADD` instructions. First, we compare the execution time of the algorithms `MUL` and `ADD` to find that these two algorithms are comparable in the sense of time consumption. The graph in Fig. 45 shows the differences in the time

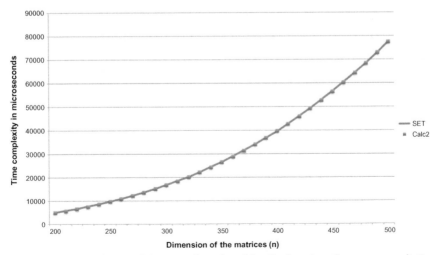

Fig. 44 Time complexity of the `SET` algorithm (blue line) and performance prediction calculated by the `Calc2` method (red dots).

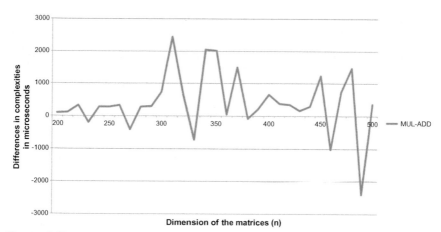

Fig. 45 Differences in the time complexities of the MUL and ADD algorithms.

complexities of the MUL and ADD algorithms. According to the oscillation of the graph we can conclude that both algorithms are equally resource-consuming, the repetitive exchange of the leadership (positive and negative values on the graph) indicates that the execution times were measured with a certain relatively small (on average <1%) error. Since the only difference between MUL and ADD is in the number of IMUL and IADD instructions used (the second one uses only IADDs, while the first one uses both) and since we proved that there is no real difference in the time consumed, we can conclude that the time consumed by the IADD and IMUL instructions are the same.

This is not just an interesting result; it also gives us an opportunity to estimate the real time consumed by both arithmetic instructions. In the MUL we have $INST(n)$ instructions, among which there are $2 \times n^3$ arithmetic instructions. Assuming that the time complexity of an arithmetic instruction (I_A) equals $I_A = I_S + \lambda$, we obtain

$$\lambda = AVG_n \left(\frac{T_{MUL}(n) - T_{SET}(n)}{2n^3} \right).$$

Using the measured times of MUL and SET and averaging them for $n = 300$, ..., 500 we obtain $\lambda = 0.22$ and $IA = 0.24$ ns. This means that the average cost of the arithmetic operations IADD and IMUL is 9.7-times larger than an average cost of a simple instruction.

To estimate the execution time of the MUL algorithm we use the following Calc3 method: given the factors I_S and I_A, calculate the estimation of the time complexity of the MUL algorithm by

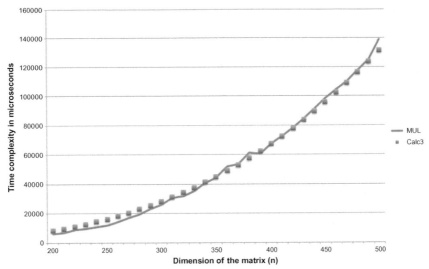

Fig. 46 Time complexity of the MUL algorithm (blue line) and performance prediction calculated by the Calc3 method (red dots).

$$T_{MUL}(n) = \left(INST(n) - 2n^3 \right) \times I_S + 2n^3 \times I_A.$$

Using this estimation we find that it much better fits the MUL algorithm than the previous ones. The graph in Fig. 46 shows the time complexity of MUL with the blue line and the Calc3 estimation with red dots. An average error of this estimation (for $n = 300, \ldots, 500$) is 2.3%.

5. Conclusions and future work

We present several aspects of a system for an automatic algorithm evaluation. We look at ALGATOR from the theoretical, implementation and the practical points of view. The abstract model presented in the theoretical part was designed to enable a simple representation of a wide range of problems. By defining the abstract parts of the model virtually any problem can be encoded into the "ALGATOR language" and the evaluation process can take place. In the second part of the article we present some of ALGATOR's implementation details. We show "how to play the game" and explain the mechanisms that drive the whole system. We expose some of the problems that can arise, especially problems with time measurement, and present our solutions. In the third part of the article we introduce the results obtained when using ALGATOR in practice. To leave the focus of the article on the algorithm-evaluation system (rather than on pure mathematical and algorithmic problems) the tests presented herein are relatively simple. Nevertheless, some of the results are quite interesting. For example, for the Array-Sorting Problem we found that the time needed to sort the array depends on the number of different elements in the array (i.e., sorting an array containing only a few different elements is much faster than sorting an array in which all the elements are different). The analysis of three recursive DFT-solving algorithms revealed the connection between the number of recursive calls and the input size. Using this result we derived the relation between the execution time and the size of the input, which fitted the future measurements with an average relative error of $<4\%$. By analyzing ALGATOR's results for the Matrix-Multiplication Problem, we derived the formulas for the number of each Java bytecode instruction used during the algorithm's execution. Combining these formulas with the time-consumption measurements, we showed that one arithmetic instruction (IADD and IMUL) is about 10 times slower than one "simple" instruction (like ICONST_0, ILOAD, ALOAD_0, ...). It is worth mentioning that ALGATOR also works well for more complex practical cases, and even though the implementation of the algorithms in those cases might be more complicated, the performance evaluation is just as easy as the one presented in this chapter.

Although ALGATOR has already been used in several practical cases and lots of problems and bugs have already been fixed, there are still many open issues and options for further development. The system could be supplemented with a better tool for data visualization. The current tool provides simple data-visualization options and works in practice only to a certain extent. For a more advanced presentation, the results should be exported and used in a third-party program. It is not only that this approach is cumbersome, but it also offers limited possibilities since the external programs are not aware of ALGATOR and thus cannot take advantage of its rich capabilities. We also miss a better web interface that would support all of ALGATOR's functionality. The current version is a kind of a garage version and needs lots of fixes for serious use by general public. Only a good online version of the system would really open up ALGATOR for the purposes for which it was originally created. In addition, ALGATOR misses a module for creating generic test cases and for performing an independent analysis based on the generated tests. This module would allow ALGATOR to more autonomously control the project and to trigger automatic tests to obtain useful information about the algorithms.

ALGATOR is an open-source program that was designed to be used by the wide community of algorithm developers. The latest version of the program with the installation instructions and usage examples is available at [12]. The author will be happy to receive any encouraging words, suggestions for improvements, or bug reports in an e-mail sent to tomaz.dobravec@fri.uni-lj.si.

References

[1] D.E. Knuth, The Art of Computer Programming, third ed. Fundamental Algorithms, vol. 1, Addison Wesley Longman Publishing Co., Inc, Redwood, USA, 1997.

[2] R. Segedwick, The analysis of quicksort programs, Acta Informatica 7 (1977) 327–355.

[3] V. Milutinovic, J. Salom, N. Trifunovic, R. Giorgi, Guide to dataflow supercomputing; basic concepts, case studies, and a detailed example, in: A.J. Sammes (Ed.), Computer Communications and Networks, Springer International Publishing, 2015.

[4] V. Milutinovic, et al., DataFlow supercomputing essentials; research, development and education, in: A.J. Sammes (Ed.), Computer Communications and Networks, Springer International Publishing, 2017.

[5] U. Čibej, J. Mihelič, Adaptation and evaluation of the simplex algorithm for a data-flow architecture, Adv. Comput. 106 (Suppl. C) (2017) 63–105 (Chapter 3).

[6] M. Muller-Hannemann, S. Schirra, Algorithm Engineering: Bridging the Gap between Algorithm Theory and Practice, Springer-Verlag, Berlin, Heidelberg, 2010.

[7] D.S. Johnson, A theoreticin's guide to the experimental analysis of algorithms, in: Proceedings of the 5th and 6th DIMACS Implementation Challenges, American Mathematical Society, 2002.

[8] C.C. McGeoch, A Guide to Experimental Algorithmics, first ed., Cambridge University Press, New York, USA, 2012.

[9] V. Blagojevic, et al., A systematic approach to generation of new ideas for PhD research in computing, Adv. Comput. 104 (2016) 1–19.

[10] J. Mihelič, B. Robič, Flexible-attribute problems, Comput. Optim. Appl. 47 (3) (2010) 553–566.

[11] J.M. Lambert, J.F. Power, Platform independent timing of java virtual machine bytecode instructions, Electron. Notes Theor. Comput. Sci. 220 (2008) 79–113.

[12] T. Dobravec. A GitHub repo of the ALGator system. 2018, Online available at github. com/ALGatorDevel/Algator, 2012–2017.

[13] N. Janko, Predelava Javanskega navideznega stroja za štetje ukazov zložne kode (A Java Virtual Machine for Counting Bytecode Instructions), bachelor thesis (mentor: T. Dobravec), UL FRI, Ljubljana, 2014.

[14] R. Lougher, JamVM—An Open-Source Java Virtual Machine, Online available at jamvm.sourceforge.net, 2014.

[15] N. Janko. A GitHub repo of VMPE. 2014, Online available at github.com/nikolai5slo/ jamvm, November 2014.

[16] U. Čibej, J. Mihelič, Improvements to Ullmann's algorithm for the subgraph isomor-phism problem, Int. J. Pattern Recognit. Artif. Intell. 29 (7) (2015) 155–180.

[17] T.H. Cormen, et al., Introduction to Algorithms, third ed., MIT Press, Massachusetts, 2009.

[18] S. Wild, M.E. Nebel, Pivot sampling in dual-pivot quicksort, in: Proceedings of the 25th International Conference on Probabilistic, Combinatorial and Asymptotic Methods for the Analysis of Algorithms, Paris, France, 2014.

[19] S. Wild, M.E. Nebel, R. Reitzig, U. Laube, Engineering Java 7's dual pivot quicksort using MaLiJAn, in: Proceedings of the 15th Workshop on Algorithm Engineering and Experiments, New Orleans, USA, 2013.

[20] P. Duhamel, H.D.L. Hollmann, Split radix FFT algorithm, Electron. Lett. 20 (1) (1982) 14–16.

[21] E. Chu, A. George, Inside the FFT black box: serial and parallel fast fourier transform algorithms, in: Computational Mathematics Series, CRC Press, LLC, New York, 2000.

[22] H.J. Nussbaumer, Fast Fourier Transform and Convolution Algorithms, Springer Series in Information Science, vol. 2, Springer, Berlin, 1982, pp. 81–94.

[23] Ž. Zorman, *Primerjava algoritmov za izračun Fourierjeve transformacije s pomočjo sistema ALGator* (Comparison of Algorithms for Calculating Fourier Transform Using the ALGator System), Bachelor Thesis (Mentor: T. Dobravec), UL FRI, Ljubljana, 2017.

[24] M. Akra, L. Bazzi, On the solution of linear recurrence equations, Comput. Optim. Appl. 10 (2) (1998) 195–210.

About the author

Tomaž Dobravec received his Dipl. Ing. degree (1996) in mathematical science and his PhD (2004) in computer science, both from the University of Ljubljana, Slovenia. He is an Assistant Professor at the Faculty of Computer and Information Science, University of Ljubljana. His main research interests are in algorithm design, analyses and evaluation, in the theory of programming languages and in networks.

CHAPTER THREE

Graph grammar induction

Luka Fürst[a], Marjan Mernik[b], Viljan Mahnič[a]
[a]University of Ljubljana, Faculty of Computer and Information Science, Ljubljana, Slovenia
[b]University of Maribor, Faculty of Electrical Engineering and Computer Science, Maribor, Slovenia

Contents

Abstract

We propose a novel approach to the graph grammar induction problem, the goal of which is to find a concise graph grammar whose language (the set of graphs which can be generated using the grammar's productions, i.e., graph substitution rules) contains all of the given "positive" graphs and none of the given "negative" graphs. Most of the existing induction methods learn graph grammars from positive examples alone, performing an iterative specific-to-general search procedure in which candidate productions for the grammar being induced are created from repeated subgraphs in the input graph set. Our approach, by contrast, couples the induction process with a

Advances in Computers, Volume 116
ISSN 0065-2458
https://doi.org/10.1016/bs.adcom.2019.07.003

graph grammar parser, thus inducing grammars from both positive and negative graphs. In addition to a graph grammar generalization operator akin to those found in the majority of existing approaches, we employ a novel generalization technique based on merging similar productions, which makes it possible to further reduce the size of the induced grammar. We applied our approach to several nontrivial graph sets, including a set of chemical structural formulas with both positive and negative examples, and obtained graph grammars which can be regarded as a concise and meaningful generalization of the input graphs.

1. Introduction

Graph grammars are systems for generating graphs using a predefined set of subgraph replacement rules (*productions*) and may thus be regarded as a generalization of better-known string grammars. A graph grammar generates all graphs which can be derived from a fixed set of initial graphs by an arbitrary number of production applications. Those graphs are said to form the grammar's *language*. Therefore, a graph grammar may be viewed as a concise representation of a (potentially infinite) set of graphs.

By analogy to string grammars and their role in defining the syntax of textual languages, graph grammars can be used to specify the syntax of graphical (visual) languages [1, 2]. However, graph grammars have also been applied to areas such as pattern recognition [3, 4], computer graphics [5, 6], software modeling [7–10], functional programming languages [11], etc.

A typical problem related to grammars is that of *membership*: given a graph grammar GG and a graph G, determine whether G belongs to the language of GG. If one is also interested in finding a sequence of production applications leading from an initial graph to G (called a *derivation* of G), we speak of the *parsing* problem. In the string grammar domain, the parsing problem has been extensively investigated, owing to the ubiquity of textual programming languages and programs [12, 13]. In the case of graph grammars, parsing has generally provoked less interest, mostly because of its NP-hardness for a vast majority of graph grammar classes. Nevertheless, the problem remains of theoretical and practical importance [1, 14, 15].

In this article, we will be dealing with what could be seen as the inverse of the parsing problem—the *induction* (or *inference* or *learning*) problem. This problem consists of finding a grammar for a language which contains a given set of graphs. Since the target language is unknown, the problem can be trivially solved by a grammar which generates only input graphs (each of them by a single production application) and nothing else. To prevent such solutions, the output grammar is typically required to generalize input graphs in a meaningful way (with "meaningful" being defined by the user) or to meet some other criteria. Furthermore, Gold [16] showed that positive examples alone do not suffice to identify infinite languages in the limit; consequently, the input to the induction problem may also include a set of "negative" graphs, which are not allowed to be generated by the target grammar. In a general case, the expected result of the induction algorithm is thus a grammar which generates all of the positive and none of the negative input graphs and satisfies all of the additional criteria, if present.

Graph grammar induction can be used to obtain a generative summary of a given set of graphs [17], to compress graphs [18], or to build a classifier to distinguish between graphs which belong to the (unknown) target concept and those which do not. Furthermore, graph grammar induction may benefit language engineers and domain experts who want to design a visual language but lack expertise or patience to construct a suitable graph grammar. Instead of performing tedious manual work, they may provide examples of graphs from the target language and let the inducer do (most of) the work [1, 2]. Alternatively, domain experts may interactively collaborate with the inducer to obtain a desirable grammar [19]. For instance, the user may prepare an initial set of positive graphs and run the inducer. The inducer constructs a grammar and generates some random graphs belonging to its language. The user may then inspect the generated graphs, label those which are not part of the target language as negative graphs, and rerun the induction algorithm. This user-inducer interplay continues until the user is satisfied with the induced grammar.

Apart from visual language design, graph grammars have been induced in domains such as pattern recognition [20], program behavior modeling [10], and website analysis [21]. The relevance of the graph grammar induction problem is expected to increase over time. Considering the growth of social networks, both in size and importance, we may be fairly certain that the demands for algorithms which find patterns in graphs and for techniques which summarize, compress, or classify graphs are going to become increasingly larger.

Inspired by Occam's razor (concise hypotheses are more likely to be correct), we seek the smallest grammar consistent with the given input set. From an algorithmic point of view, we formulate graph grammar induction as a beam search process. The induction algorithm maintains a queue with a predefined number of slots. The queue initially contains just the grammar which generates precisely the positive input graph set. Subsequently, the algorithm runs as an iterative procedure. Each iteration replaces the smallest grammar from the queue with its elementary generalizations, i.e., grammars obtained from a given base grammar by a set of predefined rules which either restructure the base grammar or slightly generalize it. Every grammar is checked against the graph grammar parser; if it is found to generate any of the negative input graphs, it is immediately discarded. The algorithm thus maintains only the grammars consistent with the input. It halts when the queue becomes empty and outputs the smallest grammar generated in the search process. Since the algorithm continuously keeps track of the smallest grammar discovered so far, it can be safely stopped at any time.

In the process of devising a novel graph grammar induction approach, we employed several systematic idea generation techniques introduced by Blagojević et al. [22]. An existing solution, namely a graph grammar parser, was *implanted* into the graph grammar induction engine. The parser itself was *revitalized*: the Rekers–Schürr graph grammar parser [1], published in 1997, had not been improved until 2011 [14], when it became clear that an efficient parsing algorithm is a necessary prerequisite for a viable induction method. While *generalization* was not used as a research technique *per se*, it nevertheless permeates our work; to induce a graph grammar means to find a graph grammar which generalizes the given positive sample graphs. If we allow ourselves to stretch the meaning of idea generation techniques to a certain extent, it could be said that our research process also involves a touch of *Mendeleyevization*. There are string grammar induction approaches which interleave the search process with parsing, and there are graph grammar induction approaches which do not use a parser and hence construct a graph grammar from positive examples only. However, hardly any graph grammar induction approach is coupled with a parser. In fact, we are only aware of the work of Ates et al. [23], but even in this case, the parser only checks the final grammar rather than being used for a continuous validation of candidate grammars.

This article is an extension of our earlier conference paper [17]. Almost every section has been considerably expanded. We now describe the algorithm in much more detail and include recent experimental results. For readers unfamiliar to graph grammars, we provide a concise, yet gradual, introduction to the topic.

In Section 2, we give an overview of published research related to our approach. Section 3 introduces the terms pertaining to graphs and graph grammars which will be used in this article. In Section 4, we present our approach. In Section 5, the induction algorithm is experimentally validated. Section 6 concludes the article.

2. Related work

The higher we climb in the hierarchy of language representation models, the fewer induction approaches can be found in literature. While relatively many[a] approaches induce regular grammars or, equivalently, finite automata, considerably less has been done in the area of context-free string grammar induction, let alone graph grammar induction [24, 25]. In this brief and by no means complete survey, we restrict ourselves to the induction of context-free string grammars and (general) graph grammars.

One of the earliest results in the area of grammar induction was discovered by Gold [16], who introduced the concept of *language identification in the limit*. A learner is continuously presented with labeled positive and optionally negative examples of the target language and asked to identify the target language after every example. If, after some finite time, the learner's guesses are all equivalent to some representation of the target language, the language is said to have been identified in the limit. Gold subsequently showed that a class containing infinite languages (this includes regular languages and everything above them) can only be identified in the limit from both positive and negative examples; positive examples by themselves do not suffice. Despite this theoretical result, many approaches induce grammars from positive examples alone. The reason might be the nonexistence of a viable parsing algorithm to validate the induced grammars or the unavailability of a suitable negative example set.

Wharton [26] induced context-free string grammars by systematic enumeration. The algorithm generates grammars in the order of increasing size and outputs the smallest grammar consistent with the input examples, i.e., the smallest grammar which can generate all of the positive and none of the negative input examples. Unfortunately, the combinatorial complexity renders the method impractical even for realistic string grammars. For graph grammars, such an approach is probably hopeless, since they have even more degrees of freedom.

VanLehn and Ball [27] adapted Mitchell's elegant *version space* concept [28] to context-free string grammar induction. Version space represents the set of hypotheses consistent with the given input examples by its upper bound $\mathcal{G}$ (the set of most general hypotheses consistent with the examples)

---

[a] "Relatively many" is still an infinitesimal fraction of the deluge we are nowadays witnessing in areas such as computer vision or "big data."

and its lower bound S (the set of most specific consistent hypotheses). Every positive example makes the set S more general (because some hypotheses in S may be too specific), and every negative example makes the set G more specific (because some hypotheses in G may be too general). The version space thus shrinks and approaches the true hypothesis in the limit. Unfortunately, VanLehn and Ball showed that the concept cannot be naturally adapted to grammar induction. In their approach, the set G can, counterintuitively, become more general as positive examples arrive. What is worse, though, is that a new example may create a large set of derivation trees, an even larger set of their products, and a still larger set of partitions of those product trees. Needless to say, a corresponding graph grammar induction approach would be prone to an even more severe combinatorial explosion.

Nevill-Manning and Witten proposed a string grammar induction approach named Sequitur [29], which could be (and was) adapted to graph grammars. To be honest, Sequitur is a compression rather than induction method; for a given string, the algorithm finds a grammar which can reconstruct precisely that string, without performing any actual generalization. Sequitur scans the input string one character at a time and maintains a grammar consisting of the sole production $S ::= \alpha$, with α being the string processed so far, until it discovers that the sequence of the previous character (say, x) and the current character (say, y) has already been observed. Now, the algorithm introduces the production $A ::= xy$ and replaces all occurrences of the digram xy in the production $S ::= \alpha$ with the symbol A. The algorithm also ensures that the output grammar contains no repeated digrams and that every production is used at least twice. Maneth and Peternek [18] employ a similar idea to compress (hyper)graphs[b] by hyperedge replacement graph grammars [15, 30]. In their algorithm, the role of digrams is played by subhypergraphs composed of two adjacent hyperedges and the incident vertices. In each iteration, the algorithm finds the most frequent digram and introduces a new production with that digram on its RHS (right-hand side) and a hyperedge with a fresh nonterminal label on the LHS (left-hand side). The authors faced numerous challenges not present in the textual case, such as counting nonoverlapping occurrences of digrams.

Mernik et al. developed a family of context-free string grammar induction approaches based on evolutionary computation. In their initial publication [31], they present a method which first constructs a random

---

[b] A hypergraph is a graph-like structure composed of hyperedges and ordinary vertices. In contrast to ordinary (binary) edges, a hyperedge may connect any number of vertices.

population of grammars and then iteratively modifies it by applying cross-over, mutation, and selected heuristic operators, using the input set of positive and negative examples as a benchmark for estimating the quality of individual grammars in the population. They later [32] discovered that the initial grammar population should not be generated at random; instead, considerably more promising results were obtained by initializing the evolutionary process by a grammar constructed from a selected set of short substrings of the input set. The initial grammar was found by enumerating all possible labeled derivation trees for those substrings. The approach was subsequently used in a metamodel inference system [33], which first converts the set of input models into a corresponding textual representation, then induces a context-free string grammar, and finally converts the induced grammar into a metamodel.

Nakamura and Matsumoto [34] took a different path, inducing a context-free grammar in an incremental fashion. The algorithm starts with a (possibly empty) grammar and processes a single input example at a time. When faced with a positive example, the algorithm tries to parse the example against the grammar constructed up to that point by running the CYK algorithm [35–37]. If it is successfully parsed, no further action is necessary; otherwise, the state of the CYK parser is examined and a new production is non-deterministically added. If the current set of productions covers any negative example, the algorithm backtracks to the previous choice point and non-deterministically adds another production. The approach was later adapted to inducing grammar productions augmented with semantic rules [38].

Javed et al. [39] proposed a somewhat similar incremental approach. Every example is parsed against the existing grammar using an LR(1) parser; in case of failure, the grammar is updated by analyzing the state of the parser stack. The main limitation of the algorithm is that the adjacent examples are only allowed to differ in one of three predefined ways. In an approach called *memetic*, Hrnčič et al. [40] overcame that limitation and also combined incremental grammar-update techniques with grammar-aware genetic operators.

Dubey et al. [41] induced a grammar for a dialect of a given programming language, a typical example of which is a language such as C or Java being augmented with a new keyword. Specifically, given an existing grammar and a set of examples, their method proposes a set of productions to be added to the grammar in order to accommodate the given examples. Every example is parsed against the existing grammar; in case of failure, a new production is suggested by analyzing both the LR(1) parser stack and the state of the CYK algorithm.

One of the earliest *graph* grammar induction methods was proposed by Jeltsch and Kreowski [42]. Their algorithm induces hyperedge replacement graph grammars from a set of positive examples. Starting with a trivial grammar which covers precisely the input graphs and nothing else, the algorithm iteratively forms new productions by decomposing the existing ones. To make the current grammar more general, productions are reverse-applied to their own RHSs whenever possible.

A family of graph grammar induction methods was derived from the graph compression approach named Subdue, proposed by Holder et al. [43, 44]. This algorithm starts with a set of single-vertex subgraphs of the input graphs, with each subgraph being equipped with a set of its occurrences in the input set. Each subgraph is assigned a compression score, computed as the number of bits required to encode the current graph set minus the sum of the number of bits to encode the subgraph and the number of bits to encode the graph set after all occurrences of the subgraph have been replaced by a single vertex. Following the beam search paradigm, Subdue maintains a queue containing the current top k subgraphs ranked by their compression score. In each iteration, the best subgraph is removed from the queue and extended in all possible ways based on the neighborhoods of its occurrences in the input graph set. Each extension is placed back into the queue, upon which the algorithm again retains only the top k subgraphs. Since the vertices which replace subgraph occurrences can themselves be part of other subgraphs, Subdue is able to build a hierarchical representation of the graph set.

As shown by Jonyer et al. [45], there is a small step from Subdue's output to a graph grammar. Indeed, when Subdue replaces a subgraph S by a vertex v, it has effectively introduced a grammar production $v ::= S$. However, a grammar obtained in this way is suitable only for parsing graphs (i.e., transforming them into a single nonterminal vertex by reverse-applying the productions), not for generating graphs belonging to the putative target language. To obtain a generative grammar, a production $v ::= S$ has to be supplemented with embedding (or context) specification rules which determine how to connect the vertices of a copy of S to the neighbors of an occurrence of v after removing the occurrence from a given host graph. This problem is remedied in subsequent incarnations of the approach [46, 47], although the authors invented their own ad hoc embedding mechanism rather than resorted to well-established formalisms [15, 48].

Under certain circumstances, all grammar induction methods in the Subdue family build recursive productions and are hence able to induce a grammar which strictly generalizes the input graph set rather than just compresses it. Jonyer et al. [45] induced recursive productions which encode sequences of isomorphic graphs connected with single edges. Kukluk et al. learned recursive productions from isomorphic graphs sharing a single vertex [46] or a single edge [47].

Ates et al. [23] employed a Subdue-like induction scheme to learn productions in which a single vertex is expanded to a graph. The induced grammars belong to the Reserved Graph Grammars [2] formalism. Zhao et al. [10] learned behavioral patterns of a given program by representing the program's execution traces using graphs and inducing a grammar belonging to the spatial graph grammars (SGG) [49] formalism. Roudaki et al. induce SGG grammars to represent typical web page design patterns; in particular, a grammar encodes an arrangement of elements such as text, menus, or images on a web page. However, in contrast to most other graph grammar induction approaches, they do not induce a graph grammar directly from the input graph set, but rather represent the input graphs by strings, learn a regular expression which covers those strings, and convert the induced regular expression to a graph grammar.

Brijder and Blockeel [50] proposed an algorithm which accepts a graph G and a set of its isomorphic disjoint subgraphs S and finds out whether there exists a node-label-controlled grammar [15] which can reconstruct the graph G from the graph obtained by replacing each occurrence of S in G by a nonterminal vertex.

Oates et al. [51] learned stochastic graph grammars. Given a set of graphs and a graph grammar with a known structure, their approach computes the probabilities of individual productions such that the likelihood of input data is maximal.

We were inspired by some of the ideas discovered by our predecessors. The overall grammar induction algorithm bears some similarities with Subdue. In our approach, the search space is also explored using the beam search algorithm, and the so-called *type-A generalization step* enumerates candidate production RHSs by extending individual subgraph occurrences by a single vertex in all possible directions. As with Jeltsch and Kreowski [42], we start with a trivial (most specific) graph grammar and then induce increasingly general grammars. However, to the best of our knowledge, the so-called *type-B generalization step*, which merges a pair of similar productions in a grammar to obtain a more general grammar, may be regarded as a novelty. Our approach learns grammars from both positive and negative graphs, which is also a rarity in the graph grammar induction world, despite being practically the norm for string grammar induction methods. One of the

reasons is undoubtedly the (both theoretical and practical) shortage of efficient graph grammar parsers for any but the most restricted graph grammar
formalisms [15, 52]. Ates et al. [23] did employ a parser, but they used it only
for validating final grammars. By contrast, our approach intertwines grammar induction and parsing. If an induced grammar is discovered (by means of
the parser) to generate any of the negative input graphs, it is immediately
discarded.

3. Preliminaries
3.1 Sets, tuples, and functions

A *set* is an unordered collection of elements without repetitions. For a set A,
let $\wp(A)$ denote its power set, i.e., $\wp(A) = \{A'|A'\subseteq A\}$.

A *tuple* is an ordered collection of elements with possible repetitions. The
elements of a tuple will be listed in parentheses. A tuple $a = (a_1, \ldots, a_n)$ is
lexicographically smaller than a tuple $b = (b_1, \ldots, b_n)$ (denoted $a<_{\text{lex}}b$) if there
exists an index $i \in\{1, \ldots, n\}$ such that $a_i < b_i$ and $a_j = b_j$ for all $j < i$.
A 2-element tuple is customarily called a *pair*, a 3-element tuple is a *triple*, etc.

Unless otherwise specified, a function $f: A \rightarrow B$ is assumed to be *total*,
i.e., defined for all elements of the set A. Given a function $f: A \rightarrow B$ and a set
$A'\subseteq A$, let $f(A') = \{f(a')|a' \in A'\}$. Likewise, if $a_1, \ldots, a_k \in A$, then $f((a_1, \ldots, a_k)) =
(f(a_1), \ldots, f(a_k))$. Let $f|_{A'}$ represent the restriction of the function $f: A \rightarrow B$ to the
domain $A'\subseteq A$, i.e., the function $f': A'\rightarrow B$ such that $f'(a) = f(a)$ for all $a \in A'$.

3.2 Graphs

A graph is usually defined as a pair (V, E), where V is the set of vertices and $E
\subseteq V \times V$ is the set of edges (connections between the vertices). However,
this definition does not cover labeled graphs and graphs with multiple
parallel edges between a given pair of vertices. We therefore prefer the
following pair of definitions:

Definition 1 (directed graph).
A *directed graph* G over a label set Σ is a tuple $(V, E, conn, label)$, where:
- V is the set of *vertices* of G.
- E is the set of *edges* of G.
- $conn: E \rightarrow V \times V$ is a function which maps an edge to its *endpoints* (the
  pair of vertices which it connects).
- $label: V \cup E \rightarrow \Sigma$ is a function which assigns a label to each vertex
  and edge.

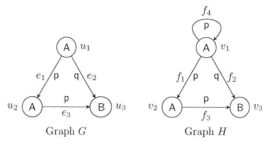

Fig. 1 Sample graphs G and H.

For example, the graph H from Fig. 1 (page 143) has $V = \{v_1, v_2, v_3\}$, $E = \{f_1, f_2, f_3, f_4\}$, $conn = \{f_1 \mapsto (v_1, v_2), f_2 \mapsto (v_1, v_3), f_3 \mapsto (v_2, v_3), f_4 \mapsto (v_1, v_1)\}$, and $label = \{v_1 \mapsto \mathsf{A}, v_2 \mapsto \mathsf{A}, v_3 \mapsto \mathsf{B}, f_1 \mapsto \mathsf{p}, f_2 \mapsto \mathsf{q}, f_3 \mapsto \mathsf{p}, f_4 \mapsto \mathsf{p}\}$.

For convenience, let us define the *source vertex* and the *target vertex* of an edge e with $conn(e) = (u, v)$ as $source(e) = u$ and $target(e) = v$. Additionally, let $connset(e) = \{v | v \in conn(e)\}$ be the set of endpoints of e.

Definition 2 (undirected graph).

An *undirected graph* G over a label set Σ is a tuple $(V, E, conn, label)$, where:

- V is the set of *vertices* of G.
- E is the set of *edges* of G.
- $conn$: $E \to \wp(V)$ is a function which maps an edge to the set of vertices which it connects. For every edge e, $1 \leq |conn(e)| \leq 2$.
- $label$: $V \cup E \to \Sigma$ is a function which assigns a label to each vertex and edge.

In an undirected graph, edges have no direction, so the notions of source and target vertices are meaningless. Therefore, the function $conn$ maps an edge to a 2-element set (or 1-element set, if the edge is a loop) rather than to a pair. For undirected graphs, $connset = conn$.

To emphasize that a set or a function pertains to a specific graph G, we will attach a subscript to its label, e.g., V_G, $conn_G$. This convention will be followed throughout the article and will not be limited to graphs.

Vertices and edges will be collectively called *(graph) elements*: $Elems = V \cup E$. To simplify notation, we will often write $x \in G$ to mean $x \in Elems_G$. A graph with $Elems = \varnothing$ will be called a *null graph* and denoted λ.

The sets $VLabels = \{label(v) | v \in V\}$ and $ELabels = \{label(e) | e \in E\}$ will be assumed to be disjoint. Let $Labels = VLabels \cup ELabels$. Unlabeled vertices will be treated as having a special label $\$_V$. Likewise, unlabeled edges will be considered to have a distinct label $\$_E$. Except for $\$_V$ and $\$_E$, literal labels will be typeset in **sans serif**.

The following concept is not part of the established graph theory but will come in handy when dealing with graph grammars:

Definition 3 (deficient subgraph).
A tuple $H = (V_H, E_H, conn_H, label_H)$ is a *deficient subgraph* of a graph G (denoted $H \sqsubseteq G$) if $V_H \subseteq V_G$, $E_H \subseteq E_G$, $conn_H = conn_G|_{E_H}$, and $label_H = label_G|_{Elems_H}$.

A deficient subgraph H of a graph G is not necessarily a proper graph, since it might contain *dangling edges*.

Definition 4 (dangling edge).
In a deficient subgraph $H \sqsubseteq G$, an edge $e \in E_H$ is *dangling* if at least one of its endpoints belongs to $V_G \backslash V_H$, i.e., $connset_H(e) \setminus V_H \neq \varnothing$.

Definition 5 (graph difference).
Given a graph G and a deficient subgraph $H \sqsubseteq G$, the *graph difference* $G \setminus H$ is a deficient subgraph D with $V_D = V_G \backslash V_H$, $E_D = E_G \backslash E_H$, $conn_D = conn_G|_{E_D}$, and $label_D = label_G|_{Elems_D}$.

A (proper) subgraph is a deficient subgraph without dangling edges:

Definition 6 (subgraph).
A deficient subgraph $H \sqsubseteq G$ is a *subgraph* of G (denoted $H \subseteq G$) if $connset_H(e) \subseteq V_H$ for all $e \in E_H$.

Definition 7 (deficient: subgraph formed by an element set).
Given a graph G, a vertex set $V' \subseteq V_G$, and an edge set $E' \subseteq E_G$, $G[V, E] = (V', E', conn_G|_{E'}, label_G|_{V' \cup E'})$ is a (deficient) subgraph formed by the sets V' and E'.

Definition 8 (induced subgraph).
Given a graph G and a set $W \subseteq V_G$, the *induced subgraph* formed by the set W is the graph $G[W, \{e \in E_G | connset_G(e) \subseteq W\}]$, i.e., the subgraph composed of the vertices in W and all edges between them.

The following nonstandard concept plays an important role in our grammar induction approach:

Definition 9 (neighborhood).
Given a graph G and a graph $H \subseteq G$, the *neighborhood* of H in G is the set of all vertices of $G \backslash H$ adjacent to a vertex in H: $Nh_G(H) = \{v \in V_{G \backslash H} | \exists w \in V_H, e \in E_{G \backslash H}: connset(e) = \{v, w\}\}$.

In the subsequent definitions, G and H are arbitrary graphs.

Definition 10 (homomorphism).
A *homomorphism* $h: G \to H$ is a pair of functions $h_V: V_G \to V_H$ and $h_E: E_G \to E_H$ such that $label_H(h_V(v)) = label_G(v)$, $label_H(h_E(e)) = label_G(e)$, and $conn_H(h_E(e)) = h_V(conn_G(e))$ for all $v \in V_G$ and $e \in E_G$.

We will write $h(x)$ to denote $h_V(x)$ if $x \in V$ and $h_E(x)$ if $x \in E$. Furthermore, if G' is a (deficient) subgraph of G, we will write $h(G')$ to denote the (deficient) subgraph of H formed by the element set $\{h(x)|x \in G'\}$.

An injective homomorphism is called a *monomorphism*; a surjective monomorphism is called an *isomorphism*.

Definition 11 (monomorphism).

A homomorphism $h: G \to H$ is a *monomorphism* if $x \neq y$ implies $h(x) \neq h(y)$ for all $x \in G$ and $y \in G$.

Definition 12 (isomorphism, isomorphic graphs).

A monomorphism $h: G \to H$ is an *isomorphism* if for each $y \in H$ there exists a unique $x \in G$ such that $h(x) = y$. A graph G is *isomorphic* to a graph G (denoted $G \simeq H$) if there exists an isomorphism $h: G \to H$.

A monomorphism $h: G \to H$ is an isomorphism between the graph G and a subgraph of H. For this reason, a monomorphism is also called a *subgraph isomorphism*. The problem of determining whether a given graph has an isomorphic subgraph in another graph is a well-known NP-complete problem [53–55] and a prominent, albeit not immediately obvious, part of the graph grammar parser by Rekers and Schürr [1].

Considering possible homomorphisms $G \to H$ in Fig. 1, $h_1 = \{u_1 \mapsto v_1, u_2 \mapsto v_1, u_3 \mapsto v_3, e_1 \mapsto f_4, e_2 \mapsto f_2, e_3 \mapsto f_2\}$ is a (plain) homomorphism, while $h_2 = \{u_1 \mapsto v_1, u_2 \mapsto v_2, u_3 \mapsto v_3, e_1 \mapsto f_1, e_2 \mapsto f_2, e_3 \mapsto f_3\}$ is a monomorphism (but not isomorphism, since the edge f_4 has no match in G).

Definition 13 (occurrence).

Given a monomorphism $h: G \to H$, an *occurrence* of G in H is the graph $H' = h(G)$.

Definition 14 (induced occurrence).

Given a monomorphism $h: G \to H$, the occurrence H' of G in H is *induced* if H' is an induced subgraph of H formed by the vertices $\{h(v)|v \in G\}$.

Fig. 2 shows an induced and a noninduced occurrence of a graph G in a graph H.

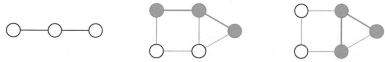

Fig. 2 A graph G (*left*) and two of its occurrences in a graph H (*middle* and *right*). Only the first occurrence is induced.

The following definition will prove to be useful when stating the goal of the graph grammar induction algorithm:

Definition 15 (size).

The *size* of a graph G is the number of its elements, i.e., $|G| = |Elems_G| = |V_G| + |E_G|$.

3.3 Graph grammars

Informally, a graph grammar is a collection of graph replacement rules called *productions*. By analogy to string grammars, each graph grammar production is a rule stating that a subgraph of a host graph may be replaced by a copy of another graph. However, while the process of deleting a substring and inserting a new string at the same place is well-defined and straightforward (e.g., given the production aB := CdE and the host string aBAaBB, the result is either CdEAaBB or aBACdEB, depending on which occurrence of the left-hand side string aB we choose to replace), this is far from being the case with graph productions. When deleting an occurrence of a production's left-hand side graph from the host graph, what should be done with the ensuing dangling edges? How should a copy of the graph on the production's right-hand side be connected with the remaining part of the host graph?

There are many possible answers to these questions, giving rise to a variety of graph grammar formalisms [56]. Two of the more prominent ones are *node replacement grammars* and *hyperedge replacement grammars*. In node replacement grammars, productions take the form $X ::= (R, \mathcal{C})$, which is interpreted in the following way: when applied to a host graph H, search for a vertex labeled X, delete it together with its incident edges, disjointly add a copy R' of the graph R to H, and connect the vertices of R' with the former neighbors of the deleted vertex as dictated by the embedding rules $\mathcal{C}$. On the other hand, hyperedge replacement grammars operate on *hypergraphs* — graph-like structures in which individual edges, called *hyperedges*, may connect any number of vertices. An ordinary edge may thus be regarded as a hyperedge with two *tentacles* (connectors). Productions of hyperedge replacement grammars take the form $X ::= R$, where X is the type of a hyperedge (determined by a label and an integer n, the number of tentacles) and R is a hypergraph with n designated vertices called "external vertices." To apply a production $X ::= R$ to a host hypergraph H, a hyperedge e of type X in H is replaced by a copy R' of the hypergraph R, where the ith vertex formerly connected to e is identified with the ith external vertex of the hypergraph R'. Hyperedge replacement grammars thus do not need embedding rules.

Graph grammars which can be produced by our induction algorithm are a subclass of a formalism called *layered graph grammars*. In grammars of this type, each production has a *context*, which implicitly determines how copies of the right-hand-side graph elements are to be embedded into the host graph. When applying a production, the part of the host graph corresponding to the context does not change but serves as the anchor for deleting and adding graph elements.

Definition 16 (layered graph grammar).

A *layered graph grammar* (LGG) is a tuple $GG = (\mathcal{N}, \mathcal{T}, \mathcal{P})$, where $\mathcal{N}$ and $\mathcal{T}$ ($\mathcal{N} \cap \mathcal{T} = \varnothing$) are the sets of *nonterminal* and *terminal* graph element labels, respectively, and $\mathcal{P}$ is the set of *productions*. Each production in $\mathcal{P}$ is a triple (*Lhs*, *Rhs*, *Common*), where the graph *Common* (the *context*) is a subgraph of both the graph *Lhs* (the *left-hand side*, or LHS) and the graph *Rhs* (the *right-hand side*, or RHS). At least one production in $\mathcal{P}$ takes the form (λ, R, λ).

Definition 17 (additional components of a production).

For a production, let *Xlhs* = *Lhs* \ *Common* (the *exclusive left-hand side*) and *Xrhs* = *Rhs* \ *Common* (the *exclusive right-hand side*). Furthermore, let *Union* be the graph formed by the union of the elements of *Xlhs*, *Common*, and *Xrhs*. Therefore, *Xlhs* = *Union* \ *Rhs* and *Xrhs* = *Union* \ *Lhs*.

We will assume that every LGG is either *directed* or *undirected*.

Definition 18 ((un)directed LGG).

An LGG is *(un)directed* if every graph in every production is (un)directed.

To illustrate these concepts, let us consider the grammar GG^0_{LHSDB} with $\mathcal{N} = \{a, b, d, s\}, \mathcal{T} = \{C, H, \$_E\}$, and the set of productions as displayed in Fig. 3. The productions are shown in the familiar LHS ::= RHS notation.

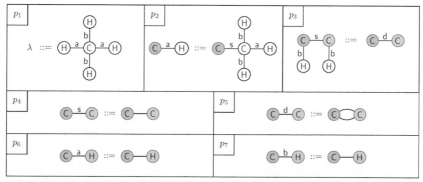

Fig. 3 GG^0_{LHSDB}: an LGG for generating structural formulas of linear hydrocarbons with single and double bonds.

The colored vertices are part of the context graphs of individual productions. Note that the context graph elements in the RHS are *the same* as those in the LHS, rather than being copies of each other. To make the relationship between production components clearer, Fig. 4 dissects the production p_3. The vertices labeled **C** are not part of the deficient subgraphs *Xlhs* and *Xrhs*; therefore, *Xrhs* consists only of a single edge, which, of course, is dangling.

Like their string counterparts, every graph grammar defines a *language*—the (possibly infinite) set of graphs which can be obtained by applying productions, starting from the null graph. Before the language can be formally defined, however, we have to introduce a few auxiliary concepts.

Definition 19 (l–homomorphism, l–occurrence).
Given a production p and a host graph H, an *l-homomorphism* $h{:}Lhs_p \to H$ is a homomorphism such that $h|_{Xlhs_p}$ is a monomorphism. The graph $h(Lhs_p)$ is an *l-occurrence* of p in H.

Definition 20 (r–homomorphism, r–occurrence).
Given a production p and a host graph H, an *r-homomorphism* $h{:}Rhs_p \to H$ is a homomorphism such that $h|_{Xrhs_p}$ is a monomorphism. The graph $h(Rhs_p)$ is an *r-occurrence* of p in H.

Definition 21 (production application).
A production p can be applied to a host graph H only if an l-homomorphism $h{:}Lhs_p \to H$ exists and if the graph difference $H \setminus h(Xlhs_p)$ does not have any dangling edges. If both conditions are met, p is applied to H by first removing $h(Xlhs_p)$ from H and then attaching copies of the elements of $Xrhs_p$ to the subgraph $h(Common_p)$ in such a way that the attached copies and the

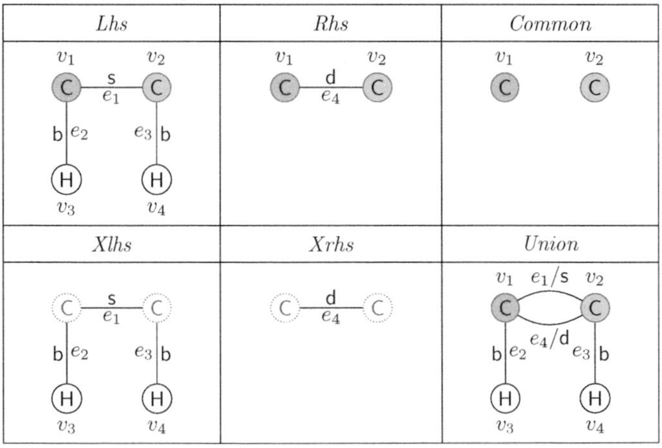

Fig. 4 Components of the production p_3 of the grammar GG^0_{LHSDB}.

subgraph $h(Common_p)$ jointly form an r-occurrence of p in H. The fact that the application of a production p to a graph H yields a graph H' will be denoted $H \overset{p}{\Rightarrow} H'$.

Definition 22 (reverse production application).
A graph H is the result of a *reverse application* of a production p to a graph H' (denoted $H' \overset{p}{\Leftarrow} H$) if $H \overset{p}{\Rightarrow} H'$.

Fig. 5 shows an application of the production p_3 of GG^0_{LHSDB} to a sample graph.

Definition 23 (derivation).
A *derivation* of a graph G in a graph grammar GG is a sequence $\lambda \overset{p_1}{\Rightarrow} G_1 \overset{p_2}{\Rightarrow} \ldots \overset{p_n}{\Rightarrow} G_n = G$, where $p_1, \ldots, p_n \in \mathcal{P}_{GG}$.

We will say that a graph G *can be derived in*, *is derivable in*, *can be generated by*, or *is covered by* a graph grammar GG if there exists a derivation of G in GG. Fig. 6 depicts a derivation of the structural formula of propene in the grammar GG^0_{LHSDB}.

Definition 24 (terminal graph).
A graph G is *terminal* with respect to a graph grammar GG if $Labels_G \subseteq \mathcal{T}_{GG}$.

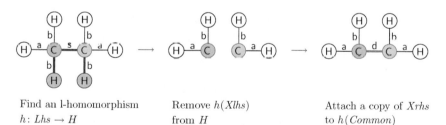

H
C
a
b
s
d
Find an l-homomorphism
h: $Lhs \to H$
Remove $h(Xlhs)$
from H
Attach a copy of $Xrhs$
to $h(Common)$

Fig. 5 An application of the production p_3 of GG^0_{LHSDB} to a sample graph.

λ
p_1
p_2
p_3
p_4
p_5
$2 \times p_6$
$4 \times p_7$
H
C
a
b
s
d

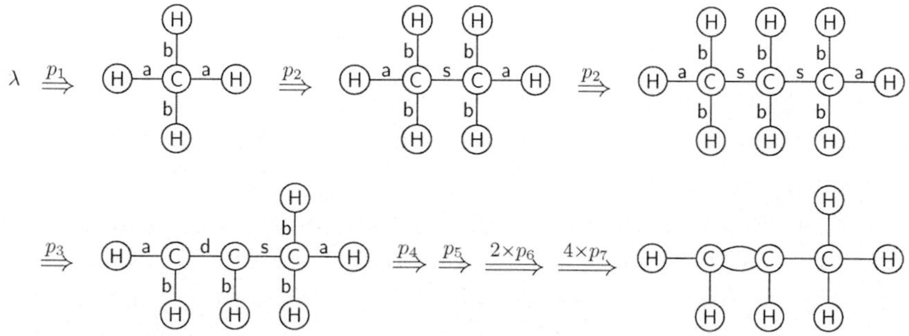

Fig. 6 A derivation of a sample graph in the grammar GG^0_{LHSDB}.

Definition 25 (language).

The *language* of a graph grammar GG (denoted $L(GG)$) is the set of all nonnull terminal graphs derivable in GG.

The language of the grammar GG^0_{LHSDB} is the set of structural formulas of linear hydrocarbons with single and double bonds (LHSDB). A hydrocarbon is a chemical compound composed of carbon (C) and hydrogen (H) atoms where each carbon atom forms exactly four bonds with other atoms and each hydrogen atom forms exactly one bond. The structural formula of a hydrocarbon is a graph in which the atoms and bonds are represented by the vertices and edges, respectively. The grammar GG^0_{LHSDB} generates all hydrocarbons where carbon atoms form a single chain (linear hydrocarbons) and can be connected by single or double bonds. (A double bond counts as two single bonds.) To see that every graph in $L(GG^0_{\text{LHSDB}})$ is a valid structural formula of a hydrocarbon, observe that the edge labels a and b represent a future C–H bond, while the labels s and d represent a future single and double C–C bond, respectively. Therefore, the production p_1 establishes a valid future LHSDB graph, and all other productions maintain this property (for example, the production p_3 removes a future bond to H from both C atoms but adds another future bond between the two C atoms, maintaining four future bonds for both C atoms). Furthermore, carbon atoms form a single chain, since a new carbon atom can be added only at the beginning or at the end of the current chain. Conversely, to see that every valid LHSDB graph can be generated by the grammar, consider that any such graph can be reduced to λ by a sequence of reverse applications of productions. First, reverse–apply the production p_6 to the two outermost C–H bonds and the production p_7 to all other such bonds. Second, reverse–apply the productions p_4 and p_5 to all single and double C–C bonds. By now, every bond has been converted to a labeled edge. Following that, reverse–apply the production p_3 to all d-labeled edges; every C vertex has thus again obtained four incident edges. Currently, all C–C edges are labeled s, whereas the C–H edges are labeled a (the two outermost ones) or b (all the others). Now, we can iteratively reverse–apply the production p_2 to an outermost C vertex and its neighbors until arriving at a copy of the RHS of the production p_1.

Definition 26 (parsing, parser).

To *parse* a terminal graph G against a graph grammar GG means to find a derivation of G in GG if one exists and to declare that $G \notin L(GG)$ in the opposite case. A *parser* is an algorithm which parses a given graph against a given graph grammar.

Definition 27 (parsability).

A layered graph grammar $(\mathcal{N}, \mathcal{T}, \mathcal{P})$ is *parsable* if there exists an integer M and a function $layer : \mathcal{N} \cup \mathcal{T} \to \{1, ..., M\}$ such that for each production $p \in \mathcal{P}$ it holds that $\ell(Lhs_p) <_{\text{lex}} \ell(Rhs_p)$, where $\ell(G) = (l_1, ..., l_M)$ with $l_m = |\{x \in G | layer(label(x)) = m\}|$ is the *layering tuple* of a graph G.

The grammar GG^0_{LHSDB} is parsable. If $M = 2$ and $layer = \{\mathsf{C} \mapsto 1,$ $\$_E \mapsto 1,$ $\mathsf{d} \mapsto 1,$ $\mathsf{H} \mapsto 2,$ $\mathsf{a} \mapsto 2,$ $\mathsf{b} \mapsto 2,$ $\mathsf{s} \mapsto 2\}$, then $\ell(Lhs_p) <_{\text{lex}} \ell(Rhs_p)$ for all $p \in \mathcal{P}$. For instance, $\ell(Lhs_{p_3}) = (2,5) <_{\text{lex}} \ell(Rhs_{p_3}) = (3,0)$ and $\ell(Lhs_{p_6}) = (1,2) <_{\text{lex}} \ell(Rhs_{p_6}) = (2,1)$.

The parsability condition for the LGG formalism plays a similar role as the LHS-smaller-than-RHS condition for context-sensitive string grammars. It ensures the existence of a parsing algorithm: to parse a graph G against a grammar GG, simply try to reduce G to λ by reverse-applying productions of GG, using backtracking if necessary. This procedure will always halt, since every reverse application strictly diminishes (in the lexicographical sense) the layering tuple of the graph being parsed.

Rekers and Schürr [1] proposed a more intelligent parsing method. In the first stage, their algorithm produces a set of candidate production applications, a subset of which is guaranteed to constitute a derivation of the input graph if the graph belongs to the language of the input graph grammar. The second stage finds that subset and the order of production applications which forms a valid derivation, if it exists. To minimize the amount of backtracking needed to find a derivation sequence, the parsing algorithm makes use of ordering and mutual exclusion relationships between candidate production applications.

Since the parsing problem is NP-hard even for highly restricted graph grammar formalisms [52], a polynomial-time LGG parser (in terms of the input graph size) is unlikely to exist. Nevertheless, the Rekers–Schürr parser, enhanced with the improvements by Fürst et al. [14], performs reasonably well for many interesting examples of graph grammars, including GG^0_{LHSDB}.

4. Our approach

4.1 Overview

As mentioned in the introduction, the goal of the induction problem is to construct a graph grammar given a set of positive graphs $(\mathcal{G}^+)$ and a possibly empty set of negative graphs $(\mathcal{G}^-)$, where $\mathcal{G}^+ \cap \mathcal{G}^- = \varnothing$. Since the problem can be trivially solved by a grammar composed of productions $\lambda ::= G$ for all graphs $G \in \mathcal{G}^+$, we seek the smallest grammar in the space of *candidate graph grammars* — the grammars which cover all positive and no negative input graphs.

The induction algorithm has been formulated as a heuristic search in the space of candidate graph grammars. Candidate grammars belong to a subclass of the LGG formalism; to explore the unrestricted LGG class would be prohibitively expensive. The *target formalism*, as we call the class of grammars which can be induced by our algorithm, is described in Section 4.2. Despite restrictions, many interesting LGGs can be rewritten as grammars in the target formalism, including GG^0_{LHSDB}.

The search space is explored in a specific-to-general fashion. The induction algorithm operates as a beam search, maintaining and iteratively manipulating a queue of candidate grammars. At the beginning, the queue contains only the trivial grammar—the grammar with productions $\lambda ::= G$ for all graphs $G \in \mathcal{G}^+$. In each iteration, the algorithm replaces the smallest grammar from the queue with a set of its generalizations and retains a fixed number of smallest grammars in the queue. The induction process finishes once the queue becomes empty. In Sections 4.3–4.6, the algorithm is explained in detail.

4.2 Target formalism

To save some space, let $u:A \overset{e:a}{\underline{\hspace{1em}}} v:B$ represent an undirected graph with $V = \{u, v\}$, $E = \{e\}$, $conn(e) = \{u, v\}$, and $label = \{u \mapsto A, v \mapsto B, e \mapsto a\}$. Likewise, $u:A \overset{e:a}{\longrightarrow} v:B$ stands for a directed graph with $V = \{u, v\}$, $E = \{e\}$, $conn(e) = (u, v)$, and $label = \{u \mapsto A, v \mapsto B, e \mapsto a\}$. Labels may be (partially) omitted, e.g., $u \overset{e:a}{\underline{\hspace{1em}}} v$ or $u \overset{e}{\underline{\hspace{1em}}} v$. The notation $L ::= R / C$ represents a graph grammar production with $Lhs = L$, $Rhs = R$ and $Elems_{Common} = C$. Productions with $Common = \lambda$ will be written as $L ::= R$.

Graph grammars inducible by our algorithm belong to a subclass of the LGG formalism, called *restricted layered graph grammars* in the rest of this article.

Definition 28 (restricted layered graph grammar).
A layered graph grammar $(\mathcal{N}, \mathcal{T}, \mathcal{P})$ is a *restricted LGG* (RLGG) if $\mathcal{N} = \{\#1, \#2, ..., \#k\}$ for some $k \in \mathbb{N}$ and if each production $p \in \mathcal{P}$ belongs to one of the following types:

Type 1: Productions of this type take the form

$$\lambda ::= R,$$

where R is a connected graph with $Labels \subseteq \mathcal{N} \cup \mathcal{T}$.

Type 2: Productions of this type take the form

$$u : A \xrightarrow{e\,:\,\#i} v : B \quad ::= \quad R/\{u, v\}$$

or the form

$$u : A \xrightarrow{e\,:\,\#i} v : B \quad ::= \quad R/\{u, v\},$$

where $\{A, B\} \subseteq \mathcal{T}$, $\#i \in \mathcal{N}$, and R is a connected graph with *Labels* $\subseteq \mathcal{N} \cup \mathcal{T}$ which contains a nonnull subgraph R' such that $Nh_R(R') = \{u, v\}$.

Type 3: Productions of this type take the form

$$u : A \xrightarrow{e\,:\,\#i} v : B \quad ::= \quad u \xrightarrow{f} v/\{u, v\}$$

or the form

$$u : A \xrightarrow{e\,:\,\#i} v : B \quad ::= \quad u \xrightarrow{f} v/\{u, v\},$$

where $\{A, B\} \subseteq \mathcal{T}$, $\#i \in \mathcal{N}$, and $label(f) \neq \#i$.

As with ordinary LGGs, an RLGG will be assumed to be either directed (all graphs of all productions are directed) or undirected (all graphs of all productions are undirected). To simplify the presentation, we will focus on the undirected case, except where the directed and undirected case are not completely analogous.

Definition 29 (guards, core, straps).

For a type-2 production $u \xrightarrow{e\,:\,\#i} v ::= R/\{u, v\}$, the vertices u and v will be called the *guards*, the graph $S \subseteq R$ with $Nh_R(S) = \{u, v\}$ will be called the *core*, and the edges with one endpoint being a vertex of S and the other endpoint being u or v will be called the *straps*.

The core graph is therefore obtained from the production's RHS graph (R) by removing the two context vertices and the straps.

Definition 30 (base graphs).

Given an RLGG, the set of *base graphs* consists of the RHSs of all type-I productions:

$$\mathcal{B}_{GG} = \{G | (\lambda ::= G) \in \mathcal{P}_{GG}\} \tag{1}$$

Definition 31 (cyclic set of productions).

A *sequence* of type-III productions $(p_1, \ldots, p_k)$ is *cyclic* if, for all $i \in \{1, \ldots, k - 1\}$, p_i takes the form $u \xrightarrow{e\,:\,\#j_i} v ::= u \xrightarrow{f\,:\,\#j_{i+1}} v$ and p_k takes the form

$u \xrightarrow{e:\#j_k} v ::= u \xrightarrow{f:\#j_1} v$. A *set* of type-III productions $\{p_1, \ldots, p_k\}$ is *cyclic* if there exists a permutation $\sigma: \{1, \ldots, k\} \to \{1, \ldots, k\}$ such that the sequence $(p_{\sigma(1)}, \ldots, p_{\sigma(k)})$ is cyclic.

Lemma 1. *An RLGG $(\mathcal{N}, \mathcal{T}, \mathcal{P})$ is parsable if the set of its type-III productions does not contain any cyclic subset.*

Proof. Let us assume that the production set satisfies the condition in the lemma. Let $\overline{G}$ be a directed graph in which vertices correspond to individual nonterminal labels and there is an edge from the vertex i to the vertex j if and only if the set $\mathcal{N}$ contains a production $u \xrightarrow{e:\#i} v ::= u \xrightarrow{f:\#j} v$. If there is no cyclic subset in the set of type-III productions, the graph $\overline{G}$ is acyclic, too. Let $order(i)$ denote the 1-based index of the vertex i in the reverse topological order of the vertex set of $\overline{G}$. (The vertex with $order = 1$ thus has no outgoing edges, and the one with $order = |\mathcal{N}|$ has no incoming edges.) Now, let $M = |\mathcal{N}| + 1$, and let $layer: \mathcal{N} \cup \mathcal{T} \to \{1, \ldots, M\}$ be a function such that $layer(a) = 1$ for all labels $a \in \mathcal{T}$ and $layer(\#i) = 1 + order(i)$ for all labels $i \in \mathcal{N}$. Let us now show that this assignment implies $\ell(Lhs_p) <_{\text{lex}} \ell(Rhs_p)$ for all productions p and hence the parsability of the grammar. For type-II productions, the first element of $\ell(Lhs)$ is 2 (the LHS always contains exactly two elements with $layer = 1$), and the first element of $\ell(Rhs)$ is strictly greater than 2, since the core graph is always nonnull and thus contains at least one (terminal-labeled) vertex. For a type-III production $u \xrightarrow{e:\#i} v ::= u \xrightarrow{f:\#j} v$, $\ell(Lhs) = (l_1, l_2, \ldots, l_M)$ with $l_1 = 2$, $l_{order(i)} = 1$, and $l_k = 0$ for all $k \notin \{1, order(i)\}$, while $\ell(Rhs) = (r_1, r_2, \ldots, r_M)$ with $r_1 = 2$, $r_{order(j)} = 1$, and $r_k = 0$ for all $k \notin \{1, order(j)\}$. Since $order(i) > order(j)$, we have $\ell(Lhs) <_{\text{lex}} \ell(Rhs)$. □

Despite limitations, it is possible to construct meaningful nontrivial grammars in the RLGG formalism. Fig. 7 shows the productions of an RLGG, henceforth called GG$_{\text{LHSDB}}$, which generates the same language

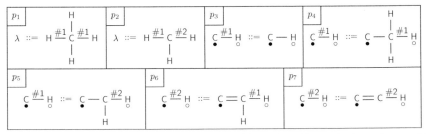

Fig. 7 GG$_{\text{LHSDB}}$: an RLGG for generating structural formulas of linear hydrocarbons with single and double bonds.

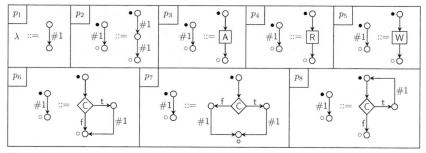

Fig. 8 GG_{FC}: an RLGG for generating flowcharts.

as GG^0_{LHSDB} (the symbols • and ∘ mark the guards). Fig. 8 displays the productions of an RLGG, denoted GG_{FC}, for generating flowcharts which consist of elementary statements (A—assign, R—read, and W—write), sequences of statements, conditional statements (both *if* and *if-else*), and loops.

Note that RLGGs can be interpreted as a special case of not only layered graph grammars but also hyperedge replacement grammars. Indeed: productions of type II and type III specify the replacement of an edge (i.e., a special case of a hyperedge) with a copy of a graph (i.e., a special case of a hypergraph), whereas the null graph, which serves as the LHS of type-I productions, can be interpreted as a hyperedge with no tentacles. The main difference between RLGGs and hyperedge replacement grammars is that the latter, at least in the standard definition, do not support labeled vertices. However, vertex labels can be appended to the labels of incident edges, so that an edge label identifies not only the edge itself but also its endpoints. Therefore, an RLGG could be rewritten as a hyperedge replacement grammar without much effort.

4.3 The big picture

The following terms have already been used in an intuitive sense, but have not yet been strictly defined:

Definition 32 (consistency).

A graph grammar GG is *consistent* with a given pair of a positive and a negative graph set, $\mathcal{G}^+$ and $\mathcal{G}^-$, if $G^+ \in L(GG)$ and $G^- \notin L(GG)$ for all graphs $G^+ \in \mathcal{G}^+$ and $G^- \in \mathcal{G}^-$.

Definition 33 (size).

The *size* of a production p is the size of its union graph, i.e., $|p| = |Union_p|$. The size of a grammar GG is the sum of the sizes of its productions, i.e., $|GG| = \sum_{p \in \mathcal{P}_{GG}} |p|$.

The size of a type-I production is thus equal to the size of its RHS. For a production of type-II or type-III, the size can be computed as $|Lhs| + |Rhs| - 2 = |Rhs| + 1$ (two vertices are common to LHS and RHS).

The induction algorithm tries to find the smallest grammar in the space of grammars consistent with the input graph sets, $\mathcal{G}^+$ and $\mathcal{G}^-$. The search space can be viewed as a graph in which vertices represent individual candidate grammars and edges represent *elementary generalizations*—operations which slightly generalize or merely restructure a given grammar. We use two types of elementary generalizations. The first type (type A) introduces a new type-II production and reverse-applies it to the base graphs (the RHSs of the type-I productions) wherever possible. The total size of the base graph set thus decreases, and if this reduction is greater than the size of the introduced production, the size of the grammar decreases as well. Under certain circumstances, the grammar also becomes more general. The second type of elementary generalizations (type B) replaces a pair of productions having isomorphic LHSs and unifiable (to be defined later) RHSs with a production which generalizes both of the original productions and a set of type-III productions which encode the differences in the original RHSs. This transformation may also make the grammar smaller.

For every grammar produced in the induction process, the algorithm employs the improved Rekers–Schürr LGG parser to determine whether the grammar covers any negative input graph. If this is the case, the grammar is immediately discarded, since it cannot be considered a candidate grammar.

Fig. 9 shows a part of the induction process starting from the structural formula of butane (the sole element of $\mathcal{G}^+$) and an empty set $\mathcal{G}^-$. The boxes contain individual candidate grammars, and the arrows represent elementary generalizations. For instance, the grammar GG_1 can be transformed by adding the production $p_{2,2}$ and reverse-applying it to the existing production $p_{1,1}$, which becomes $p_{2,1}$. The resulting grammar is even larger than the original, but it is not immediately discarded, since there is still a possibility for it to fruitfully participate in the induction process. By introducing and reverse-applying the production $p_{3,2}$, however, we arrive at a grammar strictly smaller than the source grammar. Both grammars, as well as GG_4, the result of the third possible type-A generalization of GG_1, are mere reformulations of the trivial grammar; they generate the structural formula of butane and nothing else. However, the grammar GG_5, obtained as a type-A generalization of GG_3, is strictly more general, since the production $p_{5,3}$ can be applied to occurrences of its own RHS. The language of GG_5 is the entire set of linear hydrocarbons with single bonds, which can be regarded as a

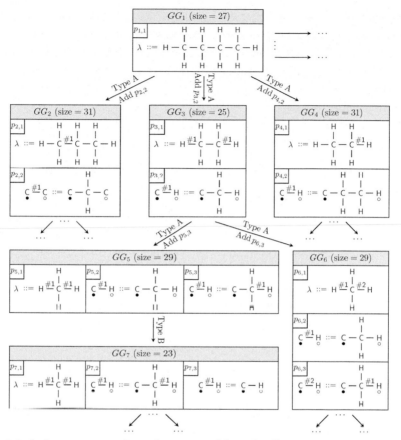

Fig. 9 Inducing a grammar from the structural formula of butane.

sensible generalization of the positive input graph set. The final result of the induction process, the grammar GG_7, is derived from GG_5 by a type-B generalization step, which replaces the productions $p_{5,2}$ and $p_{5,3}$ with $p_{7,2}$ and $p_{7,3}$.

In principle, the graph of candidate grammars could be explored by any search algorithm. Beam search, the algorithm of our choice, maintains a queue of candidate grammars, which initially contains only the trivial grammar. In each iteration, the algorithm replaces the smallest grammar in the queue with all of its elementary generalizations and retains the *beamWidth* smallest grammars in the queue. The algorithm halts when the queue becomes empty. Beam search can thus be considered to be halfway between breadth-first search (where all newly generated candidates are kept in the queue) and best-only search (where only the best candidate is kept in each iteration, effectively abolishing the need for a queue).

Algorithm 1 presents the pseudocode of the induction algorithm. The outer loop performs the beam search process, and the inner loop generates all elementary generalizations of the grammar which has just been removed from the queue. Since all candidate grammars are consistent with input graph sets, the algorithm may be interrupted any time, and the smallest grammar generated so far can be taken as the final result.

The auxiliary procedure BUILDNEWPRODUCTIONS builds a set of productions which can be used to augment the current grammar. This procedure will be explained in Section 4.4.

4.4 Type-A generalization

A type-A generalization step adds a new production to a given candidate grammar GG and reverse-applies the production to the graphs in $\mathcal{B}(GG)$ any number of times and in any order. Formally:

ALGORITHM 1 The algorithm to induce an RLGG from a positive ($\mathcal{G}^+$) and a negative ($\mathcal{G}^-$) input graph set.

1: **function** INDUCE($\mathcal{G}^+$, $\mathcal{G}^-$, *beamWidth*, *maxNCoreVertices*)
2:    $GG_1 :=$ the grammar with productions $\lambda ::= G$ for each graph $G \in \mathcal{G}^+$
3:    $GG_{\min} := GG_1$ // $GG_{\min}$: *the smallest candidate grammar so far*
4:    $Q := \{GG_1\}$
5:    **while** $Q \neq \emptyset$ **do**
6:      $GG :=$ the smallest grammar in the queue Q
7:      **if** $|GG| < |GG_{\min}|$ **then**
8:        $GG_{\min} := GG$
9:      **end if**
10:     $Q := Q \setminus \{GG\}$
11:     *NewProductions* := BUILDNEWPRODUCTIONS(GG, *maxNCoreVertices*)
12:     **for all** $p \in$ *NewProductions* **do**
13:       $GG' :=$ type-A generalization of GG using p
14:       **if** GG' does not cover any graph from $\mathcal{G}^-$ **then**
15:         $Q := Q \cup \{GG'\}$
16:         **if** type-B generalization of GG' exists **then**
17:           $GG'' :=$ type-B generalization of GG'
18:           **if** GG'' does not cover any graph from $\mathcal{G}^-$ **then**
19:             $Q := Q \setminus \{GG'\} \cup \{GG''\}$
20:           **end if**
21:         **end if**
22:       **end if**
23:     **end for**
24:     retain in Q only the *beamWidth* smallest grammars
25:    **end while**
26:    **return** $GG_{\min}$
27: **end function**

Definition 34 (type-A generalization).

A grammar $GG' = (\mathcal{N}', \mathcal{T}, \mathcal{P}')$ is a *type-A generalization* of a grammar $GG = (\mathcal{N}, \mathcal{T}, \mathcal{P})$ using a type-2 production p_{new} if $\mathcal{N}' = \mathcal{N} \cup label(Xlhs(p_{\text{new}}))$, $\mathcal{P}' = \mathcal{P} \cup \{p_{\text{new}}\} \setminus \{\lambda ::= G \,|\, G \in \mathcal{B}(GG)\} \cup \{\lambda ::= G' \,|\, G \in \mathcal{B}(GG) \wedge G \overset{p_{\text{new}}}{\Rightarrow} \cdots \overset{p_{\text{new}}}{\Rightarrow} G'\}$.

Since each reverse application of a type-2 production reduces the size of the graph, we reverse-apply the introduced production as many times as possible. The resulting grammar is hoped to be smaller than the original one and to be its meaningful generalization. However, even if none of these goals is immediately fulfilled, this may still be the case at some future moment in the induction process.

Fig. 9 shows several instances of type-A generalization. For example, the grammar GG_3 is obtained from the grammar GG_1 by introducing the production $p_{3,2}$ and reverse-applying it to the production $p_{1,1}$, which, as a result, becomes $p_{3,1}$.

A type-A generalization step might not strictly generalize the grammar, but it never shrinks its language:

Lemma 2. *A grammar GG' obtained from a grammar GG by a type-A generalization step is at least as general as GG.*

Proof. Let p_{new} be the production introduced by the generalization step. We have to show that every graph G derivable in GG is also derivable in GG'.

Let $\lambda \overset{p}{\Rightarrow} G_1 \Rightarrow \ldots \Rightarrow G_n = G$ be a derivation of G in GG. Since the set of type-II and type-III productions of GG is a subset of that of GG', the derivation $G_1 \Rightarrow \ldots \Rightarrow G_n = G$ exists in GG' too. As for the graph G_1, it can be derived in GG' either by the production $\lambda ::= G_1$ (if it has "survived" the generalization step) or by a sequence of productions $\lambda ::= G_0$ (with G_0 being the result of reverse-applying p_{new} to G_1) and p_{new}. $\qquad\square$

Type-A generalization could, in principle, add any type-II production to the given grammar. However, a reasonable strategy is to consider only those productions which can be reverse-applied at least once to at least one base graph. The following lemma specifies the necessary and sufficient condition for a production to be reverse-applicable:

Lemma 3. *Let p be a production $u \overset{e}{—} v ::= R/\{u,v\}$, and let S be the subgraph of R such that $Nh_R(S) = \{u, v\}$. The production p can be reverse-applied to a graph G if and only if all of the following holds:*

1. *G contains an r-occurrence R' of R (let h be the associated r-homomorphism).*

2. *R' contains an induced occurrence S' of S.*

3. *The only edges with one endpoint in S' and the other endpoint in $R' \setminus S'$ are the images (via h) of the edges having one endpoint in S and the other in $R \setminus S$.*

Proof. Condition (1) is necessary by virtue of how (reverse) production application is defined: to be able to reverse-apply a production p to a host graph G, a copy of Rhs_p has to appear in G, although the subgraph $Common_p$ need not be mapped injectively. In the first step, a reverse application removes the image of $Xrhs_p$ from G, leaving only the image of the guards ($Common_p$). The removal must not create any dangling edges. This is only possible if S' is an *induced* occurrence of S (condition (2)) and if there are no surplus edges between S' and $R' \setminus S'$ (condition (3)). If these conditions were not met, at least one edge (either between the vertices of S' or between those of S' and R') would be left dangling upon removing $h(Xrhs_p)$. Therefore, all three conditions are necessary.

Conversely, if all three conditions are satisfied, the production can be reverse-applied to R'.   □

Corollary 1. *Let W be a subset of the vertex set of a graph G, and let S be the induced subgraph of G formed by W. If $|Nh_G(S)| = 2$, then let $Nh_G(S) = \{u,v\}$ and let R be the subgraph of G made up of the elements of S, the vertices u and v, and all edges with one endpoint being a vertex of S and the other endpoint being either u or v. Now, the production $u\overset{e}{\text{---}}v ::= R'/\{u,v\}$, where R' is a copy of R, can be reverse-applied to the graph R.*

In other words, every induced subgraph with exactly two vertices in its neighborhood can form the core of some type-II production. This observation leads to a simple strategy for finding eligible type-II productions: in each graph G in the set of base graphs, enumerate all induced subgraphs S with $|Nh_G(S)| = 2$ and construct a production with a copy of S as its core, copies of the vertices $Nh_G(S) = \{u, v\}$ as its guards, and copies of the edges between S and $Nh_G(S)$ as its straps. The RHS of the production is thus completely determined. However, there is still a degree of freedom as to the label of the edge on the LHS. In principle, we could use any nonterminal label, but the productions $u : A\overset{e:\#i}{\text{---}}v : B ::= R/\{u,v\}$ and $u : A\overset{e:\#j}{\text{---}}v : B ::= R/\{u,v\}$ have the same effect on the grammar if neither $u : A\overset{e:\#i}{\text{---}}v : B$ nor $u : A\overset{e:\#j}{\text{---}}v : B$ occurs as a subgraph in the grammar. For this reason, we construct a production $u : A\overset{e:\#i}{\text{---}}v : B ::= R/\{u,v\}$ for each i such that the graph $u : A\overset{e:\#i}{\text{---}}v : B$ occurs as a subgraph somewhere in the grammar and for a single i without this property. In the latter case, the production will restructure the grammar rather than actually generalize it, but if a copy of the production's LHS is already part of an existing production, the language of the grammar might become strictly larger.

The procedure for constructing a set of productions suitable for augmenting a given grammar GG is shown as Algorithm 2. To save resources, the algorithm considers only the subgraphs with at most *maxNCoreVertices* vertices to be potential cores of future productions. The auxiliary function FINDPATTSOCCS (Section 4.6) returns a complete set of pairs $(Patt_i, Occs_i)$ with $Occs_i = \{(Host_{i1}, Occ_{i1}), ..., (Host_{ik_i}, Occ_{ik_i})\}$, where $Patt_i$ is a graph with at most *maxNCoreVertices* vertices, $Host_{ij}$ is one of the

---

ALGORITHM 2 The algorithm to construct a set of productions to augment a given grammar within the type-A generalization procedure.

```
 1: function BUILDNEWPRODUCTIONS(GG, maxNCoreVertices)
 2:     PattsOccs := FINDPATTSOCCS(B_GG, maxNCoreVertices)
 3:     Productions := ∅
 4:     for all (Patt, Occs) ∈ PattsOccs do
 5:         for all (Host, Occ) ∈ Occs do
 6:             if |Nh_Host(Occ)| = 2 then
 7:                 let u and v be the two vertices in Nh_Host(Occ)
 8:                 A := label(u);
 9:                 B := label(v);
10:                 I := {i | (u: A ──e: #i── v: B) occurs in GG}
11:                 if I = ∅ then
12:                     k := 1
13:                 else
14:                     k := max(I) + 1
15:                 end if
16:                 R := BUILDRHS(Host, Occ)
17:                 for all i ∈ I ∪ {k} do
18:                     Productions := Productions ∪ {(u: A ──e: #i── v: B) ::= R}
19:                 end for
20:             end if
21:         end for
22:     end for
23:     return Productions
24: end function
25: function BUILDRHS(Host, Occ)
26:     N := Nh_Host(Occ)
27:     V := V_Occ ∪ N
28:     E := E_Occ ∪ {e ∈ E_Host | (conn(e) ∩ V_Occ ≠ ∅ ∧ conn(e) ∩ N ≠ ∅}
29:     return a copy of Host[V, E]
30: end function
```

base graphs of GG, and Occ_{ij} is an induced occurrence of $Patt_i$ in $Host_{ij}$. For each occurrence with a 2-vertex neighborhood, the function then builds a set of productions as described in the previous paragraph. As we saw in Algorithm 1, each of those productions is then separately added to the grammar and reverse-applied to the base graphs.

For directed graphs, the procedure for finding candidate productions is analogous to the undirected case. However, in the directed case, there are two degrees of freedom when constructing the LHS of a production. Besides the label, there are two possible directions of the edge. For each label $\#i$, we thus create the productions

$$u : A \xrightarrow{e:\#i} v : B := R/\{u, v\},$$

$$u : A \xleftarrow{e:\#i} v : B := R/\{u, v\}$$

instead of the production

$$u : A \xrightarrow{\quad e:\#i \quad} v : B := R/\{u, v\}.$$

4.5 Type-B generalization

Type-B generalization is based on the notion of *unifiability*:

Definition 35 (unifiability and unification of labels).
Labels a and b are *unifiable* (denoted $a \sim b$) if $a = b$ or $a \in \mathcal{N} \vee b \in \mathcal{N}$. The *unification* of unifiable labels a and b is defined as follows:

$$unif(a, b) = \begin{cases} a & \text{if } a \in \mathcal{N}, \\ b & \text{otherwise.} \end{cases} \tag{2}$$

This concept can be extended to graphs:

Definition 36 (unifiability of graphs, unifying isomorphism).
Graphs G and H are *unifiable* (denoted $G \sim H$) if there exists a bijective mapping $h: G \to H$ such that $label(h(x)) \sim label(x)$ for all $x \in G$ and $h(conn(e)) = conn(h(e))$ for all $e \in E_G$. Such a mapping is called a *unifying isomorphism*.

While an ordinary isomorphism can map an element $x \in G$ to an element $x' \in H$ only if they have the same label, a unifying isomorphism can also match pairs of elements with unifiable labels.

Definition 37 (unification of graphs).
Given a unifiable pair of graphs G and H and a unifying isomorphism $h: G \to H$, the *unification* of the graphs G and H is a graph K obtained by copying the graph G and setting $label_K(e') = unif(label_G(e), label_H(h(e)))$ for all edges $e \in E_G$ and their copies $e' \in E_K$.

An example of graph unification is shown in Fig. 10. Graphs G and H are unifiable because the function which maps the vertex labeled A in G to the vertex labeled A in H, the vertex labeled B in G to the vertex labeled B in H, etc. is a unifying isomorphism. The graph K is the unification of graphs G and H.

A type-B generalization step replaces a pair of type-1 or type-2 productions having isomorphic LHSs and unifiable RHSs with a type-1 or type-2 production which generalizes the two original productions and with a set of type-3 productions which represent the differences between the original RHSs. Formally:

Definition 38 (type-B generalization).

Let p: $L ::= R$ and q: $L ::= R'$ be productions of a grammar $GG = (\mathcal{N}, \mathcal{T}, \mathcal{P})$ such that the graphs R and R' are unifiable. Furthermore, let $R'' = \textit{unif}(R, R')$, and let h: $R'' \to R$ and h': $R'' \to R'$ be unifying isomorphisms. A grammar $GG' = (\mathcal{N}, \mathcal{T}, \mathcal{P}')$ is a *type-B generalization* of GG if $\mathcal{P}' = \mathcal{P} \setminus \{p, q\} \cup \{r\} \cup S$, where r is the production $L ::= R''$ and the set S contains a type III production for each edge $e \in E_{R''}$ where either $\textit{label}(h(e)) \neq \textit{label}(e)$ or $\textit{label}(h'(e)) \neq \textit{label}(e)$. In the first case, the production in S corresponding to the edge $e \in E_{R''}$ is $u : A \xrightarrow{\ e_1 : label(e)\ } v : B ::= u \xrightarrow{\ e_2 : label(h(e))\ } v$, where A and B are the labels of the endpoints of e. In the second case, the production in S corresponding to the edge $e \in E_{R''}$ is $u : A \xrightarrow{\ e_1 : label(e)\ } v : B ::= u \xrightarrow{\ e_2 : label(h'(e))\ } v$, where A and B are the labels of the endpoints of e.

Fig. 9 presents a single example of type-B generalization: the grammar GG_7 is derived from GG_5 by replacing the productions $p_{5,2}$ and $p_{5,3}$ with the generalizing production $p_{7,2}$ ($\textit{unif}(\#1, \$_E) = \#1$) and the difference representation set $\{p_{7,3}\}$.

As with type-A generalization, type-B generalization never removes anything from the language of the grammar:

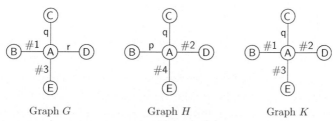

Graph G Graph H Graph K

Fig. 10 Graph K is the unification of graphs G and H.

Lemma 4. *A grammar* GG' *obtained from a grammar* GG *by a type-B general-ization step is at least as general as* GG.

Proof. Let us assume that a type-B generalization step replaced type-I (or type-II) productions p and q with a type-I (or type-II) production r and type-III productions s_1, ..., s_k, ..., s_l for some value of k and l. Without loss of gen-erality, we may assume that s_1, ..., s_k correspond to the differences between r and p and s_{k+1}, ..., s_l correspond to the differences between r and q. Now, a derivation step $G \overset{p}{\Rightarrow} H$ can be simulated by a sequence $G \overset{r}{\Rightarrow} \overset{s_1}{\Rightarrow} ... \overset{s_k}{\Rightarrow} H$, and a derivation step $G \overset{q}{\Rightarrow} H$ can be simulated by a sequence $G \overset{r}{\Rightarrow} \overset{s_{k+1}}{\Rightarrow} ... \overset{s_l}{\Rightarrow} H$. Consequently, any graph derivable in the original grammar is also derivable in its type-B generalization. $\square$

Generalization steps of both types may create a grammar with duplicate productions. If that happens, the induction algorithm simply removes all redundant productions.

4.6 Candidate subgraph search

The procedure for finding candidate productions to augment a grammar in a type-A generalization step relies on an algorithm to enumerate subgraph pat-terns and their induced occurrences in the grammar's base graph set. There are multiple approaches which serve this purpose; see [44] for a detailed review. Our algorithm, shown as Algorithm 3, was inspired by the Subdue [43] and VSiGram [57] graph data mining approaches.

Algorithm 3 starts with enumerating all single-vertex subgraphs in the base graph set of the given grammar. The result of this step is a set of single-vertex pattern graphs, $Patts_1$, and a set $Occs(P)$ for each graph P $\in Patts_1$, where $Occs(P) = \{(H_1, O_1), (H_2, O_2), ...\}$ indicates that the graph P occurs as the subgraph O_1 in the graph H_1, as the subgraph O_2 in the graph H_2, etc. The ith iteration of the algorithm starts with the set of $(i-1)$-vertex pattern graphs, $Patts_{i-1}$. For each graph $P_{i-1} \in Patts_{i-1}$ and for each induced occurrence $O \subseteq H$ of P_{i-1}, the algorithm collects all vertices $v \in H$ adjacent to O and, for each such vertex v, constructs a new subgraph O' by extending O by v and all edges $e \in H$ which connect v and one of the existing vertices in O. If the subgraph O' has already been generated (i.e., if its vertices have been visited in a different order), it can safely be discarded. If the subgraph O' has not yet been visited, but we have already constructed a pattern graph P isomorphic to O', then O' is recorded as another occurrence of P. The third possibility is that no graph P isomorphic to O' has been produced so far. In this case, we have found a new pattern graph. A copy P' of O' is added to the set $Patts_i$, and O' is recorded as one of its occurrences.

ALGORITHM 3 The algorithm to build a set of pattern graphs and their occurrences in a given graph set $\mathcal{G}$.

1: **function** FINDPATTSOCCS($\mathcal{G}$, $maxNCoreVertices$)
2:     $Patts_1 := \{P \mid |V_P| = 1 \land P$ has at least one induced occurrence in $\mathcal{G}\}$
3:     **for all** $P \in Patts_1$ **do**
4:         $Occs(P) := \{(H, O) \mid H \in \mathcal{G} \land O$ is an ind. occurrence of P in $H\}$
5:     **end for**
6:     **for all** $i \in \{2, \ldots, maxNCoreVertices\}$ **do**
7:         $Patts_i := \emptyset;$
8:         **for all** $P_{i-1} \in Patts_{i-1}$ **do**
9:             **for all** $(H, O) \in Occs(P_{i-1})$ **do**
10:                 **for all** $v \in Nh_H(O)$ **do**
11:                     $O' := O \cup \{v\}$
12:                     $O' := O' \cup \{e \in E_H \setminus E_{O'} \mid \exists w \in V_{O'}: conn(e) = \{v, w\}\}$
13:                     **if** $\neg\exists(H, O'') \in Occs(P_{i-1}): O'' = O'$ **then**
14:                         **if** $\exists P \in Patts_i: P$ and O' are isomorphic **then**
15:                             // O' is a new occurrence of an existing pattern P
16:                             $Occs(P) := Occs(P) \cup \{(H, O')\}$
17:                         **else**
18:                             // O' is the first occurrence of a new pattern P'
19:                             $P' :=$ a copy of O'
20:                             $Patts_i := Patts_i \cup \{P'\}$
21:                             $Occs(P') := \{(H, O')\}$
22:                         **end if**
23:                     **end if**
24:                 **end for**
25:             **end for**
26:         **end for**
27:     **end for**
28:     **return** $\{(P, Occs(P)) \mid P \in \bigcup_{i=1}^{maxNCoreVertices} Patts_i\}$
29: **end function**

The final result of the algorithm is a set of pairs $(P, Occs(P))$, where P is a graph with at most $maxNCoreVertices$ vertices and at least one induced occurrence in the given grammar's base graph set, and $Occs(P)$ is a set of pairs (H, O), which determine the induced occurrences of P.

Note that we are interested solely in *induced* occurrences (Corollary 1). For this reason, we augment a given occurrence O not only by a single host graph vertex but also by all host graph edges between that vertex and the existing vertices of the occurrence.

For every generated subgraph, the algorithm has to determine whether it is isomorphic to some pattern graph. The efficiency of the algorithm is thus heavily influenced by the efficiency of isomorphism checks. A potentially efficient solution is to convert each graph to the so-called *canonical string* [58], which is a string invariant to different numberings of vertices and edges. As a consequence, two graphs are isomorphic if and only if their canonical strings are equal. In this way, the problem of checking whether two graphs are isomorphic is reduced to that of checking whether two strings are equal.

Unfortunately, we should not be surprised that creating a canonical string is, in terms of computational complexity, no less hard than solving the graph isomorphism problem itself. Therefore, we content ourselves with *pseudo-canonical strings* which have to fulfill a weaker requirement: if two graphs have the same pseudo-canonical string, they are isomorphic. The reverse is, of course, strongly desirable, but not required; if two isomorphic graphs have different pseudo-canonical strings, the pattern-occurrence set will contain two copies of the same pattern, each with its own set of occurrences, but no real harm will be done.

In practice, the following method achieves fairly good results. The overall idea is to build a robust pseudo-canonical string on the basis of the robust *pseudo-canonical numbering* of graph vertices. The pseudo-canonical numbering and, in turn, the pseudo-canonical string should be as insensitive as possible to the initial numbering of the vertices. For each vertex in a given n-vertex graph, we first build its *local-features string* from the label of the vertex itself, the lexicographically ordered list of labels of the edges incident to that vertex, and, for each incident edge, the label of the neighbor at the other end of that edge. The list of local-features strings is then lexicographically sorted, and the index of a vertex's local-feature string in the ordered list is taken as the vertex's pseudo-canonical number. To break at least some ties, the neighbors' labels in the local-features strings of the vertices having the same pseudo-canonical number are augmented with the neighbors' pseudo-canonical numbers, and the strings are sorted again. After reassigning pseudo-canonical numbers, the pseudo-canonical string of the graph is constructed as $label(v_{\sigma(1)}) \circ / \circ label(v_{\sigma(2)}) \circ / \circ \ldots / \circ label (v_{\sigma(n)}) \circ / / \circ \alpha_{11} \circ / \circ \alpha_{12} \circ / \circ \ldots / \circ \alpha_{1n} \circ / / \circ \ldots \circ \alpha_{n1} \circ / \circ \alpha_{n2} \circ / \circ \ldots / \circ \alpha_{nn}$, where $\circ$ denotes string concatenation, $v_{\sigma(i)}$ is the vertex whose pseudo-canonical number is i, $/$ is a unique separator (a label which does not belong to the universal set of graph element labels), and α_{ij} is a string representing the relationship between the vertices $v_{\sigma(i)}$ and $v_{\sigma(j)}$. In particular, α_{ij} is an lexicographically ordered list of labels of the parallel edges connecting the vertices $v_{\sigma(i)}$ and $v_{\sigma(j)}$, and $-$ (another unique marker) if the two vertices are not connected.

To illustrate these concepts, consider the graph in Fig. 11. Column LFS_1 in Table 1 shows the initial local-features strings for individual vertices. The local-features string of a vertex v takes the form $L(l_1/L_1, \ldots, l_k/L_k)$, where $L = label(v)$, l_i is the label of the ith edge incident to v, and L_i is the label of the vertex on the opposite end of that edge. The indices are assigned in such a way that the list $l_1/L_1, \ldots, l_k/L_k$ is ordered lexicographically. In the next step (column PCN_1), the vertices receive initial pseudo-canonical numbers corresponding to the lexicographical ordering of local-features strings. The vertices v_3, v_4, and v_5, which have the lexicographically smallest local-features strings, receive a pseudo-canonical number of 1. The vertices v_2 and v_1 are assigned numbers 4 and 5, respectively. In the next step (column LFS_2), the neighbors' labels in the local-features strings of the vertices v_3, v_4, and v_5 are augmented with their pseudo-canonical numbers. After that (column PCN_2), the pseudo-canonical numbers of those vertices are computed again (note that the vertices v_3 and v_4 still have equal local-features strings, and so we could also have swapped their pseudo-canonical numbers). On the basis of the pseudo-canonical ordering v_5, v_3, v_4, v_2, v_1, we construct the following pseudo-canonical string for our graph:

$$A/A/A/A/B//-/\$_E/\$_E/-/-//\$_E/-/-/\$_E/-//\$_E/-/-/\$_E/-//$$
$$-/\$_E/\$_E/-/\$_E//-/-/-/-/\$_E$$

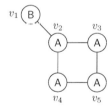

Fig. 11 A sample graph.

Table 1 Local-features strings (LFS) and pseudo-canonical numbers (PCN) for the vertices of the graph from Fig. 11.

	LFS_1	PCN_1	LFS_2	PCN_2
v_1	B($\$_E$/A)	5		
v_2	A($\$_E$/A, $\$_E$/A, $\$_E$/B)	4		
v_3	A($\$_E$/A, $\$_E$/A)	1	A($\$_E$/A/1, $\$_E$/A/4)	2
v_4	A($\$_E$/A, $\$_E$/A)	1	A($\$_E$/A/1, $\$_E$/A/4)	3
v_5	A($\$_E$/A, $\$_E$/A)	1	A($\$_E$/A/1, $\$_E$/A/1)	1

To understand why a pair of equal pseudo-canonical strings signifies a pair of isomorphic graphs, consider that a pseudo-canonical string can be viewed as a (generalized) adjacency matrix rewritten in a row-by-row fashion. Conversely, a pseudo-canonical string relies on the labels and neighborhoods of individual vertices, and these are, at least to a certain degree, independent of the initial numbering of the graph.

5. Experimental results

By virtue of its design, the induction algorithm always produces a graph grammar consistent with the given input. However, the output grammar is not necessarily a meaningful generalization of the positive input graph set. The grammar may cover just the set $\mathcal{G}^+$ and nothing else, or the set $\mathcal{G}^+$ might be generalized in an undesirable way.

In this section, we show the performance of the induction algorithm on four meaningful test suites. We applied the algorithm to cycles (Section 5.1), binary trees (Section 5.2), flowcharts (Section 5.3), and LHSDB graphs (Section 5.4). In the first three experiments, the grammar is induced solely from positive graphs, whereas in the fourth one, the input set contains both positive and negative graphs. Cycles, binary trees, and LHSDB graphs are undirected, while flowcharts are directed.

Unless noted otherwise, the parameters *beamWidth* and *maxNCoreVertices* were set to 10 and 5, respectively. Nevertheless, the algorithm proved to be fairly insensitive to changes in these parameters, as long as they had sensible values. For example, since the productions of the reference grammar for LHSDB graphs (Fig. 7) have at most three vertices in their *Xrhs* sets, the value of *maxNCoreVertices* has to be at least 3 to make it possible to recreate the reference grammar from a set of input graphs. A larger value increases the computational demands, but is not likely to produce a significantly smaller grammar, as grammars with larger *Xrhs* sets tend to have greater overall sizes.

5.1 Induction from cycles

In the first set of experiments, we induced graph grammars from cycles. The n-cycle (denoted C_n) is the undirected graph with vertices $v_1, \ldots, v_n$ and edges $v_1 - v_2, \ldots, v_{n-1} - v_n, v_n - n_1$. Fig. 12 shows the grammars induced from positive sets $\{C_6\}, \{C_8\}, \{C_{10}\}$, and $\{C_6, C_8\}$. In all four experiments, the negative input set was empty.

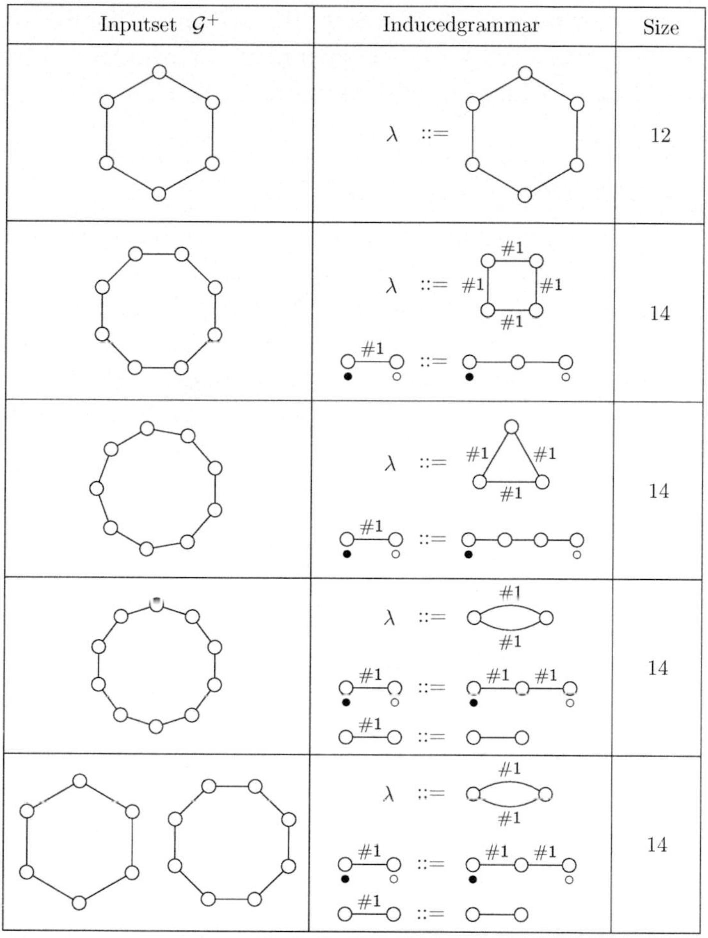

Fig. 12 Graph grammars induced from sets of cycles.

If the algorithm is fed with sets $\mathcal{G}^+ = \{C_6\}$ and $\mathcal{G}^- = \varnothing$, the result is the trivial grammar. The same goes for the set $\{C_7\}$. The positive input set $\{C_8\}$ leads to a grammar which is not trivial but still fails to cover anything more than the 8-cycle itself. A similar result is obtained from the set $\{C_9\}$. However, the set $\{C_{10}\}$ results in a grammar $GG^\star$ which covers all cycles with at least two vertices. This grammar, which is, in fact, induced from any set $\{C_n\}$ with $n \geq 10$ and also from sets such as $\{C_6, C_8\}$, can definitely be considered a meaningful generalization of the input set.

Actually, the grammar $GG^\star$ is created in the induction process even if the algorithm is applied to a set $\{C_n\}$ with $n < 10$. However, in such cases the grammar is not small enough to compete with the trivial grammar or its restructurings. For example, the size of the trivial grammar for the set $\{C_6\}$ is 12, while the size of GG^* is 14. The grammar GG^* also fails to be smaller than the grammars induced from the sets $\{C_8\}$ and $\{C_9\}$.

5.2 Induction from binary trees

For the purpose of this section, a *binary tree* is a full binary tree (i.e., each vertex has either zero or two children) in which the leaves are labeled **B** and the interior vertices are labeled **A**. A binary tree is assumed to have at least three vertices; a single vertex will thus not be considered a binary tree.

Fig. 13 presents induction from sample singleton sets of binary trees. Again, the sets $\mathcal{G}^-$ were empty in all three cases. In the case of the input set $\mathcal{G}_1^+$, the algorithm produced the most favorable output: the induced grammar (GG_1) covers the entire set of binary trees. In fact, this grammar was induced whenever the tree in the input set had at least three complete levels.

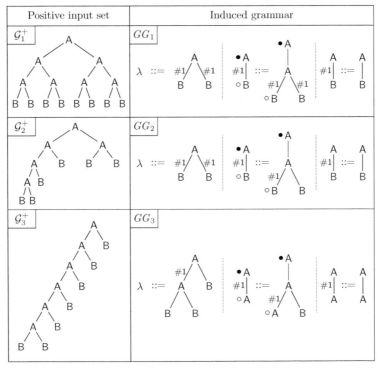

Fig. 13 Graph grammars induced from singleton sets of binary trees.

The tree in the set $\mathcal{G}_2^+$ has only two complete levels, and so the induction algorithm could not carry out all the type-A generalization steps necessary to produce a full grammar for binary trees. On the other hand, the set $\mathcal{G}_3^+$ contains a degenerate tree, and the algorithm induces a grammar for such trees.

5.3 Induction from flowcharts

In the third set of experiments, we induced a grammar from sets of flowcharts. Each input set consisted of 10–50 random graphs belonging to the language of GG_{FC} (Fig. 8). The individual graphs, constructed by sequences of random applications of GG_{FC} productions, comprised up to 25 vertices. In most cases, the induction algorithm produced the grammar of Fig. 14 (called GG_{FC}^I in the following text) or a slight variation thereof.

Superficially, the induced grammar may appear fairly different from the reference grammar. However, a closer inspection reveals that the productions p_1–p_8 of GG_{FC}^I actually correspond to the productions p_1–p_8 of GG_{FC}, despite the fact that some arrows are reversed. Since the productions p_9–p_{13} make it possible to generate a graph which does not belong to $L(GG_{FC})$, it holds that $L(GG_{FC}^I) \supseteq L(GG_{FC})$.

5.4 Induction from LHSDB graphs

In our last experimental setup, we tried to induce a grammar of structural formulas of linear hydrocarbons with single and double bonds. It soon became apparent that a grammar for the target language (or at least something similar) is unlikely to be inducible from a positive set alone. For this

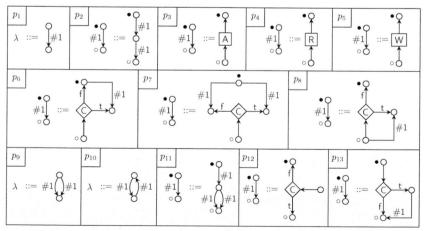

Fig. 14 GG_{FC}^I: the grammar induced from a set of flowcharts.

reason, we prepared both a positive and a negative set of input graphs. The positive input set ($\mathcal{G}_{\text{LHSDB}}^+$) consisted of all LHSDB graphs with up to 6 vertices labeled C (carbon atoms), which means 42 graphs in total. The negative input set ($\mathcal{G}_{\text{LHSDB}}^-$) comprised 200 graphs obtained by randomly choosing LHSDB graphs with at most 4 vertices C and randomly removing one or two vertices H from them.

Having prepared the input sets, we induced a grammar using Algorithm 4. The algorithm validates the grammars as they are being induced. It tries to find small subsets $\mathcal{S}^+ \subseteq \mathcal{G}_{\text{LHSDB}}^+$ and $\mathcal{S}^- \subseteq \mathcal{G}_{\text{LHSDB}}^-$ such that the grammar induced from them is consistent with the "universal" sets $\mathcal{G}_{\text{LHSDB}}^+$ and $\mathcal{G}_{\text{LHSDB}}^-$. At the beginning, the set $\mathcal{S}^+$ contains only the smallest graph from the set $\mathcal{G}_{\text{LHSDB}}^+$ (i.e., methane), while the set $\mathcal{S}^-$ is empty. The algorithm then induces a grammar GG from the sets $\mathcal{S}^+$ and $\mathcal{S}^-$ and builds a set ($\mathcal{Z}^+$) of all graphs from $\mathcal{G}_{\text{LHSDB}}^+$ which are not covered by GG and a set ($\mathcal{Z}^-$) of all graphs from $\mathcal{G}_{\text{LHSDB}}^-$ covered by GG. In other words, the graphs from the sets $\mathcal{Z}^+$ and $\mathcal{Z}^-$ are those misclassified by the grammar GG. If both sets are empty, the algorithm halts, since it has successfully produced a grammar consistent with the sets $\mathcal{G}_{\text{LHSDB}}^+$ and $\mathcal{G}_{\text{LHSDB}}^-$. Otherwise, the algorithm

ALGORITHM 4 The algorithm to induce a grammar with simultaneous validation.

```
 1: function INDUCEANDVALIDATE(𝒢₀⁺, 𝒢₀⁻, beamWidth, maxNCoreVertices)
 2:     𝒮⁺ := {the smallest graph from 𝒢₀⁺}
 3:     𝒮⁻ := ∅
 4:     GG := INDUCE(𝒮⁺, 𝒮⁻, beamWidth, maxNCoreVertices)
 5:     𝒵⁺ := {G ∈ 𝒢₀⁺ | GG does not cover G}
 6:     𝒵⁻ := {G ∈ 𝒢₀⁻ | GG covers G}
 7:     while (𝒵⁺ ≠ ∅) ∨ (𝒵⁻ ≠ ∅) do
 8:         if 𝒵⁻ ≠ ∅ then
 9:             𝒮⁻ := 𝒮⁻ ∪ {the smallest graph from 𝒵⁻}
10:         else
11:             𝒮⁺ := 𝒮⁺ ∪ {the smallest graph from 𝒵⁺}
12:         end if
13:         GG := INDUCE(𝒮⁺, 𝒮⁻, beamWidth, maxNCoreVertices)
14:         𝒵⁺ := {G ∈ 𝒢₀⁺ | GG does not cover G}
15:         𝒵⁻ := {G ∈ 𝒢₀⁻ | GG covers G}
16:     end while
17:     return (𝒮⁺, 𝒮⁻, GG)
18: end function
```

repeats the following procedure as long as $\mathcal{Z}^+ \cup \mathcal{Z}^- \neq \varnothing$: add the smallest graph from $\mathcal{Z}^-$ to $\mathcal{S}^-$ (if $\mathcal{Z}^- \neq \varnothing$) or the smallest graph from $\mathcal{Z}^+$ to $\mathcal{S}^+$ (in the opposite case), re-induce the grammar GG from the updated sets $\mathcal{S}^+$ and $\mathcal{S}^-$, and rebuild the sets $\mathcal{Z}^+$ and $\mathcal{Z}^-$.

Fig. 15 displays the sets $\mathcal{S}^+$ and $\mathcal{S}^-$ at the end of the algorithm. The grammar induced from these sets (let us call it GG^I_{LHSDB}) is shown in Fig. 16. As with the reference grammar (GG_{LHSDB} of Fig. 7), the label #1 represents an edge which will ultimately develop into a C–H bond or a single C–C bond, while the label #2 represents a future double C–C bond. With this observation in mind, it is easy to see that the grammar GG_{LHSDB} generates only valid hydrocarbon graphs and that any LHSDB graph is a member of $L(GG_{LHSDB})$. However, the language of the induced grammar is a superset of the LHSDB graph set; the grammar GG^I_{LHSDB} is capable of producing structural formulas of some nonlinear hydrocarbon graphs with single and double bonds. The reason for this (slight) overgeneralization can be traced to type-B generalization. Ideally, a type-B generalization step should find a unifying isomorphism which minimizes the number of productions to be added to the grammar. For the pair of isomorphic RHSs shown in Fig. 17, for example, an "optimal" unifying isomorphism would map the vertex u_3 to v_4 and the vertex u_4 to v_3. However, our implementation of type-B generalization does not

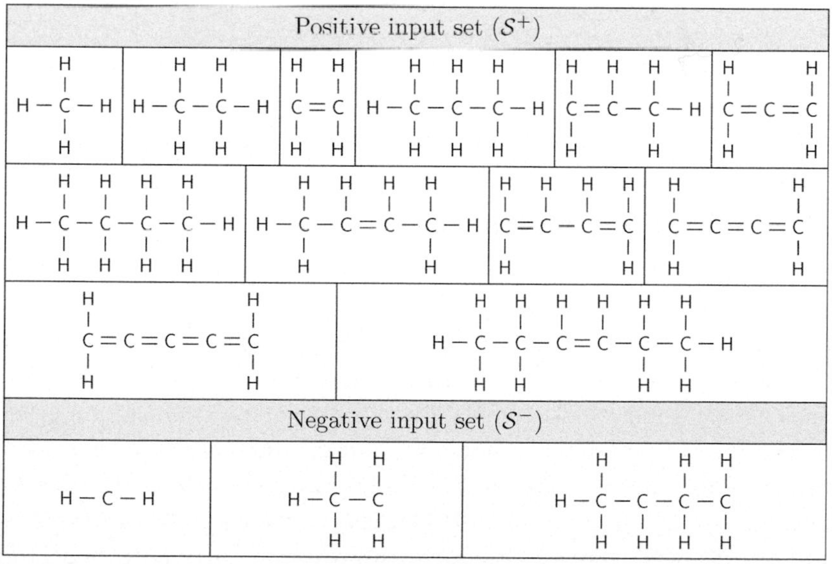

Fig. 15 The sets $\mathcal{S}^+$ and $\mathcal{S}^-$ created by Algorithm 4 in the LHSDB induction setup.

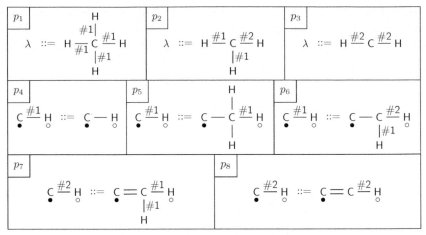

Fig. 16 The grammar induced from the sets $\mathcal{S}^+$ and $\mathcal{S}^-$ from Fig. 15.

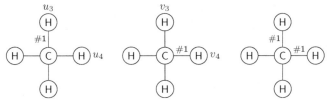

Fig. 17 A pair of isomorphic graphs (*left* and *middle*) and a possible result of their unification (*right*).

perform any optimizations and might instead produce a unifying isomorphism which maps u_3 to v_3 and u_4 to v_4. In this case, we obtain an RHS with two #1-labeled edges, and the grammar consequently becomes more general.

5.5 Computational performance

The time required to induce a grammar may be exponential in terms of the size of input sets. The potentially computationally expensive parts of the algorithm are the procedure for enumerating subgraphs (a graph may have exponentially many induced subgraphs) and the (improved) Rekers–Schürr parser. The computational complexity of the subgraph enumeration procedure can be brought down to polynomial at the price of missing some subgraphs; in our case, this is the role of the parameter *maxNCoreVertices*. On the other hand, in view of the NP-hardness of the graph grammar parsing problem, the parser probably cannot be improved to run in polynomial time in the worst case.

To give an impression of the *actual* performance of the algorithm, Table 2 shows the number of generated grammars (including those inconsistent with input sets) and the total time consumption when inducing a grammar from the sets S^+ and S^- from Fig. 15. The experiment was repeated for different values of the parameters *beamWidth* (with *maxNCoreVertices* being fixed at 5) and *maxNCoreVertices* (with *beamWidth* = 10). For the default values of *beamWidth* and *maxNCoreVertices* (10 and 5, respectively), the algorithm took 7.2 s to complete.[c] Naturally, a greater value of *beamWidth* implies a larger number of generated grammars and, in turn, longer duration of the algorithm, since more grammars are retained in the beam search queue. As for the *maxNCoreVertices* parameter, a greater value implies a longer search for subgraphs and, once again, a larger number of generated grammars.

Fig. 18 displays the running time for inducing grammars from subsets of the sets $\mathcal{G}^+_{\text{LHSDB}}$ and $\mathcal{G}^-_{\text{LHSDB}}$. We conducted two series of experiments. In the *i*th experiment (for $i \in \{1, \dots, 42\}$) of the first series, we induced a grammar

Table 2 The total number of grammars created during the induction from the sets S^+ and S^- from Fig. 15, and the total time spent by the induction process.

	beamWidth				*maxNCoreVertices*			
	1	10	100	1000	3	5	7	9
Number of grammars	116	148	590	24 435	95	148	195	224
Duration (in seconds)	6.8	7.2	12.2	370	3.4	7.2	12.4	17.5

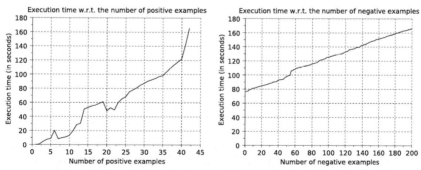

Fig. 18 The dependence of execution time on the number of positive input graphs (*left*) and the number of negative input graphs (*right*) in the LHSDB induction scenario.

[c] All experiments were conducted on an 8-core 3.40 GHz Intel Core i7 machine.

from a set composed of the i smallest graphs in $\mathcal{G}^+_{\text{LHSDB}}$, which served as the positive input set, and the entire set $\mathcal{G}^-_{\text{LHSDB}}$, which served as the negative input set. In the second series of experiments, grammars were induced from the entire set $\mathcal{G}^+_{\text{LHSDB}}$ and from sets composed of the i smallest graphs from the set $\mathcal{G}^-_{\text{LHSDB}}$. As we can see, in the first series the execution time does not grow monotonically with the increasing number of input graphs. The reason is that, all other things being equal, adding one more (positive) graph to the input set may result in a grammar which is more "parser-friendly" than the original one. The time required to parse a graph against a grammar does not necessarily grow with the size of the grammar.

6. Conclusion

In this paper, we have dealt with the problem of graph grammar induction, the goal of which is to find a concise graph grammar which can generate all of the given positive graphs and, optionally, none of the given negative graphs. Most of the existing approaches learn grammars only from positive graphs, and they pursue their objective by finding and reverse-applying productions which compress the given graph set to the greatest extent. Having intertwined the induction process with a graph grammar parser, we induce graph grammars from both positive and negative graphs, and we use not only a generalization approach based on reverse applications of candidate productions but also a technique which merges productions with isomorphic LHSs and similar RHSs and can be, to the best of our knowledge, regarded as a novelty. We have demonstrated the capabilities of our approach by applying it to several nontrivial graph sets, one of which had to contain both positive and negative graphs in order to give rise to a suitable graph grammar.

Currently, our algorithm induces only the grammars where the LHSs of noninitial productions consist of a pair of context vertices connected with a directed or undirected edge. Therefore, a natural extension is to build productions with any number of context vertices, with the idea to induce a sub-class of LGG grammars equivalent to the class of hyperedge replacement grammars. A hyperedge could be represented by a hub vertex with ordered "tentacles" connecting it with each of the context vertices. However, while the idea might seem simple to implement (just remove the restriction that a core graph needs to have exactly two neighbors), the challenge lies in the combinatorial explosion of the number of ways the tentacles can be ordered.

To arrive at a feasible solution, the stipulation requiring the equivalence to hyperedge replacement grammars should almost certainly be relaxed, and even in this case we could hardly do without advanced heuristic techniques.

In the type-A generalization step, productions used to augment the given grammar are (indirectly) evaluated by their immediate effect on the size of the grammar. However, some productions might have a limited short-term impact but a potentially large long-term influence. In particular, the production $L ::= R$ might only have a single occurrence of R in the current grammar, but its reverse application at that occurrence might result in an occurrence of an RHS unifiable with R, giving rise to a recursive production. (For instance, when the production $p_{3,2}$ of the grammar GG_3 in Fig. 9 is reverse-applied to one of its occurrences in the input graph (the RHS of GG_1), the RHS of the production $p_{5,3}$ arises at that occurrence, leading to the recursive production $p_{5,3}$ and subsequently (after merging $p_{5,2}$ and $p_{5,3}$) to $p_{7,2}$.) Since no true generalization is possible without directly or indirectly recursive productions, they should be given priority over nonrecursive ones. To prevent overlooking a potentially recursive production with a low static score, productions should be evaluated by considering not only their static compression capabilities but also the potential effects of their reverse applications.

To make the intricate details of the matter presented herein more accessible to students, we could build a tool which makes it possible to perform graph grammar induction in a step-by-step manner. As shown by Nikolić et al. [59], the pedagogical benefits of such tools cannot be overestimated.

References

[1] J. Rekers, A. Schürr, Defining and parsing visual languages with layered graph grammars, J. Vis. Lang. Comput. 8 (1) (1997) 27–55.
[2] D.Q. Zhang, K. Zhang, J. Cao, A context-sensitive graph grammar formalism for the specification of visual languages, Comput. J. 44 (3) (2001) 186–200.
[3] M. Flasiński, S. Myśliński, On the use of graph parsing for recognition of isolated hand postures of Polish Sign Language, Pattern Recognit. 43 (6) (2010) 2249–2264.
[4] L. Lin, T. Wu, J. Porway, Z. Xu, A stochastic graph grammar for compositional object representation and recognition, Pattern Recogn. 42 (7) (2009) 1297–1307.
[5] N. Aschenbrenner, L. Geiger, Transforming scene graphs using Triple Graph Grammars—a practice report, in: A. Schürr, M. Nagl, A. Zündorf (Eds.), Applications of Graph Transformations with Industrial Relevance, LNCS, vol. 5088, Springer, Heidelberg, 2007, pp. 32–43.
[6] F. Schöler, V. Steinhage, Towards an automated 3D reconstruction of plant architecture, in: A. Schürr, D. Varró, G. Varró (Eds.), Applications of Graph Transformations with Industrial Relevance, LNCS, vol. 7233, Springer, Heidelberg, 2011, pp. 51–64.
[7] K. Ehrig, J.M. Küster, G. Taentzer, Generating instance models from meta models, Softw. Syst. Model. 8 (4) (2009) 479–500.

[8] L. Fürst, M. Mernik, V. Mahnič, Converting metamodels to graph grammars: doing without advanced graph grammar features, Softw. Syst. Model. 14 (3) (2015) 1297–1317.

[9] P. Gómez-Abajo, E. Guerra, J de Lara, A domain-specific language for model mutation and its application to the automated generation of exercises, Comput. Lang. Syst. Struct. 49 (2017) 152–173.

[10] C. Zhao, K. Ates, J. Kong, K. Zhang, Discovering program's behavioral patterns by inferring graph-grammars from execution traces, in: 20th International Conference on Tools with Artificial Intelligence, ICTAI 2008, November 3–5, 2008, Dayton, Ohio, USA, vol. 2, IEEE Computer Society, 2008, pp. 395–402.

[11] M.J. Plasmeijer, M.C.J.D. van Eekelen, Term graph rewriting and mobile expressions in functional languages, in: M. Nagl, A. Schürr, M. Münch (Eds.), Applications of Graph Transformations with Industrial Relevance, LNCS, vol. 1779, Springer, Heidelberg, 1999, pp. 1–13.

[12] S. Sippu, E. Soisalon-Soininen, Parsing theory, Volume I Languages and Parsing, EATCS Monographs on Theoretical Computer Science. Springer-Verlag, 1988.

[13] B. Slivnik, On different LL and LR parsers used in LLLR parsing, Comput. Lang. Syst. Struct. 50 (2017) 108–126.

[14] L. Fürst, M. Mernik, V. Mahnič, Improving the graph grammar parser of Rekers and Schürr, IET Softw. 5 (2) (2011) 246–261.

[15] G. Rozenberg, Handbook of Graph Grammars and Computing by Graph Transformation: Volume 1 (Foundations), World Scientific, 1997.

[16] E.M. Gold, Language identification in the limit, Inf. Control. 10 (5) (1967) 447–474.

[17] L. Fürst, M. Mernik, V. Mahnič, Graph grammar induction as a parser-controlled heuristic search process, in: A. Schürr, D. Varró, G. Varró (Eds.), Applications of Graph Transformations With Industrial Relevance (AGTIVE), LNCS, vol. 7233, Springer, Budapest, Hungary, 2011, pp. 121–136.

[18] S. Maneth, F. Peternek, Compressing graphs by grammars, in: 32nd IEEE International Conference on Data Engineering, ICDE 2016, Helsinki, Finland, May 16–20, 2016, IEEE Computer Society, 2016, pp. 109–120.

[19] L. Fürst, Interactive graph grammar induction, World Usability Day, Ljubljana, Slovenia, 2013, pp. 25–28.

[20] M. Flasiński, Inference of parsable graph grammars for syntactic pattern recognition, Fund. Inform. 80 (4) (2007) 379–413.

[21] A. Roudaki, J. Kong, K. Zhang, Specification and discovery of web patterns: a graph grammar approach, Inform. Sci. 328 (2016) 528–545.

[22] V. Blagojević, D. Bojić, M. Bojović, M. Cvetanović, J. Đorđević, D. Đurđević, B. Furlan, S. Gajin, Z. Jovanović, D. Milićev, V. Milutinović, B. Nikolić, J. Protić, M. Punt, Z. Radivojević, ž. Stanisavljević, S. Stojanović, I. Tartalja, M. Tomašević, P. Vuletić, A systematic approach to generation of new ideas for PhD research in computing, in: A.R. Hurson, V. Milutinović (Eds.), Creativity in Computing and DataFlow SuperComputing, Advances in Computers, vol. 104, 2017, pp. 1–31 (Chapter 1).

[23] K. Ates, J.P. Kukluk, L.B. Holder, D.J. Cook, K. Zhang, Graph grammar induction on structural data for visual programming, in: 18th IEEE International Conference on Tools with Artificial Intelligence, IEEE Computer Society, Washington, DC, 2006, pp. 232–242.

[24] R. Parekh, V. Honavar, Grammar inference, automata induction, and language acquisition, in: R. Dale, H.L. Somers, H. Moisl (Eds.), Handbook of Natural Language Processing, Marcel Dekker, New York, 2000, pp. 727–764.

[25] A. Stevenson, J.R. Cordy, Grammatical inference in software engineering: an overview of the state of the art, in: K. Czarnecki, G. Hedin (Eds.), Int. Conf. on Software Language Engineering, LNCS, vol. 7745, Springer, Dresden, Germany, 2012, pp. 204–223.

[26] R.M. Wharton, Grammar enumeration and inference, Inf. Control. 33 (3) (1977) 253–272.

[27] K. VanLehn, W. Ball, A version space approach to learning context-free grammars, Mach. Learn. 2 (1) (1987) 39–74.

[28] T.M. Mitchell, Generalization as search, Artif. Intell. 18 (2) (1982) 203–226.

[29] C.G. Nevill-Manning, I.H. Witten, Identifying hierarchical structure in sequences: a linear-time algorithm, J. Artif. Intell. Res. 7 (1997) 67–82.

[30] A. Habel, Hyperedge Replacement: Grammars and Languages, LNCS, vol. 643, Springer, 1992.

[31] M. Mernik, G. Gerlič, V. žumer, B.R. Bryant, Can a parser be generated from examples? in: ACM Symposium on Applied Computing (SAC), March 9–12, 2003, Melbourne, FL, USA, 2003, pp. 1063–1067.

[32] M. Črepinšek, M. Mernik, B.R. Bryant, F. Javed, A. Sprague, Context-free grammar inference for domain-specific languages, Technical Report UABCIS-TR-2006-0301-1, University of Alabama at Birmingham, 2006.

[33] F. Javed, M. Mernik, J. Gray, B.R. Bryant, MARS: A metamodel recovery system using grammar inference, Inf. Softw. Technol. 50 (9–10) (2008) 948–968.

[34] K. Nakamura, M. Matsumoto, Incremental learning of context free grammars based on bottom-up parsing and search, Pattern Recogn. 38 (9) (2005) 1384–1392.

[35] D.H. Younger, Recognition and parsing of context-free languages in time n^3, Inf. Control. 10 (2) (1967) 189–208.

[36] S.L. Graham, M.A. Harrison, Parsing of general context-free languages, Adv. Comput. 14 (1976) 77–185.

[37] D. Bojić, M. Bojović, Chapter five—A streaming dataflow implementation of parallel Cocke-Younger-Kasami parser, in: A.R. Hurson, V. Milutinović (Eds.), Creativity in Computing and DataFlow SuperComputing, Advances in Computers, vol. 104, Elsevier, 2017, pp. 159–199.

[38] K. Imada, K. Nakamura, Towards machine learning of grammars and compilers of programming languages, in: W. Daelemans, B. Goethals, K. Morik (Eds.), Machine Learning and Knowledge Discovery in Databases, Springer, Berlin, Heidelberg, 2008, pp. 98–112.

[39] F. Javed, M. Mernik, B.R. Bryant, A. Sprague, An unsupervised incremental learning algorithm for domain-specific language development, Appl. Artif. Intell. 22 (7–8) (2008) 707–729.

[40] D. Hrnčič, M. Mernik, B.R. Bryant, F. Javed, A memetic grammar inference algorithm for language learning, Appl. Soft Comput. 12 (3) (2012) 1006–1020.

[41] A. Dubey, P. Jalote, S.K. Aggarwal, Learning context-free grammar rules from a set of programs, IET Softw. 2 (3) (2008) 223–240.

[42] E. Jeltsch, H.J. Kreowski, Grammatical inference based on hyperedge replacement, Springer, 1990, pp. 461–474.

[43] L.B. Holder, D.J. Cook, S. Djoko, Substructure discovery in the SUBDUE system, in: - U.M. Fayyad, R. Uthurusamy (Eds.), Knowledge Discovery in Databases (Workshop), AAAI Press, Seattle, Washington, USA, 1994, pp. 169–180.

[44] D.J. Cook, L.B. Holder, Mining Graph Data, John Wiley & Sons, 2006.

[45] I. Jonyer, L.B. Holder, D.J. Cook, MDL-based context-free graph grammar induction and applications, Int. J. Artif. Intell. Tools 13 (1) (2004) 65–79.

[46] J.P. Kukluk, L.B. Holder, D.J. Cook, Inference of node replacement graph grammars, Intell. Data Anal. 11 (4) (2007) 377–400.

[47] J. Kukluk, L. Holder, D. Cook, Inferring graph grammars by detecting overlap in frequent subgraphs, Int. J. Appl. Math. Comput. Sci. 18 (2) (2008) 241–250.

[48] H. Ehrig, G. Engels, G. Rozenberg, Handbook of Graph Grammars and Computing by Graph Transformation, Applications, Languages, and Tools, vol. 2, World Scientific, 1999.

[49] J. Kong, K. Zhang, X. Zeng, Spatial graph grammars for graphical user interfaces, ACM Trans. Comput.-Human Interaction 13 (2) (2006) 268–307.

[50] R. Brijder, H. Blockeel, On the inference of non-confluent NLC graph grammars, J. Log. Comput. 23 (4) (2013) 799–814.

[51] T. Oates, S. Doshi, F. Huang, Estimating maximum likelihood parameters for stochastic context-free graph grammars, in: T. Horváth (Ed.), Proc. 13th Int. Conf. on Inductive Logic Programming, LNCS, vol. 2835, Springer, Szeged, Hungary, 2003, pp. 281–298.

[52] G. Rozenberg, E. Welzl, Boundary NLC graph grammars—basic definitions, normal forms, and complexity, Inf. Control 69 (1–3) (1986) 136–167.

[53] J.R. Ullmann, An algorithm for subgraph isomorphism, J. ACM 23 (1) (1976) 31–42.

[54] U. Čibej, J. Mihelič, Search strategies for subgraph isomorphism algorithms, in: P. Gupta, C.D. Zaroliagis (Eds.), International Conference on Applied Algorithms, ICAA 2014, Kolkata, India, January 13–15, 2014, 2014, pp. 77–88.

[55] U. Čibej, J. Mihelič, Improvements to Ullmann's algorithm for the subgraph isomorphism problem, Int. J. Pattern Recognit. Artif. Intell. 29 (7) (2015) 28.

[56] H. Ehrig, G. Engels, H.J. Kreowski, G. Rozenberg, U. Montanari, Handbook of Graph Grammars and Computing by Graph Transformation, vol. 1–3, World Scientific, 1997–1999.

[57] M. Kuramochi, G. Karypis, Finding frequent patterns in a large sparse graph, Data Min. Knowl. Disc. 11 (3) (2005) 243–271.

[58] L. Babai, E.M. Luks, Canonical labeling of graphs, in: D.S. Johnson, R. Fagin, M.L. Fredman, D. Harel, R.M. Karp, N.A. Lynch, …J.I. Seiferas (Eds.), Proc. 15th Annual ACM Symposium on Theory of Computing, ACM, Boston, Massachusetts, USA, 1983, pp. 171–183.

[59] B. Nikolić, Z. Radivojević, J. Djordjević, V. Milutinović, A survey and evaluation of simulators suitable for teaching courses in computer architecture and organization, IEEE Trans. Educ. 52 (4) (2009) 449–458.

About the authors

Luka Fürst received his M.Sc. and Ph.D. in computer science from the University of Ljubljana in 2007 and 2013, respectively. He works as a teaching assistant at the University of Ljubljana, Faculty of Computer and Information Science. His research areas include graph grammars, general graph theory (with a particular emphasis on graph symmetries), software engineering, image analysis, and computer programming education.

Marjan Mernik received the M.Sc. and Ph. D. in computer science from the University of Maribor in 1994 and 1998, respectively. He is a professor at the University of Maribor, Faculty of Electrical Engineering and Computer Science. He is also a visiting professor at the University of Alabama at Birmingham, Department of Computer and Information Sciences, and at the University of Novi Sad, Faculty of Technical Sciences. His research interests include programming languages, compilers, domain-specific (modeling) languages, grammar-based systems, grammatical inference, and evolutionary computations. He is a member of the IEEE, ACM, and EAPLS. Dr. Mernik is the editor-in-chief of Computer Languages, Systems and Structures journal, as well as associate editor of Applied Soft Computing journal.

Viljan Mahnič received the M.Sc. and Ph.D. degrees in computer science from the University of Ljubljana in 1981 and 1990, respectively. He is a professor and the head of the Software Engineering Laboratory at the Faculty of Computer and Information Science of the University of Ljubljana, Slovenia. His teaching and research interests include programming languages, software development methods, software process improvement, empirical software engineering, and software measurement.

CHAPTER FOUR

Asymmetric windows in digital signal processing

Robert Rozman

University of Ljubljana, The Faculty of Computer and Information Science, Ljubljana, Slovenia

Contents

Abstract

Symmetric windows are widely used in the field of digital signal processing due to their easy design and linear phase property. Nevertheless, symmetry also implies a few potential drawbacks like longer time delay in short-time frequency analysis and some limitations in frequency response. The removal of the symmetry constraint can therefore lead to asymmetric windows better in certain respects. In signal processing, better signal representations and related improved processing performance can be accomplished. In addition, shorter time delay can be achieved with asymmetric windows. This feature

Advances in Computers, Volume 116
ISSN 0065-2458
https://doi.org/10.1016/bs.adcom.2019.07.004

is important for contemporary spoken communications in the Internet or mobile networks and all other real-time signal processing applications.

The article gives a comprehensive review of the past and current work in the field of asymmetric windows. We elaborate on our work and related efforts of other researchers inspired by the idea of asymmetry. Shorter time delay and some better spectral properties are the most prominent potential of asymmetric windows. However, there are also some other more subtle properties which can improve the performance in specific application contexts (e.g., frequency estimation and detection of closely spaced components in frequency analysis). Several examples of interesting effects of asymmetric windows are presented, followed by empirical evaluations in the fields of pitch modification, shorter time delay audio processing (e.g., speech coding), frequency analysis, speech processing, and FIR filter design. In addition, a detailed comparison of various asymmetric windows found in the literature to widely known symmetric windows is made taking into account several practical and theoretical aspects. Finally, all presented achievements are summarized in a table which provides a complete overview of the current state of this interesting research and application field.

Abbreviations

AMFCC	autocorrelation mel frequency cepstral coefficient
ASR	automatic speech recognition
DCT	discrete cosine transform
DFT	discrete Fourier transform
DNN	deep neural network
DSP	digital signal processing
FBANK	filter bank feature
FIR	finite impulse response
FT	Fourier transform
HMM	hidden Markov model
IIR	infinite impulse response
LP	linear prediction
MFCC	mel frequency cepstral coefficient
MLP	multi-layer perceptron
OLA	overlap and add
PSOLA	pitch-synchronous overlap and add
RMSE	root mean square error
SNR	signal-to-noise ratio
SR(S)	speech recognition (system)
STFT	short-time Fourier transform
TF	time-frequency
TTS	text to speech
UBM	universal background model

1. Introduction

Digital signals are becoming a ubiquitous part of our life. Signal processing platforms are more and more compact, low-priced and capable of performing complex computing operations. Nowadays, we acquire and process many signals which were some years ago either not available or too complex to be processed. We are able to solve complex problems with low-priced desktop systems, and to produce small ubiquitous devices which can interact with the environment, other devices, and users via signals and other communication channels.

Despite the ongoing evolution, some concepts stay more or less constant and unchanged in the field of Digital Signal Processing (DSP). Signals have to be acquired through an analog-to-digital conversion process denoted as sampling, after which signals are processed in a digital form as a series of samples in the time domain or in the transformed domain (mostly frequency domain).

When processing digital signals, two tasks are generally performed. The first one is the information extraction task (e.g., speech recognition, frequency analysis) which tries to gain specific information from the signal's content. The second task is the transformation of the signal into a new form which is in certain aspects better than the original one (e.g., filtering, speech enhancement, denoising). Sometimes the time domain representation of the signal suffices for both tasks. However, in the majority of nontrivial cases and more complex signal processing procedures, the signal is transformed into some other domain (mostly frequency domain). Here the desired properties are better recognized and distinguished from other less important ones.

By far the most commonly used concept for the transformation of digital signals into the frequency domain is the Discrete Fourier Transform (DFT), which reveals the frequency content of the signal more informative for a series of signal processing tasks. We denote such procedure of determining the frequency content of a signal as frequency analysis and we will put a lot of focus on it in this article. It is particularly important because it serves as a starting point for various methods from the fields of speech, image and audio processing. Those are among the most widespread research and practical application areas of DSP.

Several examples of such applications will be presented in this article. They will demonstrate how improvements in frequency analysis can enhance the performance of systems which build on the time-frequency presentation of the signals. As we have said, there are many such systems which could easily benefit from the improved results of frequency analysis. Therefore, we could achieve multiplicative effects on the widespread area of related techniques and systems. This is the main reason to acknowledge the high importance of frequency analysis as an elementary procedure in these fields.

Regarding the variability of a signal's frequency content, we can classify signals into two common groups. If a signal's frequency content is stationary, then it is commonly analyzed in a single segment with a finite number of samples. In this simple case, we do not have any time dimension in frequency presentation; it is mostly not needed anyway.

If the signal's content is nonstationary over time, then it is generally analyzed in a series of finite, overlapped segments (commonly denoted as frames) with the assumption of the stationarity in each frame (Fig. 2). We denote this procedure as a short-time frequency analysis which usually results in common three-dimensional (time, frequency, magnitude) signal representation denoted as a spectrogram (Fig. 3). In both cases, the "average" frequency content of the signal is calculated in each frame. In the second case of multiple frames, those are generally shorter and stationarity is assumed for a single frame only. Frame length is frequently determined in the context of the application, and is generally a compromise between the desired time and frequency resolutions of the signal's representation.

In both variations of frequency analysis, we perform analysis or processing tasks on frames with a finite number of samples. Samples outside of a particular frame are assumed to be equal to zero. Theoretically, this step is equivalent to a product between the infinite discrete signal being sampled and another infinite sequence of samples with non-zero values only inside the same frame and all zero values elsewhere. This sequence is denoted as a window and is a built-in part of any frame-based processing procedure on digital signals.

The emerging digital evolution has generated more and more powerful computing platforms available for digital signal processing tasks. However, there are two permanent aspects which require proper attention even in modern signal processing systems: the processing time delay and the undesired variability of signals. No matter how powerful and efficient our systems are, we always aspire to perform better, faster, and do more complex computations in the same time unit. Consequently, the two aspects will probably stay in focus in the future as well.

In this section, we further elaborate on the two highlighted aspects (time delay in processing and undesired signal variability) in separate subsections. We also present the concept of asymmetric windows as a potential approach to deal with negative effects of those aspects. In the last subsection, a few important general thoughts are highlighted, explaining the research path we follow in our work.

The rest of the chapter is organized as follows. Section 2 is a general overview of windows in signal processing with explanation of their influence in both approaches to frequency analysis and the Finite Impulse Response (FIR) filter design. A lot of practical applications in the signal processing area are based on these elementary tasks. Section 3 represents a more detailed introduction to asymmetric windows, their potentials, and design methods. The following section, which is the most important part of the article, gives a general review of related work on asymmetric windows. Various approaches to the design of asymmetric windows from various application areas are here presented and analyzed in more detail with the aim of evaluating the effect of presented asymmetric windows and explaining their advantages and possible pitfalls. Section 5 contains discussion and conclusions. The presented asymmetric windows and their utilization in various DSP application areas are summarized in a table. In addition, we identify the related promising research topics for the future.

1.1 Time delay in signal processing

In practice, we are only able to process signals using finite sequences of samples, which means dividing longer signals into shorter frames. Such approach is particularly important for real-time performance. We process signals in finite frames, on the one-by-one basis, and perform information extraction and/or signal transformation tasks on each frame separately. Consequently, it takes time to acquire all samples in the frame and carry out additional processing steps. Therefore, we cannot avoid the intrinsic time delay between the input signal and output results. We usually want to make this time delay as short as possible as most real-time signal processing applications severely depend on a short processing time delay.

Besides traditional signal processing tasks which depend on a short time delay, such as analysis, coding, distributed processing, recognition and communications, the time delay is also important for several emerging, interactive and user-centered tasks such as [1]:
• Hearing aids
• Augmented reality

- Computer-aided music practicing tools
- Brain-computer interface
- Patient monitoring systems
- Musical pitch tracking
- Real-time onset detection

For all the mentioned applications, the short time delay of processing is crucial for their performance. For most others, it is also a desired feature. Therefore, the requirement for a short processing time delay is basically built into almost every real-time application in the field of DSP.

1.2 Undesired variability of signals

Another important aspect of modern signal processing paradigms is the variability of signals. From the pure signal processing viewpoint, an ideal signal would be constant and without any noise. However, the signal's variability normally contains useful information and related variability is usually denoted as a desired variability. On the other hand, quite often there are various unwanted distortions present in signals, which are denoted as undesired variability. In practice, we always try to distinguish between those two types of variability in signals. This is usually a complex task, at least for nontrivial problems.

Generally, we want to focus on useful variability in signals. However, undesired variability commonly interferes with its useful counterpart in different ways. Therefore, we create signal presentations which merge both domains. In short-time frequency analysis, we denote such representation as Time-Frequency (TF). One axis represents time and the other frequency. We try to utilize both dimensions when focusing only on useful variability in signals.

The most common interfering variability is noise. It is always present in real signals at various power levels. The relation of power between the useful part of the signal and the unwanted noise is denoted as the Signal-to-Noise Ratio (SNR). However, the measure itself does not help much without the processing techniques which enhance signals achieving higher SNR ratios. In some cases, SNR loss is inevitable due to the processing of signals, and we only try to minimize that SNR loss; this situation is denoted a passive approach. On the other side, if we succeed to increase the SNR ratio (e.g., filtering, speech enhancement, denoising), those methods are usually recognized as an active approach.

1.3 Can asymmetric windows help?

The solution to these two major issues of modern signal processing (namely, time delay and variability of signals) seems easy at first sight. We simply need to design and implement systems which will perform with a short time delay and will be able to distinguish between the useful and the undesired variability in signals. Unfortunately, this is easier to describe in theory than achieve in practice. Still, there are various ways to improve existing systems in both directions.

One possibility is the use of asymmetric windows which can help tackle the stated problems in practice. As we will show in the continuation of the article, asymmetric windows can in certain respects be better than the commonly used symmetric windows. They can produce better signal presentations with a shorter time delay in processing. Despite being considered a passive approach, their integration is usually a simple replacement with no additional computing cost.

As already stated, asymmetric windows can have better spectral and time delay properties compared to the commonly used symmetric windows. As a simple, but a built in step in a frame-based frequency analysis procedure, the influence on final results is generally expected not to be very significant. Since asymmetric windowing can be considered as a passive method, it tries to minimize the loss of the SNR ratio during frequency analysis. However, as we will show in the article, it can contribute to better results in most procedures based on frequency analysis. This fact represents the main motivation for this article.

There is some more interesting potential for the enhancement of frequency estimation results. Certain properties of asymmetric windows which result in better frequency representation could also contribute to a better performance of entire systems based on frequency analysis. In addition, we could empirically evaluate the hypothesis that passive asymmetric window improvement to a certain level also has an effect on the performance of complex systems which already operate in real conditions using symmetric windows. Some of our preliminary tests, carried out in a research work related to speech recognition, confirmed that exciting finding [2].

Finally, we would like to highlight some general thoughts on the evolution of the signal processing research area in the next subsection.

1.4 Knowledge vs. data, and inherent robustness

The first thought is the dilemma whether it is better to enhance the performance of systems by incorporating more and more knowledge or to rely on emerging machine learning concepts (e.g., deep learning models). The latter can learn from data even without initial domain knowledge and are capable of quite an outstanding performance. However, they also exhibit certain shortcomings (e.g., limited ability of explanation or presentation of gained knowledge).

During our research work, we have tested both approaches. However, we have always preferred the knowledge-based approach with proper empirical evaluation. We have focused on existing knowledge and constantly tried to expand it with actual research achievements. Consequently, we have always wanted to have a full understanding of a common workflow from the input, processing and output of various systems.

On the other hand, we feel that deep learning networks are a promising concept and deserve proper research attention. In addition, we can gain a lot of new knowledge from such systems. However, we want to foster knowledge-based approaches to maintain a proper level of understanding of the systems which perform more and more complex tasks. We also believe that we can enhance the performance of knowledge-based systems by inclusion of knowledge gained from machine learning concepts.

The second dilemma is related to the importance of system robustness to signal distortions not present in the learning phase. In our work, we denote this property as "inherent robustness." This situation is quite common in many cases. Systems are usually designed and/or trained in "learning" conditions and it is practically impossible to include all "real" environment conditions in a system's learning phase. Therefore, systems' performance is often substantially degraded in a "real" environment. Actually, the systems which are able to perform successfully in different conditions are very rare. This is particularly true in the fields of speech processing and recognition. Consequently, we strongly believe that it is always important to enhance the system's inherent robustness and get better performance in this way although, this is often much harder and takes more time. However, if we perform this task successfully, even the systems which are trained in "laboratory" (clean) conditions can perform better in unseen real environmental conditions. In our opinion, this is the key point for further development of such systems and deserves a proper research attention.

As we have explained in the above subsections, we believe that asymmetric windows are a promising concept which needs proper research attention and can generate a lot of interesting new research ideas [3]. However, their implementations in signal processing systems require background knowledge in computer architecture and organization areas.

Therefore, a significant effort is also needed in computer engineering education [4]. Both arguments contribute to the motivation for the work in the field of asymmetric windows presented in this article.

2. Windows in signal processing

In practice, signals are often acquired and processed using finite time intervals. By doing this, we multiply infinite signals with windows which are non-zero only inside a particular time interval. Therefore, windows can be essentially seen as weighting functions applied in time or frequency domains in various situations with certain purposes. As classified in Ref. [5], windows are used in four common situations and are characterized accordingly:

- Data window
  - ○ Time domain weighting function for operations on data (e.g., averaging, smoothing, PSOLA pitch alteration [6])
- Lag window
  - ○ Time domain weighting function for the covariance of the observed process
- Frequency window
  - ○ Frequency domain weighting function equivalent to the operation of convolution in the time domain (e.g., Mel-Frequency Filter bank)
- Spectral window
  - ○ Time domain weighting function with related effects in the frequency domain

For our purpose, windows will be treated primarily as spectral windows, i.e., weighting functions with certain effects in the frequency domain presentation of the signal. The element-wise product of a signal's frame and a window corresponds to a dual operation in the frequency domain—a convolution integral (Eq. 3). Because of the expressed duality relation between both domains, windows can also be implemented in the frequency domain using a convolution operation; for some windows, this implementation can be even more efficient [5]. However, the time domain implementation is frequently in the primary focus in the area of signal processing, which is also the case in this article.

Symmetric windows represent the vast majority of windows used in signal processing, which can be attributed to several reasons. They can mostly be easily calculated, understood and mathematically analyzed. However, the property of symmetry can also be recognized as a restriction. In a symmetric window, half of the values represent a symmetrical pair of the other half. From an optimization standpoint, only half of the values can be freely changed or optimized.

If we relax the symmetry restriction, we can get windows with some better properties [2,7–9]. However, to be able to create better windows consistently, we need to understand their influence in each application and propose optimized solutions, often according to application-specific criteria. When an application problem is fairly complex, this is not an easy task. Therefore, we often use general knowledge and experiences from the frequency analysis field as an alternative source. This is often useful because frequency analysis is the first step in most signal processing applications.

On the other hand, windows belong to a group of "passive" approaches, and their influence might not be so pronounced. This is probably the strongest argument why little research attention was devoted to this field in the past. The main purpose of this article and presented related work from other authors is to expose the field of asymmetric windows, highlight its potential advantages and show its empirically confirmed importance in various practical cases.

In this section, we will present a more detailed insight into windows and their role in general signal processing. We will elaborate on three prominent fields in more detail: both variants of the frequency analysis procedure and FIR filter design methods. In all of them, windows play an important role.

2.1 Frequency analysis of stationary signals

Whenever we process discrete signals using finite frames, theoretically speaking, we perform truncation from an infinite duration signal to a finite non-zero sequence of N samples. This step is denoted as windowing and is defined with

$$s_w(n) = w(n)\, s(n), \ -\infty < n < \infty \tag{1}$$

where $s(n)$ represents discrete samples of the infinite signal, $w(n)$ is the window sequence and $s_w(n)$ is the resulting discrete signal as a sequence of samples which can be non-zero only within the finite range of N samples. Theoretically, $s_w(n)$ is still an infinite discrete sequence, but its samples are equal to 0 outside the finite frame of N samples. Therefore, only non-zero samples which form a finite sequence are kept for further processing.

After windowing, Fourier Transform (FT) is applied to $s_w(n)$, determining its frequency response

$$S_w\left(e^{j\omega}\right) = \sum_{n=-\infty}^{\infty} s_w(n)\, e^{-j\omega n} \tag{2}$$

Practically, the operation of windowing in Eq. (1) is simple to understand in the time domain. However, it is more complicated from the frequency domain viewpoint. According to Fourier Transform (FT) properties, multiplication of two signals in the time domain is dual to a convolution integral between the corresponding transforms in the frequency domain. As a consequence, the resulting transform in the frequency domain differs from the transform of the infinite signal $s(n)$. The transform of a windowed sequence $s_w(n)$ is defined with

$$S_w\left(e^{j\omega}\right) = \frac{1}{2\pi} \int_{\jmath_l}^{\pi} S\left(e^{j\theta}\right) W\left(e^{j(\omega-\theta)}\right) d\theta \tag{3}$$

where $S(e^{j\omega})$ is the Fourier Transform of the infinite signal, and $W(e^{j\omega})$ is the Fourier Transform of the window. This distortion of the resulting transform cannot be avoided in practice when using frame-based analysis methods. We therefore need to understand it properly and take it into account in the context of a particular application as much as possible.

In order to understand the difference between the "ideal" and the calculated "real" transform better, further explanation is needed. Fourier Transform in Eq. (2) equals the frequency response of the windowed sequence. By definition, it shows the frequency content of the windowed signal as a function of frequency ω and comprises two components

$$S_w\left(e^{j\omega}\right) = \left|S_w\left(e^{j\omega}\right)\right| e^{j \sphericalangle S_w(e^{j\omega})} \tag{4}$$

where $\left|S_w(e^{j\omega})\right|$ is a magnitude response and $\sphericalangle S_w(e^{j\omega})$ is a phase response of the windowed signal. In most applications, we focus on the magnitude response more, because it represents magnitudes for frequency components contained in the signal; these are often more important in practical applications. However, the frequency response gives us complex information about signal frequency content: for each component with frequency ω_0, the values of its magnitude and phase response can be determined.

To extend the understanding, we would like to present the difference between the "ideal" and the gained "real" frequency response visually.

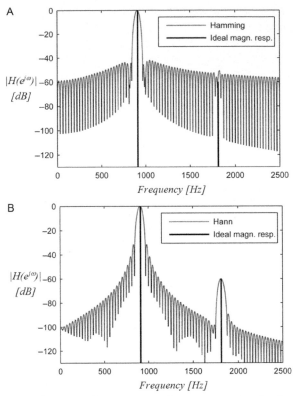

Fig. 1 Comparison of magnitude responses of a two-tone signal using (A) Hamming and (B) Hann windows.

We will focus on the effect of windows on the magnitude response only. In Fig. 1 we show magnitude responses of the same sequence which contains two sinusoidal components (i.e., tones) with unequal amplitudes. Two different windows were applied in the calculation of magnitude responses. The Hamming window is symmetric with a narrower main lobe and a higher level of side lobes in the magnitude response. Hann window is also symmetric with a similar main-lobe width, but a more rapidly decaying level of side lobes compared to the Hamming window. Examples of the main lobe and side lobes in the magnitude response of a window are shown in Fig. 4A and explained in Section 2.3.

The "ideal" magnitude response in Fig. 1 is illustrated with two straight vertical bold lines which show the frequencies and magnitudes of both sinusoidal components in the signal. The magnitude responses obtained from the finite windowed frame of the signal are presented with dotted lines.

We can clearly see how windowing affects ideal magnitude response. Two major differences between ideal and real magnitude responses are spectral smearing and spectral leakage. The first effect "smears" the narrower region around the spectral component (vertical bold lines) and is caused by the main lobe in the window's magnitude response. The second effect shows energy leaking to other parts of the magnitude response. This effect is caused by the side lobes in the magnitude response of the window.

From Fig. 1 we can also see that the second component with lower amplitude can be detected only when Hann window is used (Fig. 1B); in the case of the Hamming window (Fig. 1A), the weaker component is hidden by the spectral leakage effect caused by the side lobes of the stronger component.

In the ideal case, we would like to have a window with the narrowest main lobe and lowest side lobes, but we cannot have both minimized. Therefore, there are a lot of different windows with various relations between those two distortions. Some windows are better in one or both properties than others, but the ideal (the best of all) window does not exist. Therefore, we have to select a suitable window as a compromise in the context of a specific application. In some cases, this is rather easy (e.g., single tone detection) and in more complex applications it is much harder to do. In addition, the direct influence of the window on final results is often not so clear and pronounced (e.g., speech recognition).

2.2 Frequency analysis of nonstationary signals

It is important to note that the magnitude response shows the average spectrum across the whole frame. Therefore, the calculated magnitude response is consistent only if the signal's content is stationary across the frame, i.e., if its frequency content does not change in time. In practice, this is a rare situation and we have many more signals whose frequency content varies in time. This is particularly true in the area of audio, speech, video, and other signal processing applications (e.g., data communication). In these cases, we use the time-dependent frequency analysis method denoted as the Short-Time Fourier Transform (STFT), shown in Fig. 2. We simply divide the signal into successive frames and make an assumption of stationarity in a single frame only. The length of the frame is determined in the context of each application. For speech recognition, frame duration is typically 20–30 ms.

In STFT, the input signal is divided into finite (usually overlapping) shorter frames, in which the Fourier transform is calculated (Fig. 2). Formally, the segmentation operation is defined as the product of samples

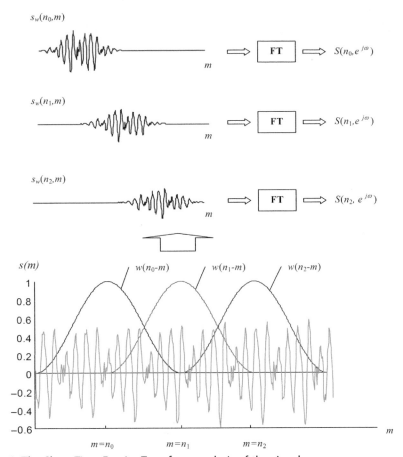

Fig. 2 The Short-Time Fourier Transform analysis of the signal.

of an infinite signal and a window sequence with non-zero values only within the selected frame.

Sliding the window on the input signal creates a sequence of individual frames

$$s_w(n, m) = w(n - m) \, s(m), \; -\infty < m < \infty, n = n_t, n_t = n_0 + tM, t \\ = 0, 1, 2, \ldots, \tag{5}$$

where $s(m)$ represents samples of the discrete infinite signal, $w(n)$ is the window sequence, n is the center point of the individual frame, and M is the distance in samples between adjacent frames. The length of all frames is equal to the length of the window. In each frame, frequency response is calculated and added to the two-dimensional, time–frequency (TF) presentation of the signal $S(n, e^{j\omega})$:

$$S\left(n, e^{j\omega}\right) = \sum_{m=-\infty}^{\infty} s_w\left(n, m\right) e^{-j\omega m}. \qquad (6)$$

As we have already defined in Eq. (3), due to the inevitable use of the window, the resulting frequency spectrum in each frame is now influenced by the spectral properties of the window in the same way as in Fig. 1. Consequently, only a better or worse approximation of the ideal frequency response can be estimated. The choice of the window, therefore, has the same significant influence on the obtained frequency response.

To choose a suitable window in the context of each application, we first have to define the desired parameters or properties of the window. There are two options: either we find an existing window which matches our specifications, or we can design a window to suit our needs. In this case, we can design a new asymmetric window by combining existing windows using various operations (e.g., convolution, product, concatenation); the properties of a new window are some sort of combination of original properties. If we still do not have an acceptable solution, we can specify an error function and an optimality criterion, and perform the optimization to produce the window with desired optimal properties.

An example of the resulting STFT magnitude response of a speech signal is shown in Fig. 3; this time–frequency (TF) presentation of the signal is denoted as a spectrogram.

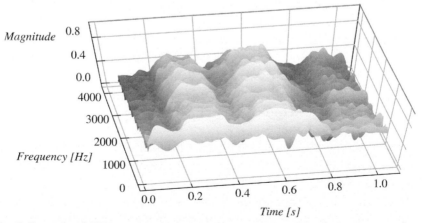

Fig. 3 Example of STFT magnitude response (i.e., spectrogram) of a spoken word.

2.3 FIR filter design

In frequency analysis, windows are used for truncation of infinite signals into finite sequences of samples, i.e., frames. We also know that tapering the values of the window to zero on both ends decreases the level of side lobes and therefore the amount of spectral leakage distortion in the magnitude response. However, this effect can only be achieved at the expense of the wider main lobe and the related greater spectral smearing effect. Similar behavior can also be noticed when it comes to using windows in the design of FIR filters.

There are three common methods for the design of FIR filters:
- Windowing method [10,11]
    - The window is applied for truncation of an infinite (ideal) discrete impulse response, determined by the Inverse Fourier Transform (IFT) of the desired frequency response.
    - This method is the simplest FIR filter design method, also suitable for real-time or adaptive applications.
- Frequency sampling method [11–13]
    - It is based on the DFT transform pair in frequency (desired frequency response) and time domains (finite impulse response).
    - It is often denoted as "naive filtering" because it needs correction for a better control of error in regions between DFT points.
- Optimal window design method (known also as "Parks and Mc Clellan method") [14,15]
    - Produces an optimal solution according to Chebyshev MiniMax criterion—Eq. (12).
    - It is a more complex method with optimal results; however, it is less suitable for adaptive applications.

The FIR filter design procedure commonly starts with the definition of the desired frequency response. Filters are typical DSP systems whose behavior is more important and defined in the frequency domain. However, they are usually implemented in the time domain. In most cases, the linear phase response is desired, and this property implies the symmetry of the impulse response in the time domain.

The Window design method for the FIR filter is the simplest and at the same time the only one which uses windows. Therefore, it will be presented here in more detail.

Generally, filter design methods start with the definition of the desired frequency or magnitude response. In this method, the infinite impulse response $h_{IIR}(n)$ in the time domain is determined by the Inverse Fourier Transform from the desired frequency response $H_d(e^{j\omega})$

$$h_{IIR}(n) = \frac{1}{2\pi} \int_{-\pi}^{\pi} H_d\left(e^{j\omega}\right) e^{j\omega n} \, d\omega \qquad (7)$$

Impulse response in Eq. (7) is a discrete and infinite sequence. It can therefore not be implemented in practice; it needs to become finite. In this case, a window $w(n)$ is used to truncate the infinite discrete impulse response $h_{IIR}(n)$

$$h_{FIR}(n) = w(n)\, h_{IIR}(n),\ -\infty < n < \infty \qquad (8)$$

The expression in Eq. (7) is solved analytically and usually results in a generic expression which can be used to produce impulse responses of FIR filters with various characteristics. Since the frequency response of the window resembles the frequency response of the low–pass FIR filter (Fig. 4), we will present the generic expression for the infinite impulse response of the low-pass type of FIR filter

$$h_{IIR}(n) = \begin{cases} \dfrac{\omega_c}{\pi} & n = 0 \\[2mm] \dfrac{\sin(\omega_c\, n)}{\pi n} & n \neq 0 \end{cases},\ -\infty < n < \infty, 0 < \omega_c < \pi, \qquad (9)$$

where ω_c determines the arbitrary frequency point at the junction of the filter's pass and stop bands. The application of the window to the infinite impulse response $h_{IIR}(n)$ from Eq. (9) as defined in Eq. (8) has an influence on filter characteristics in the frequency domain. The frequency response of the resulting filter $H(e^{j\omega})$ after windowing differs from the desired similarly as defined in Eq. (3). The resulting frequency response can be expressed as a convolution integral between the desired frequency response $H_d(e^{j\omega})$ and the frequency response of the window $W(e^{j\omega})$

$$H\left(e^{j\omega}\right) = \frac{1}{2\pi} \int_{-\pi}^{\pi} H_d\left(e^{j\theta}\right) W\left(e^{j(\omega-\theta)}\right) d\theta \qquad (10)$$

It can be seen in Fig. 4 that window properties are actually reflected in the corresponding properties of the filter. The main-lobe width in the window's magnitude response (Fig. 4A, near 0 Hz) is proportional to the width of the transition band in the magnitude response of the filter (Fig. 4B, near 0.13 Hz). The height of side lobes (Fig. 4A) is reflected in the attenuation

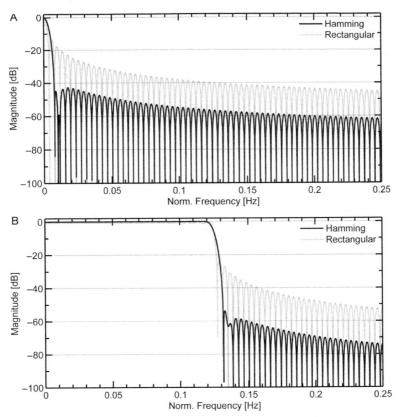

Fig. 4 Comparison of (A) magnitude responses of rectangular and Hamming windows and (B) corresponding low-pass FIR filter's magnitude responses.

of the filter (Fig. 4B, defined as the difference between the magnitude value in the passband and the highest side lobe value) and ripples in the passband. In practice, we frequently give more attention to the filter's attenuation and therefore often apply windows with lower side lobes at the expense of the wider main lobe. In this manner, according to Fig. 4, the Hamming window would be commonly preferred practical choice over the rectangular window. Consequently, the transition band is wider, but we have a better attenuation in the stopband of the filter's magnitude response.

FIR filters, designed by the Windowing method, can be optimal according to the Mean Square Error criterion (MSE), generally defined as

$$\delta_{MSE}(\boldsymbol{h}) = \frac{1}{2\pi} \int_{\omega \in \Phi} A\left(e^{j\omega}\right) \left| H_d\left(e^{j\omega}\right) - H\left(e^{j\omega}\right) \right|^2 d\omega, \tag{11}$$

where $A(e^{j\omega})$ is a positive error weighting function, $\delta_{MSE}(h)$ is the Mean Square Error (MSE) of the impulse response h, and Φ is the union of all non–overlapping frequency regions where the difference between desired $H_d(e^{j\omega})$ and actual $H(e^{j\omega})$ frequency responses is evaluated (usually transition bands are excluded from the error calculation). The designed filter is optimal according to MSE criteria only when the rectangular window is used. Despite the MSE optimality in this case, the resulting filter has a generally lower and variable attenuation rate in stopbands (Fig. 4B). In other words, the error is distributed non–uniformly across the stop and pass bands, which is frequently not compatible with the requirements for practical applications of FIR filters. Therefore, the MSE criterion is rather rarely used in the design of FIR filters.

Chebyshev MiniMax error criterion is more common for assessing frequency characteristics of digital filters and is generally defined as

$$\delta_{Cheb}(h) = max_{\omega \in \Phi} A\left(e^{j\omega}\right) \left| H_d\left(e^{j\theta}\right) - H\left(e^{j\omega}\right)\right|, \tag{12}$$

where $\delta_{Cheb}(h)$ is the maximum value of error for the impulse response h, across the union of frequency regions Φ. The optimal solution according to this criterion has a minimal value of the error defined in Eq. (12). Therefore, Parks–McClellan FIR filter design method [14] is known to produce optimal solutions according to Chebyshev MiniMax criterion.

FIR filters designed with the Windowing method are, according to Chebyshev criterion, suboptimal solutions. However, their simplicity and better compatibility with real-time and adaptive applications are their main strengths and the reason they are used in many practical cases (e.g., design of Variable Fractional Delay (VFD) FIR filters [16]).

3. Can asymmetric windows perform better?

Symmetric windows are a common choice in signal processing applications, which is probably the correct decision in most cases. However, there are situations where the symmetry could exhibit certain limitations.

If we consider a window as a sequence of N samples, it is quite obvious that symmetry actually implies that only half of the samples can be determined arbitrarily. The second half is determined by the symmetry constraint. From the optimization viewpoint, we can perform the optimization under various criteria better if we can expose all samples to the process of searching for an optimal solution. This is the reason why asymmetric windows have a promising optimization potential.

However, to achieve asymmetry, the symmetry property has to be relaxed. This can cause some positive and some negative effects. The decision on what prevails probably depends on the application context. Consequently, we can here draw a similar conclusion to the one we already reached for symmetric windows. We believe that there is room for a whole family of asymmetric windows which will have some better properties and few other worse properties than symmetric windows. Therefore, different types of asymmetric windows are expected to exist and their utilization will sometimes be more general and sometimes more application-specific.

The following subsections will present a few important topics related to asymmetric windows. After a brief overview of various asymmetric windows, some common potential benefits will be outlined. In the final subsection, we will present several approaches to the design of asymmetric windows. All presented approaches will be also elaborated in more detail with their practical applications in Section 4.

3.1 Overview of asymmetric windows

Asymmetry is a quite common concept across various research and application fields. Generally speaking, both symmetry and asymmetry have their pros and cons. However, for some applications of windowing asymmetry can exhibit certain potential advantages.

On the other hand, it seems that symmetry is much more common in the field of frequency analysis. To a certain level of understanding, this seems reasonable. Symmetric windows perform quite well for the majority of applications in this field. In addition, the influence of the windows is usually not so pronounced, which is why less research attention has been devoted to asymmetric windows. However, asymmetry can sometimes bring substantial enhancements which can be important for certain applications. We strongly believe that asymmetric windows deserve more research attention than they did in the recent past.

Hereinafter, we will present a brief overview of various ideas related to asymmetric windows in approximate chronological order. An overview is done to the best of our knowledge and hopefully all the most important approaches are included. More details about some of the presented ideas can be found in Section 4.

According to our sources, Barnwell brought the idea of asymmetry to the field of frequency analysis in 1981 [17]. Although he did not define any particular finite asymmetric window, his ideas of asymmetric processing caught the attention of a few researchers that implemented this idea into a more formal finite window context: Florencio [18,19] and Rozman et al. [2,8,9,20]. Florencio showed the advantages of asymmetric windowing in the context of shorter time delay speech coding in years 1991 and 1993. Rozman et al. confirmed improvements in the speech recognition system's performance when asymmetric windows were used in the frequency analysis stage, in the period from 2003 to 2013.

In 1995, ITU proposed the asymmetric Hamming-Cosine window to achieve a shorter time delay in the speech coding application field [21]. The Hamming-Cosine asymmetric window was simply formed as a concatenation of two symmetric windows with the center point moved more toward the edge of the window. Consequently, ITU achieved a shorter time delay in the speech coding application known as G.729 codec, while preserving the quality of the speech signal. This work certainly attracted attention and demonstrated that a shorter time delay is a common potential advantage of asymmetric windows.

A few years later (in 2002 and 2006), Zivanovic et al. published articles which considered asymmetry from a completely different viewpoint [22,23]. It was empirically confirmed that the nonlinear phase response of asymmetric windows can bring some important advantages to the detection of closely spaced components in frequency analysis. More precisely, the authors focused on detection of the components located nearer than one frequency (DFT) bin.

In the years after 2012, a lot of various asymmetry ideas were implemented and published. Alam et al. [24] and Morales et al. [25] introduced asymmetric shifted DDR windows. Alam also evaluated phase modified windows and the ITU Hamming-Cosine window. Several other authors confirmed similar improvements when asymmetric windows were used [26–28].

In 2016, Su et al. [1] proposed the Minimum–Phase (MP) transform, which can be applied to any symmetric window under certain constraints. Since the transform moves all zeros of the system function inside the unit circle in the Z-plane, it seems to be a general approach applicable to any symmetric window. In practice, the method is successful if zeros in the Z-plane are not located too close to the unit circle. Therefore, it can be successfully applied only to a certain subset of symmetric windows. Nonetheless, the main advantage of the method is that the magnitude response of the transformed window is retained, while the group delay can generally be reduced.

We strongly believe that asymmetric windows will appear more often in the field of signal processing in the future. Since previous research efforts were less intensive, more activity and appearance of various approaches to asymmetric windows can be expected.

3.2 Potential benefits of asymmetric windows

Asymmetric windows generally exhibit two potential types of benefits. One is related to better spectral properties and to the shorter processing time delay. In addition, there are some other specific properties of asymmetric windows which can be utilized in application-specific contexts. Hereinafter, both groups of potential benefits are presented in more detail, including practical examples.

3.2.1 General spectral and time delay related benefits

In frequency analysis, symmetric windows are a common choice. However, it is not clear whether this is necessary in all cases. Symmetry is related to the property of linear phase in frequency response. Similarly, it is not so obvious whether this property is always needed. This particularly refers to applications where only the magnitude response is retained for further analysis.

A typical example of "magnitude-only" signal presentation is the frequency analysis stage of speech recognition systems. This approach is a consequence of the common consensus from the past that the accuracy of human speech recognition is practically independent of phase distortions of the signal [29]. Consequently, there is no apparent reason not to use an asymmetric window in this case. A similar argument could be more generally applied to other similar applications of the "magnitude-only" frequency analysis procedure (e.g., spectrogram-based methods).

From the design of digital FIR filters it is known that some properties can improve with asymmetry, or better said, with the relaxation of symmetry constraint. With sufficiently wide pass and transition bands, filters with better magnitude response can be designed [2,8]. Two asymmetric FIR filter designs ("Nearly linear phase" and "Arbitrary phase") are compared to the symmetric case (Linear phase) in Fig. 5. It can be noticed in Fig. 5B that loosening the criterion from exact linear to approximately linear and arbitrary nonlinear phase can lead to consistently better attenuation in the stopband of the magnitude response. Filters in Fig. 5 were designed according to three different criteria:

- "Linear phase"
  - ○ Filter exhibits exact symmetry (or linear phase) property
- "Nearly-Linear phase"
  - ○ Filter exhibits approximate symmetry (or linear phase) in the passband

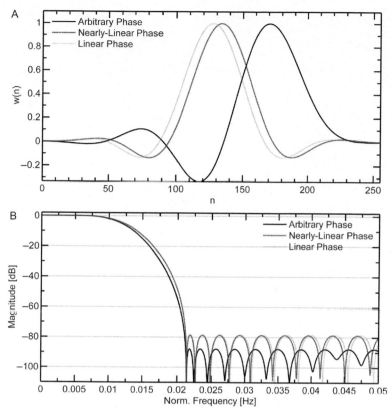

Fig. 5 Time domain (A) and magnitude response (B) comparison of three FIR filter designs at different optimality criteria: Linear, Nearly-Linear and Arbitrary phase.

○ Complex error function is defined as

$$E\left(e^{j\omega}\right) = H_d\left(e^{j\omega}\right) - H\left(e^{j\omega}\right) \tag{13}$$

○ Error function includes the magnitude and phase response difference
• "Arbitrary phase"
○ Filter exhibits arbitrary nonlinear phase response
○ Magnitude-only error function is defined as

$$E\left(e^{j\omega}\right) = \left|H_d\left(e^{j\omega}\right)\right| - \left|H\left(e^{j\omega}\right)\right| \tag{14}$$

○ Error function includes only magnitude response difference
It should be pointed out that the phase response in the two latter cases becomes more and more nonlinear (Fig. 6). It is not clear how important

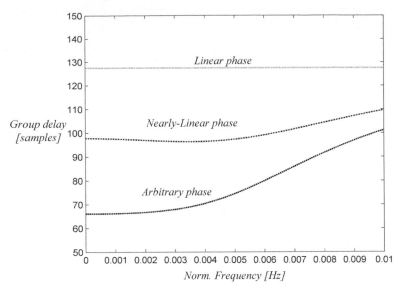

Fig. 6 Group delay comparison of three FIR filter designs at various optimality criteria.

that degradation generally is for the operation of windows. This property in the FIR filter context actually means that time delay is not the same for all components and depends on the frequency of the component. In this case, the filter can substantially change the signal's envelope even if it fully passes all components (i.e., all-pass filter). This is usually not acceptable for applications where the signal's envelope is important (e.g., communications) or variable time delays are unacceptable (e.g., audio processing). On the other hand, it seems appropriate for speech processing applications where nonlinear phase distortions do not seem to be so prominent, particularly for the intelligibility of speech.

While interpreting various effects, we have to be aware of the crucial difference between windows and impulse responses of FIR filters. Windows are applied to a signal as a weighting function in the time domain while the impulse response of the FIR filter is used in the convolution operation with the input signal. We can share a lot of knowledge between these two research and application fields. However, we need to be permanently aware of this important distinction.

As already explained, in the past there was a strong consensus against the importance of phase response in speech recognition; consequently, only the magnitude response of the window was of primary concern. Recent research on the importance of phase for speech intelligibility [30,31] has shown that this topic should be reconsidered. Although not of primary importance for this article, some related contributions to this topic will also be presented in this work.

There is another important potential benefit of asymmetric windows, related to a shorter time delay in frame-based signal processing. Symmetric windows usually have a global maximum value at the center point of a sequence. This also means that processing delay is the same for all symmetric windows and is approximately equal to a half of samples in frames. On the other side, the center of asymmetric windows can be moved more toward one or another end of the sequence. In this way, we can achieve a shorter time delay of processing, as illustrated in Fig. 7. To highlight the effect more clearly, a sliding type of analysis was performed. It can be noticed that the tones in this particular case of the spectrogram appear approximately 5 ms earlier for an asymmetric window than in the spectrogram obtained using a symmetric window.

3.2.2 Application-specific benefits

Besides the presented general advantages related to better spectral properties and a shorter time delay, there are also some other properties which are interesting and more pronounced in application-specific contexts.

The first example of that sort is the decorrelation effect important for systems which use common variants of statistical models (e.g., HMM, GMM-UBM). In those systems, full covariance matrices are often replaced by a more computationally efficient vector which keeps only diagonal

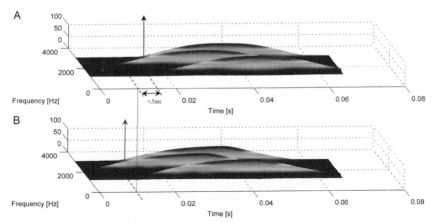

Fig. 7 Time delay comparison of two spectrograms using symmetric (A) and asymmetric (B) window.

elements instead of the whole matrix. If the features which are being modeled are less correlated, the degradation introduced by diagonal simplification is smaller.

At first sight, it is not obvious how to relate asymmetric windows and the correlation of features. However, detailed knowledge of a specific application context gives the answer.

In speech recognition and speaker related applications, we usually transform frames of the signal into vectors with features which form a compact and representative description of the signal in each frame. This process is denoted as parameterization. During this process, a Mel-frequency filter bank is applied to a short-time magnitude response.

The filter bank is essentially a weighting function which merges detailed magnitude response in several substantially wider critical bands which are most important for speech intelligibility [29]. This way we get a more general signal representation which is also more robust to variation in individual speaker properties (e.g., fundamental speech frequency). At the same time, the filter bank consists of heavily overlapped triangle weights which increase the correlation between values in neighbor bands. We can therefore say that a higher level of correlation is built into the described concept.

However, there is also a correlation expressed between non-neighbor frequency bands. This type of correlation cannot be attributed to overlapped triangle functions and it is probably more related to the spectral leakage effect of the frequency analysis procedure and the related window. This statement is illustrated in Fig. 8 and explained in more detail in Refs. [2,8]. It can be seen that the window with lower spectral leakage (asymmetric window "Solvopt3_10") can reduce average correlation in features and consequently

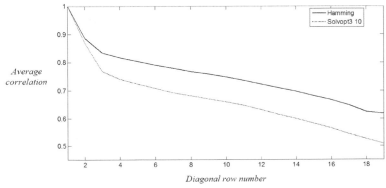

Fig. 8 Average correlation of FBANK features when using Hamming and asymmetric Solvopt3_10 window on a speech corrupted with white noise at 6 dB.

cause better performance of the modeling concept. This effect has been reported as important for HMM in Refs. [2,8] and GMM–UBM models in Ref. [32]. As we know, asymmetric windows can exhibit better spectral properties and can be an interesting concept from this practical application's viewpoint.

There are also some other situations where certain properties of asymmetric windows can contribute significantly to the solution of an application-specific problem. Sometimes a property may not seem so important from a more general viewpoint, however, it can attribute to a significant positive change in an application-specific context.

A representative example of this argumentation are $-class asymmetric windows, introduced by Zivanovic et al. [22,23]. Those windows have a great discrimination property for closely spaced components although their magnitude response does not indicate such behavior, particularly if it is considered using the knowledge of frequency analysis. Following this perspective, the discrimination property is generally more attributed to the main-lobe width in the magnitude response of the window. However, in this particular application-specific context of the sub-DFT-bin spaced components detection, authors claim that the phase response plays a more important role. This fact could be easily overlooked from the mentioned general frequency analysis viewpoint. As a particularly interesting case, it is elaborated in more detail in Section 4.3.2.

3.3 Design of asymmetric windows

When selecting windows for specific applications, a general overview needs to be done first. A list of desired window properties is usually a good starting point, after which the application context has to be considered. After that, a window can be selected.

There are generally three possible ways to select or produce the asymmetric window with desired properties:
- Selection of an existing asymmetric window with matched properties
- Simple construction methods applied to existing windows
- Optimization-based design of the desired asymmetric window

The first scenario is the easiest and also the most common one. We just need to find and select an existing window with matched properties. In this article, a lot of various examples will be presented and there is a fair chance, that the presented windows will exhibit the desired properties for various applications.

The second scenario is commonly used when there is no single existing solution to match the list of requirements. The list of possible construction methods is quite long and it will be briefly presented in the next subsection.

If the former two scenarios fail, then we are forced to apply the optimization process to the window design task. However, the main problem is the definition of the optimality criterion. After that, if the optimization method is successful, we get the optimal solution according to the defined criterion. Optimization-based methods are outlined in Section 3.3.2.

3.3.1 Simple design methods

As we have already stated, symmetry and asymmetry come both with pros and cons. We often want to create a combination and take advantage of pros from both cases, which is why we try to create a new window combining the inherited desired properties from existing windows. Sometimes, this is a fairly trivial task while sometimes it is quite the opposite. However, the methods of construction are simple and fast; therefore, one can try and evaluate several approaches before making the final selection.

Since all methods are presented in more detail in Section 4, only a brief overview will be provided here:

- Concatenation
  - Two windows are usually concatenated in the time domain to form an asymmetric window; the central point is often headed toward the edge to achieve the reduction of the processing time delay or follow some other goal.
  - The known compromise between spectral and time delay properties must be taken into account.
  - Typical examples: ITU Hamming-Cosine window [21], AAC window [40].
- Composite function or modulation $w(p(t))$
  - Function $p(t)$ as an argument modulates the window function $w(t)$.
  - Various interesting properties are gained under moderate time delay reduction.
  - Typical examples: Composed Hann windows [19].
- Truncation of (a)symmetric functions
  - A longer window is truncated to the asymmetric part.
  - Typical examples: $-class windows (Zivanovic et al. [22, 23]), windows by Luo et al. [7, 41]
- Convolution
  - Shorter windows can be combined into a new asymmetric window by convolution.

- o Frequency response of the resulting window is the product of existing responses.
- o Typical examples: DDR (asymmetric) windows [24–26].
- Stretching
  - o A symmetric window is stretched asymmetrically on both sides of the central point.
  - o Typical examples: Stretch windows (Luo et al. [7])
- Shift or translation of peak
  - o The central peak of the symmetric window is shifted more toward the edge.
  - o Typical examples: Shift windows (Luo et al. [7])
- Minimum Phase (MP) transform
  - o All zeros in Z-plane that are outside the unit circle are moved inside the unit circle.
  - o It is applicable only to certain windows.
  - o Typical example: MP Flat-top window (Su et al. [1]).
- Phase modified windows
  - o The phase response of the system is modified in a way which is useful.
  - o Typical examples: PhMod windows (Alam et al. [26], Asbai et al. [27, 32, 47, 48]).

3.3.2 Optimization-based methods

If we cannot find or construct a window with desired properties, then we need to perform an optimization procedure to find an acceptable solution. However, proper optimality criteria should be defined. This is often not a trivial task to do because it is not clear which window properties will lead to the best results. There is always a way to use general knowledge from the field of frequency analysis. However, we also know that this knowledge might in some cases not be tightly related to success in an application-specific context. Consequently, optimal solutions according to the wrongly selected criteria may perform even worse than an existing non-optimal solution. Therefore, we normally need an iterative approach and more experimentation with different criteria and optimization methods. If this is done properly, then we frequently get promising results.

To illustrate the problem of criteria selection, let us show the relation between the two commonly used criteria in the field of FIR filter design methods: Chebyshev MiniMax and MSE criteria. The relation between errors according to both criteria is shown in Fig. 9 [33]. It can be clearly seen that reaching the optimal point according to one criterion at the same time means more error according to other criteria. It can therefore be

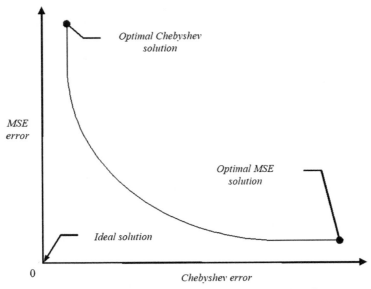

Fig. 9 Relation of MSE and Chebyshev MiniMax errors in FIR Filter design.

questioned whether reaching optimal points is always a good decision. A possible approach might be to satisfy both criteria to a certain compromised level. This conclusion might also be generalized to other similarly related criteria. Intensive optimization according to one criterion can lead to a specific optimal point, however, the solution can be suboptimal in a more general sense.

Let us elaborate further on these both criteria. From their definitions, we can conclude that the Chebyshev criterion produces optimal solutions with a uniformly distributed error. However, according to MSE, this is a clearly suboptimal solution because an aggregated error in this case certainly increases. In the field of FIR filter design, the Chebyshev criterion is a more common choice because we usually want to have uniform attenuation across filter stopbands.

On the other hand, in the field of general frequency analysis, there are other more important properties. For instance, in the detection of closely spaced components, windows with a narrow main lobe and consequently higher side lobes (e.g., rectangular window) can produce better discrimination results. However, if we use the same window in the detection of distant components of unequal amplitude, then the rectangular window will generally produce worse detection results. This fact supports the statement that each property can play a different role in a specific application context.

We just need to have proper knowledge of window properties and a specific application context and match those two to produce effective solutions.

The other similar and quite common relation is the already mentioned compromise between spectral and time delay properties of the asymmetric window. It is generally known that excess minimization regarding one property means worse performance regarding the other. This fact should be always considered in the optimization process.

As we have explained, the selection of optimization criteria is not a trivial task. In addition, "overfitting" to one criterion might lead to suboptimal solutions according to other criteria. However, this is just the first part of the window design procedure.

In the second part, we need to perform optimization according to the selected criterion. Since we are witnessing the rapid evolution of processing power and optimization algorithms, this part is much easier to perform today than it was in the past. Over the years of development, a lot of various optimization procedures appeared in this field (e.g., Refs. [34–37]). However, most of them are focused on certain specific criteria or applications.

During our work on asymmetric windows we have encountered various optimization problems and criteria. As we have done research mostly in the area of speech recognition, we needed a lot of experimentation on various criteria and optimization procedures. We have always valued the general optimization methods which allowed us to test various criteria more freely, usually at the expense of longer execution time.

With general optimization methods we were able to test various design criteria without specific programming of optimization algorithms. We probably did not achieve exact optimal solutions but nonetheless, according to the analysis of resulting windows, we could conclude that even the theoretically suboptimal solutions were good enough to perform our research task successfully. For our purposes, we used the optimization package SOLVOPT [34,38] which showed acceptable behavior. Today, when the range of accessible general optimization methods is much wider and computing platforms are more powerful (e.g., Ref. [39]), we are certain that experimentation with various criteria is substantially easier.

4. Review of related work on asymmetric windows

Asymmetric windows have several interesting properties which can result in certain advantages over symmetric windows in various areas of signal processing. The most prominent ones are a shorter processing time delay and better spectrum properties (e.g., lower side lobes, monotonicity). In addition, there are some more specific properties which can lead to better results in certain application contexts (e.g., nonlinear phase response).

This section gives a broad review of the use of asymmetric windows in various application areas. Each application area is presented in a separate subsection. The review includes the sources which we have identified throughout our work and exhaustive search for all sources in these fields. The selection and representation in the following subsections are done to the best of our knowledge.

4.1 Pitch modification in time domain

In this subsection we present a typical example of an asymmetric window, used as a data window in the time domain.

There are a variety of different methods available for the pitch modification task. The "Pitch–Synchronous Overlap and Add" (PSOLA) is one of the more popular methods. Originating from text-to-speech (TTS) synthesis systems, PSOLA also gained recognition as a pitch or duration modification method. The essence of the PSOLA pitch modification procedure can be described in three steps. The first step is to divide the speech signal into shorter overlapping segments. In the second step, segments are moved more closely together or more apart, depending on the pitch modification goal. In the last step, the signal is resynthesized using the well-known overlap-add synthesis (OLA) method. At the end of the procedure, the pitch is increased or decreased, depending on the movement of the segments in the second step.

Jung et al. [6] constructed an asymmetric window which is applied in the synthesis phase of the PSOLA method. It is shown that the proposed asymmetric window gives better results in the PSOLA based pitch modification method. The results were assessed using the Signal-to-Noise Ratio (SNR) of the method's output signal with the proposed asymmetric window (denoted as PSOLA data window) and the symmetric Hann window. The authors measured SNR in both cases of speech modification: reduction to 70% and increase to 130% of the original pitch. The results on two speakers confirmed that SNR after the PSOLA pitch modification with the proposed asymmetric window was in both cases approximately 3 dB better than with the symmetric Hann window. This means that in this particular method better speech quality was retained after pitch modification using the asymmetric window.

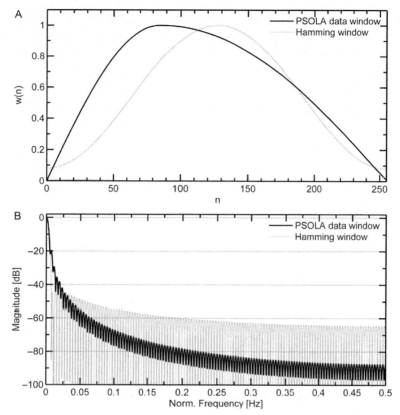

Fig. 10 Time domain presentations (A) and magnitude responses (B) of the PSOLA (asymmetric) data window and the Hann symmetric window.

If we analyze the properties of both windows (shown in Fig. 10), we can observe that the asymmetric window demonstrates better asymptotic behavior than the symmetric one and a higher level of the first few side lobes. We believe that the empirically shown effect of the asymmetric window [6] can be more contributed to the time domain presentation of the window. However, more exhaustive testing is needed for a more decisive conclusion. To conclude, it seems that the proposed asymmetric window is better suited for an application in the PSOLA pitch modification algorithm.

4.2 Shorter time delay processing in time domain

Time delay is a particularly important parameter for almost every real-time application in the field of signal processing. Symmetric windows have a constant time delay which cannot be reduced without losing symmetry.

Therefore, quite a few asymmetric windows were introduced in this field. With the main goal of a shorter time delay, frequency response should also be under careful consideration; it is desired to be at least similar to the existing responses of symmetric windows, if not better.

In this section we present the work on asymmetric windows which is closely related to the processing of signals with the aim of shortening processing time delay as much as possible. On the other hand, windowed signals are usually transformed into the frequency domain and the windows presented in this section can be attributed to the category of spectral windows.

4.2.1 Asymmetric windows with minimum phase transform

The theory of minimum phase systems in digital signal processing offers the interesting possibility to construct an asymmetric window for processing signals with a shorter time delay while keeping the same magnitude response. We can therefore achieve a shorter overall processing delay at practically no cost and without magnitude response distortion.

The theory says that systems with minimum phase have all the poles and zeros of their system function inside the unit circle in the Z-plane. Since windows have only nontrivial zeros, for minimum phase property, all zeros must lie inside the unit circle. If a window has any zeros outside unit circle, they can be moved inside the unit circle by replacing them with the corresponding conjugate reciprocal pairs. In this way the magnitude response of the window remains the same; only the phase response changes in the way that the group delay becomes smaller, and consequently, the processing time delay becomes shorter. This procedure is denoted as "Minimum Phase" (MP) transform and was proposed by Su et al. [1].

A visual confirmation of the MP transform effect is presented in Fig. 11. In the time domain, the asymmetric MP transformed Flat-top window has clearly moved the central point with the maximum value toward the beginning of the frame, in comparison to the symmetric Flat-top window (Fig. 11A). In the frequency domain, the magnitude response is the same for both windows by theoretical definition and visual confirmation in Fig. 11B.

To be able to apply the MP transform, the window must have zeros of the system function located outside the unit circle in the Z-plane. If no such zeros exist, then the window is already considered as minimum phase. It has been shown in Ref. [1] that some well-known windows already have the MP property and their group delays cannot be efficiently lowered using

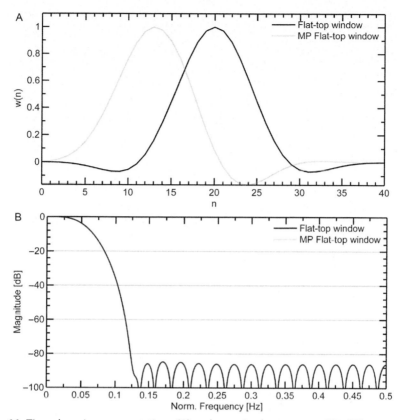

Fig. 11 Time domain representations (A) and magnitude responses (B) of Flat-top symmetric and MP Flat-top asymmetric window ($N=41$).

this method (e.g., Hamming, Hann). On the other hand, the method can be applied to some other windows with great success (e.g., Flat-top, Cosine family of windows). An example with zero locations of a symmetric and an MP transformed asymmetric Flat-top window can be seen in Fig. 12; all zeros are clearly located inside the unit circle after MP transform on the right side of Fig. 12.

4.2.2 ITU Hamming-Cosine window in speech coding

One of the major advantages of symmetric windows, besides having simple definitions, is that they have predictable characteristic features. However, this also means that sometimes they cannot be dynamically adapted to specific run–time requirements. In practice, symmetric windows are often

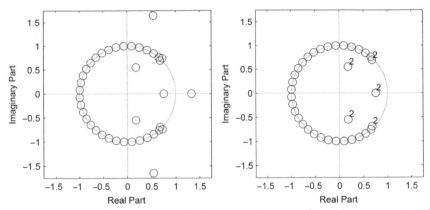

Fig. 12 Pole-zero plot of symmetric Flat-top window (left) and asymmetric MP transformed Flat-top window (right), N=41.

combined to form new asymmetric windows using a variety of simple and fast methods. Consequently, the resulting solutions combine the predefined properties of multiple windows into a new window with some better properties.

Asymmetric windows have a very important role in the speech coding area. Among these, the most well-known hybrid "Hamming-Cosine" window was introduced and recommended by ITU for use in low-delay speech coding systems [21]. The ITU window is a typical example of an asymmetrical concatenation of two well-known symmetric windows; it comprises a longer segment with the first half of Hamming and a shorter segment with the second half of the cosine window:

$$w_{ITU}(n) = G \begin{cases} 0.54 - 0.46\cos\left(\dfrac{2\pi n}{415}\right) & n = 0,\dots,207, \\ \cos\left(\dfrac{2\pi(n-208)}{191}\right) & n = 208,\dots,255. \end{cases} \quad (15)$$

The ITU Hamming-Cosine window is compared to the traditional symmetric Hamming window in Fig. 13. From both parts of Fig. 13 it is evident that the window was presumably selected on the basis of the time domain asymmetry and related shorter time delay when encoding the voice signal. The ITU window emphasizes more recent samples of the voice signal in the frame while the symmetric window (in this case Hamming) emphasizes older samples around the middle of the frame (Fig. 13A).

From the magnitude response of the ITU window (Fig. 13, right) it is evident that some of its properties (especially moderate asymptotic attenuation)

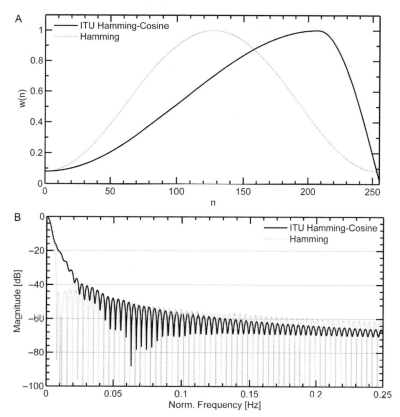

Fig. 13 Time domain presentations (A) and magnitude responses (B) of ITU Hamming-Cosine asymmetric window and Hamming symmetric window.

are not entirely in line with the knowledge of the required window properties for robust speech recognition [2,8]. Nonetheless, it should be noted that the primary purpose of using the proposed ITU window is in a speech coding process based on the LPC analysis; this case is considered less sensitive to magnitude response changes. Nevertheless, at least one alternative window can be proposed for application in both areas. It is denoted as ITU Hann-Cosine window ($w_{ITUHann}$), obtained by replacing the Hamming window with the Hann one in the former definition in Eq. (15). With this substitution, we define the ITU Hann-Cosine window as follows:

$$w_{ITUHann}(n) = G \begin{cases} 0.5 - 0.5\cos\left(\dfrac{2\pi n}{415}\right) & n = 0, \dots, 207, \\ \cos\left(\dfrac{2\pi(n-208)}{191}\right) & n = 208, \dots, 255. \end{cases} \quad (16)$$

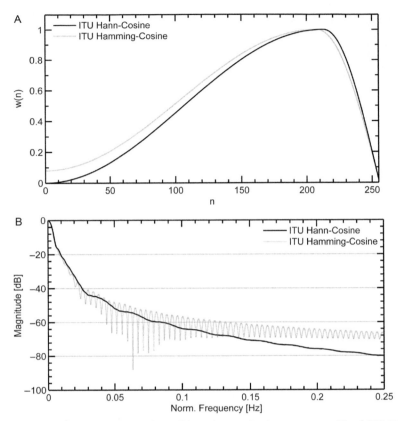

Fig. 14 Time domain presentations (A) and magnitude responses (B) of ITU Hann-Cosine window and ITU Hamming-Cosine window.

The ITU Hann-Cosine window more closely corresponds to the declared properties of the desired magnitude response in both areas and can represent an alternative to the widely used Hamming-Cosine ITU window (see the comparison in Fig. 14). Several tests in speech recognition confirmed this statement [2]. However, it is yet to be evaluated in the speech coding area.

There are also other similar asymmetric functions for short delay processing. A typical example is the AAC Low Delay filter bank using the asymmetric window [40].

4.2.3 Parabolic exponential windows in speech coding
As explained in Section 3.3, there are several possibilities for the design of asymmetric windows. One of the first asymmetric windows was proposed in Ref. [17] as an impulse response of a simple recursive system with two poles.

This work continued in several directions (Rozman et al. [2], Florencio [19]). Despite this, the common ground for all related forms of asymmetric windows as an exponential function is defined as

$$w_{exp}(n) = n\,e^{-an}. \qquad (17)$$

Since exponential windows are theoretically infinite but close to zero for large values of n, they must be truncated to a finite sequence where all samples outside of the finite range are equaled to zero. This is achieved using several methods:

- Multiplication with rectangular window [19], whereby the resulting window is denoted as the Truncated Sweep-Exponential (TSE) window;
- Multiplication by simple parabolic function $w_{par}(n) = n\,(N-n)$, whereby the resulting window is denoted as the Parabolic Exponential (PE) window.

Since the PE window has much better spectral properties, it is compared to the Hamming window in Fig. 15. We can see that what is achieved is a shorter time delay and a better spectral asymptotic side-lobe level decay rate. However, the first few side lobes are higher than those of the Hamming window. Consequently, the discrimination property in that area is obviously worse.

The PE type of asymmetric window has also been evaluated in speech coding application (FS1016 CELP coder) with promising results [18]. A time delay shorter by 5.6 ms with no loss in speech quality was empirically proven. The only modification applied was the replacement of a symmetric with an asymmetric window function. This is simple effort for such noticeable time delay reduction.

Two other simple techniques for the design of asymmetric windows, the modulation and convolution of two shorter windows, were proposed in Ref. [19]. However, little attention was devoted to those windows. These types of asymmetrical windows are presented in more detail in the next subsection.

4.3 Asymmetric windows in frequency analysis

Windows often have an important role in frequency analysis. As already explained in Sections 2.1 and 2.2, windows influence the frequency presentation of a signal obtained from a finite frame of samples. Frequency analysis is a generic concept and various applications are based on this procedure. Therefore, it is important to understand the influence of windows on the outcome of frequency analysis. For this purpose, we will put more focus on the spectral properties of windows.

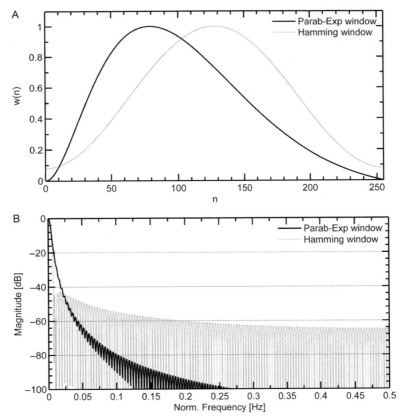

Fig. 15 Time domain presentation (A) and magnitude responses (B) of Parabolic Exponential asymmetric window and symmetric Hamming window.

As we have already mentioned, there are two major features of asymmetric windows which can lead to better results in frequency analysis: potentially better magnitude response and phase nonlinearity. In the following subsections, we present two selected application fields which coincide with both above-mentioned features. In both cases, asymmetric windows gained attention and showed better performance than symmetric windows.

4.3.1 Asymmetric windows in frequency estimation

We have already mentioned that asymmetric windows can have a shorter processing time delay and some better spectral properties. In addition, Luo et al. [7,41] empirically showed that asymmetric windows with their properties could enhance some frequency estimation methods. But before going into details, let us represent several proposed approaches to asymmetric window design which were summarized in Ref. [7]:

- Modulation method

$$w_{modwin}(t) = w_{win}\left(f_{mod}(t)\right) \tag{18}$$

- Truncation method

$$w_{truncwin}(t) = w_{win}(t)f_{trunc}(t) \tag{19}$$

- Convolution method

$$w_{convwin}(t) = w_{win}(t)*f_{conv}(t) \tag{20}$$

- Shift or Translation of the peak method

$$w_{shiftwin}(t) = \begin{cases} w_{win}\left(\dfrac{t}{2\varepsilon}\right), & t \le \varepsilon, \\[2ex] w_{win}\left(\dfrac{t}{2(1-\varepsilon)}\right), & t > \varepsilon. \end{cases} \tag{21}$$

In Eqs. (18)–(21), $w_{win}(t)$ represents a symmetric window (e.g., Hann, Hamming, Kaiser), $f(t)$ is an asymmetric modifying function, and $w_{namewin}(t)$ is the resulting window, where *name* is replaced with the corresponding method's name.

As shown in Fig. 16, the modulation method produces windows with generally worse spectral behavior. Nevertheless, the magnitude response is smooth and monotone. It is not yet fully clear, but there might be applications which could take advantage of this phenomenon; one such candidate is audio processing [42]. The practical evaluation in Ref. [7] confirms that this asymmetric window shows the greatest robustness to noise in the frequency estimation task using the Phase-difference method [41]. In addition, several other proposed asymmetric windows (one modulated and one truncated) perform better than the common symmetric Hann window. It was also confirmed that asymmetric windows can be used in time-reversed pairs to enhance the results of the phase difference method in the estimation of a single component in the presence of white noise [7].

The truncation method produces asymmetric windows which exhibit quite similar spectral properties as their symmetric counterparts. In Ref. [7], a simple truncation function

$$f_{trunc}(t) = t \tag{22}$$

is used. The resulting window is shown in Fig. 17.

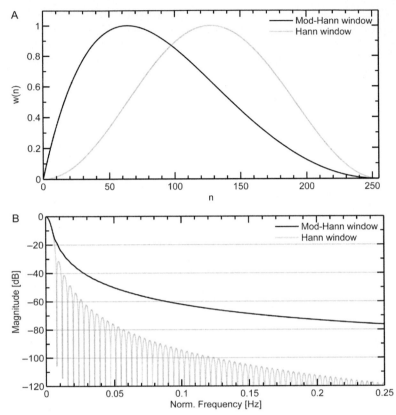

Fig. 16 Time domain presentations (A) and magnitude responses (B) of Modulated Hann asymmetric window and Hann symmetric window.

The Convolution method in Eq. (20) is interesting because it represents the dual operation of the product in the frequency domain. We can therefore determine spectral properties of the resulting window in a much simpler way, as a product of both frequency responses. Luo et al. [7] proposed the same function for convolution and truncation as in Eq. (22). The resulting windows have even lower side lobes and better asymptotic attenuation behavior, obviously at the cost of the wider main lobe.

An approach using the shift or translation of the peak method in Eq. (21) focuses on the control of the time delay property; a single parameter ε can determine the relative position of the window's center point and consequently the corresponding time delay. If the parameter ε is equal to 0.5, then the window is symmetric; with other values, asymmetric windows are obtained. These windows do not show any spectral advantages, particularly

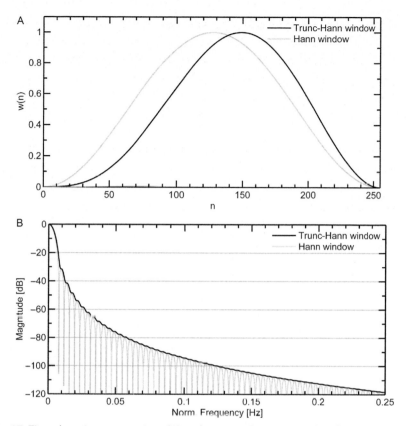

Fig. 17 Time domain presentations (A) and magnitude responses (B) of Truncated Hann asymmetric window and symmetric Hann window.

with increasing the difference between ε and 0.5. These windows are therefore not analyzed and presented in more detail here. However, more details can be found in Ref. [7].

On the basis of the presented examples we can conclude that time delay and spectral behavior are usually related. In general, shorter time delay causes worse spectral behavior and vice versa. Therefore, we must find a correct compromise between these two to obtain the best possible results.

There is also another possible conclusion, which is not so apparent and which is why we have to be more careful with the explanation. From the practical evaluations, presented in Refs. [22,41] we can see that windows with apparently worse spectral behavior in the magnitude response give better results in the spectral estimation task. This can be attributed to another component of their frequency response—the phase response which is in these cases obviously more important than expected.

All the above-mentioned findings also contribute to another conclusion: the selection of a proper window for an arbitrary task is more a trial and error iterative procedure before the final decision can be made. What is needed in order to design successful solutions is a proper combination of general knowledge, the application-specific domain, and appropriate amount of practical experimentation.

4.3.2 Detection of closely spaced components

Traditionally, the ability to detect the spectral components of a signal is related to several properties of a window's magnitude response:

- Main-lobe width
- Peak side-lobe level
- Asymptotic rate of fall for side lobes

In addition, the distance between components and the difference of their amplitudes can also be an important parameter.

The assessment which parameters are more important than others is specific to each situation. If we want to detect the distant components of similar amplitudes, then usually we do not expect any problems. If distant components are of unequal amplitudes, then the amount of spectral leakage plays the main role. Generally, in this case the lower side-lobe level and, perhaps even more importantly, higher asymptotic rate of fall for side lobes can lead to better detection results.

If the components are closer together, i.e., located at the distance comparable to the main lobe, then detection depends on the main-lobe width. On the other hand, detection also depends on the first few side lobes levels if weaker components are located close to but outside the range of the main lobe. However, none of the methods using classical spectral windows can help in the case where components are closer than one DFT bin (distance between two adjacent DFT points). Fortunately, several techniques are able to cope with this problem, and one of them, introduced by Zivanovic et al., uses asymmetric windows extensively [22,23]. A typical example of such asymmetric window (denoted as the Hann-asymmetric window) is shown and compared to the symmetric Hann window in Fig. 18. The presented window is denoted in literature as a "$-class" window, obtained from the symmetric window just by taking the first and the second half to form a pair of asymmetric windows. The spectral behavior of obtained windows in the magnitude response is apparently worse from the classical DSP viewpoint; however, the main advantage obviously comes from the nonlinear phase response.

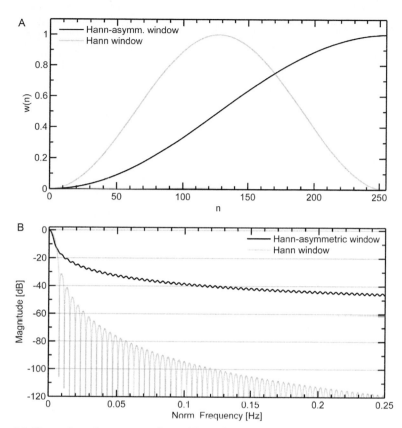

Fig. 18 Time domain presentations (A) and magnitude responses (B) of Hann-asymmetric window and Hann symmetric window.

As the authors concluded in Ref. [22], when components are closer than one DFT bin, then the phase response of the window plays a more important role than the magnitude response. This is surely surprising at first, but it has been analyzed and evaluated on several hypothetical and practical examples. Interesting results of a short practical experiment are shown in Fig. 19 and explained below.

In a more precise conclusion from the authors, the slope of the phase response is proportional to the discrimination ability of the window in this case [22]. Therefore, windows with a greater slope have better potential to discriminate among closely spaced components. This finding seems quite surprising, especially if we look at the magnitude response of the asymmetric Hann window shown in Fig. 18. Although the main lobe is a bit narrower, overall common spectral behavior does not indicate such discrimination property.

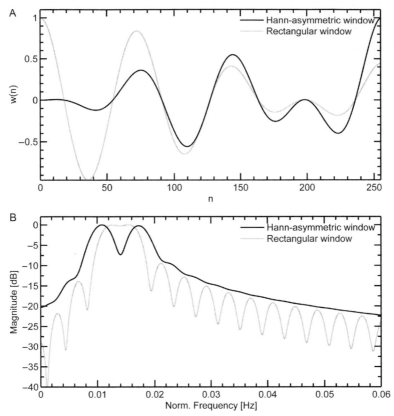

Fig. 19 Time domain presentations of a windowed signal (A) and its magnitude responses (B) using Hann-asymmetric and Rectangular symmetric windows.

To evaluate this surprising fact, additional evaluation was performed on a short test signal, consisting of two closely spaced tones. We have generated the test signal as follows

$$x(n) = \cos\left(2\pi\frac{F_1}{F_S}n\right) + \cos\left(2\pi\frac{F_2}{F_S}n\right) \quad n = 0, 1, \dots, 255, \quad (23)$$

where sampling frequency $F_S = 8000\,\text{Hz}$, $F_1 = 100\,\text{Hz}$ and $F_2 = 120.3125\,\text{Hz}$ (0.65 of DFT bin higher than F_1). For reference comparison, the rectangular window was selected because of its narrowest possible main lobe and therefore the best known discrimination property related to magnitude response among all windows. In both cases, we interpolated the magnitude response by zero padding signal to the length of 8192 samples. The results are shown in Fig. 19. Surprisingly, the two tones can be clearly distinguished in the case of the asymmetric Hann window, but to a much lesser extent in the case of the rectangular window. It is confirmed in Refs. [22,23] that for the detection of closely spaced components (less than 1 DFT bin apart), the phase response of the window is of primary importance.

A similar approach to frequency analysis by using asymmetric windows and phase difference is also elaborated in Ref. [41], where Luo et al. improved the correction procedure for frequency, phase, and amplitude estimation in discrete spectra by using asymmetric windows. Similar utilization of the difference in the phase response of asymmetric windows is noticed in this work, and interesting improvements are achieved, particularly in the presence of noise.

4.4 Asymmetric windows in speech processing

Speech processing systems are generally more complex compared to common frequency analysis tasks, which is the reason why many more factors can influence final results. Since windows are considered to be a passive technique, we normally do not expect a very pronounced influence on the final result. We can conclude that the assessment of a window's influence in complex systems is more subtle.

On the other hand, if we can observe certain progress in relation to the replacement of existing windows with better ones, then this is achieved with no additional time or space complexity. It is therefore tempting to explore such approaches. It is even more interesting if window replacement can contribute to improved inherent system robustness.

In the following subsections we present selected speech processing application fields in which asymmetric windows attracted research attention and showed promising results.

4.4.1 Speaker recognition

Various asymmetric windows were proposed in this field of research. We will represent them in the chronological order of appearance.

First, we will analyze a family of asymmetric windows proposed by Sahidullah et al. [28]. Their generation is quite simple

$$w_{diff}(n) = n^k\, w(n)\, k = 0, 1, \ldots, \tag{24}$$

where $w(n)$ is a symmetric window (Hamming in this case) and $w_{diff}(n)$ is the resulting asymmetric window whose center point is moved to the right of the symmetric center. Members of the defined family are denoted as "differentiation windows of order k."

The authors in Ref. [28] show that power spectrum, obtained by using differentiation windows, includes a wider range of information about:

- power spectrum,
- derivative of the power spectrum,
- phase response.

Besides the power spectrum alone, there are two important additions which could enhance speaker recognition performance. It was shown earlier in Ref. [43] that the slope of the power spectrum also contains speaker discrimination attributes; something similar was also confirmed for the phase response by other authors in Refs. [44,45]. An example of a differentiation window is compared to the Hamming window in Fig. 20. We can see that

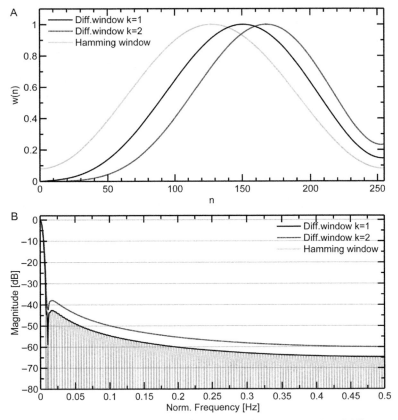

Fig. 20 Time domain presentations (A) and magnitude responses (B) of differentiation asymmetric windows of orders $k=1$, $k=2$, and the Hamming symmetric window.

the spectral behavior alone is slightly worse than the reference symmetric window (Hamming in this case). However, the inclusion of additional, practically invisible speaker discriminating attributes into signal presentation obviously makes the difference.

Empirically, speaker recognition performs consistently better with increasing the order k ($k=0$ denotes a symmetric window and for higher k we get a corresponding k-th order differentiation window) on two different speaker recognition platforms [28]. The results were also confirmed by Bakshi et al. using a smaller, slightly different scenario of speaker identification [46].

4.4.2 Speaker verification

Asymmetric windows have also been successfully applied to speaker verification. Two approaches to the design of an asymmetric window in this research field will be presented. The first one is a simple phase modification of an existing symmetric window, which has been applied to the standard procedure for MFCC features extraction. Second, an asymmetric modification of the family of symmetric windows with the Double Dynamic Range (DDR) property will be evaluated.

4.4.2.1 Phase-modified asymmetric windows

Phase-modified asymmetric windows were proposed and evaluated in many sources [26,27,32,47,48]; they are defined as follows

$$w_{phmod}(n) = c\, w(n)\, e^{k\,\theta(n)}, 0 \leq n \leq N - 1, \tag{25}$$

where $w(n)$ is a symmetric window (Hamming in this case), c the normalization constant, $\theta(n)$ instantaneous phase of $w(n)$, and k the asymmetry control parameter (higher value of k means more asymmetry). The resulting asymmetric window w_{phmod} is shown and compared to the standard symmetric Hamming window in Fig. 21. We can notice substantially lower spectral leakage, but at the expense of the wider main lobe. Other differences are more subtle.

In the first typical approach using this concept, Alam et al. proposed an asymmetric window obtained by applying phase modification to a selected symmetric window (Hamming in this case) for use in speech recognition (more details follow in Section 4.4.3). These asymmetric windows were also applied to speaker verification tasks. Various improvements were explored (e.g., front-end processing [32,47,48], a voice activity detector (VAD) [27]) but we will focus only on those related to the use of windows.

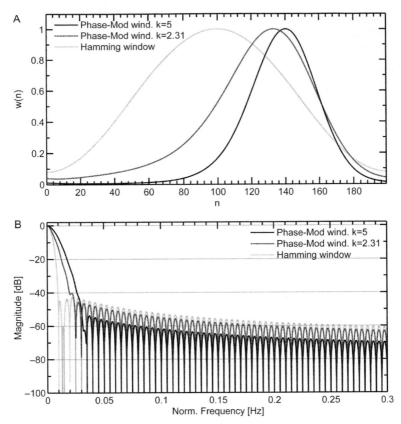

Fig. 21 Time domain presentations (A) and magnitude responses (B) of Phase-Modified asymmetric windows with values of the asymmetry control parameter $k=2.31$, $k=5$, and the Hamming symmetric window.

In all mentioned systems, asymmetric windows were applied to the standard MFCC feature extraction procedure. The resulting features are denoted as "Asym-MFCC" and were used in two speaker verification subsystems.

Asym-MFCC features were first used to improve the performance of a voice activity detector (VAD) subsystem [27]. Its functionality is of particular importance in the presence of noise because it tries to identify speech and non-speech segments in the signal. Non-speech segments are usually discarded or used to estimate noise properties, while processing focuses on speech segments, where the SNR ratio is generally higher and speech quality and intelligibility are substantially better. Therefore, better performance can be expected if we are able to identify regions with a higher SNR more reliably.

Asym–MFCC features were also used as a general acoustic presentation concept of the signal. This approach was inspired by the successful application of these features in speech recognition systems [26], and a similar achievement was also tried for speaker verification [48]. More Asym–MFCC features were used in the speaker verification task (23 features) than in the speech recognition task (13 features), which can be attributed to the importance of general (speaker independent) properties of speech signals which are generally present in MFCC features with lower indexes. However, more speaker-dependent attributes are contained in MFCC features with higher indexes. Therefore, it is quite obvious that a longer MFCC features vector is used in biometrics related applications where the focus is more on individual characteristics of the speaker (e.g., speaker verification and detection). The results of practical evaluations for this case show that MFCC is better in clean conditions while Asym–MFCC is better in noisier conditions [48].

4.4.2.2 Asymmetric DDR windows

In the second approach, asymmetric windows were designed as a simple modification of symmetric windows with the Double Dynamic Range (DDR) property. DDR windows were proposed by Shannon et al. [49,50]. The double dynamic range property of those windows was needed in the autocorrelation-based feature extraction method denoted as Higher-lag Autocorrelation Spectrum Estimation (HASE).

Asymmetric DDR windows in this field are constructed from symmetric DDR windows with a simple shift in the time domain. The center point of symmetry is moved to a non-centered location. By doing this, we get better time delay behavior of the window, probably at the expense of spectral properties. Two procedures for shift modification were proposed by Alam et al. and Morales et al. [24–26]. We were able to reproduce only the asymmetric modification from the second source and the resulting window (DDRasym) is analyzed in Fig. 22. We can conclude that with the asymmetric modification of the original DDR window, the dynamic range is substantially lowered compared to the symmetric DDR window. However, that is not crucial in this case because the window is applied to the signal only once (double range property is important in HASE method because the window is applied twice).

Alam et al. evaluated the ITU G.729 Hamming-Cosine window and the asymmetric DDR window in the speaker verification task [24]. Both are shown and compared to the symmetric Hamming window in Fig. 22.

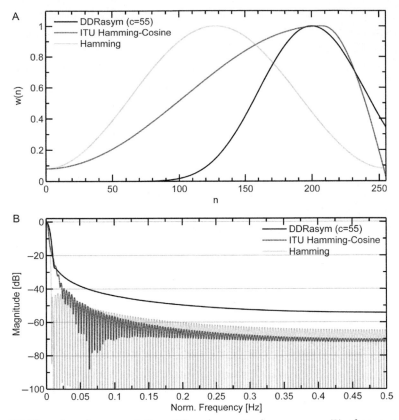

Fig. 22 Time domain presentations (A) and magnitude responses (B) of asymmetric DDR and ITU windows in comparison to the Hamming symmetric window.

In the evaluation special attention was devoted to speaker recognition robustness. Additive noise was artificially added to the test set only and it was not present during training. Both types of asymmetrical windows showed better robustness than the symmetric Hamming window.

4.4.3 Speech recognition
Speech recognition is still considered an unsolved problem. The state of the art recognition systems encompass a lot of various methods and concepts, but they still lack acceptable recognition rates in more complex usage scenarios. According to speech researchers, the most influential factor for recognition performance is the mismatch between learning and testing environments [51]—performance degrades in noisy or, more generally said, mismatched environments.

A possible solution to this problem is to incorporate real environmental conditions into a training set as much as possible. This is by no means an easy task. In addition, it can be quite expensive to create a proper test and train sets.

The other solution, which in our opinion more promising, is to create systems with better inherent robustness. Windows can contribute to those efforts and despite the fact that their influence is not so pronounced, they surely deserve proper research attention.

Window replacement is a trivial task which implies no additional processing burden. Moreover, some preliminary tests have indicated that certain windows can contribute to the robustness of the system even if window modification is applied after the learning phase is completed [2]. This shows great potential for further research, and asymmetric windows could be significant for that research direction.

To the best of our knowledge, three different types of asymmetric windows have been applied to speech recognition systems so far:
- Phase modified asymmetric windows
- Asymmetric DDR windows
- Asymmetric FIR windows

All evaluations and practical applications of asymmetric windows in speech recognition are presented in more detail in the following subsections.

4.4.3.1 Phase modified asymmetric windows

Phase modified asymmetric windows were proposed in Refs. [26,27] and presented already in Section 4.4.2 and Fig. 21. Using the definition in Eq. (25), we get the whole family of windows by varying the value of the asymmetry parameter k. In Ref. [26], four members of the family were analyzed and evaluated in the Aurora2 speech recognition task with parameter values $k = -1.41, -2.21, 2.21, 2.31$. These windows are shown in Fig. 23. For better visibility, the window with a value of 2.21 is excluded. The Hamming window is also shown for reference comparison since it represents the most common symmetric window in speech recognition.

Windows were evaluated on a speech recognition task on Aurora2 database, which is one of the most commonly used platforms for evaluation of speech recognition robustness [52]. A clean training mode was chosen for this evaluation. In this mode, the SRS was trained only on clean speech recordings without any additional noise or other degradation. After that, SRS performance was also evaluated on test sets which contained several noises and convolutional degradations at various SNR levels. This is a typical situation where the inherent robustness of the SRS is put to a proper test.

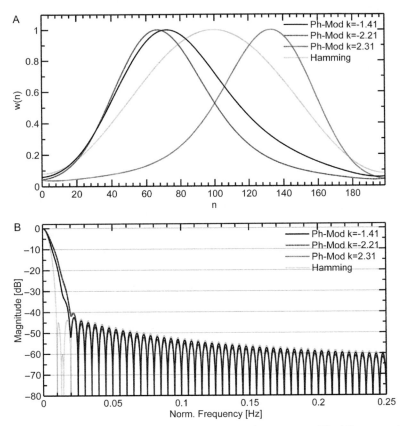

Fig. 23 Time domain presentations (A) and magnitude responses (B) of Phase modified asymmetric windows and the Hamming symmetric window.

The Aurora2 also allows another type of training mode—a multi training mode which presents a subset of noises to the system already in the training phase. In this case, more focus is placed on the system's ability to learn from and adapt to different noises and degradations. However, the influence of inherent robustness is not so prominent in this case. Consequently, we prefer the use of the clean training mode instead.

The results of the practical evaluation with the clean training mode show better system recognition robustness, particularly in noisier conditions with a lower SNR (0 dB, 5 dB) [26].

4.4.3.2 Asymmetric DDR windows
The second approach using asymmetric windows in speech recognition is presented in Ref. [25]. In this case, the window is not used in a direct

calculation of DFT, but in a slightly different way. The signal's Power Spectral Density (PSD, also denoted as power spectrum) often serves as a starting point for further and more specific feature extraction methods.

The calculation of PSD can generally be approached using two groups of methods: non-parametric and parametric. In the non-parametric approach, PSD can be estimated directly from the signal or signal's autocorrelation sequence using DFT. Windows are involved in non-parametric methods, which is why only this group will be considered here.

On the basis of a PSD estimate, MFCC features are usually calculated as the final, compact signal presentation in speech recognition. Standard features gained from DFT calculation are generally denoted as MFCC. On the other hand, features based on autocorrelation are denoted as AMFCC [49] and the method of their extraction as Higher-lag Autocorrelation Spectrum Estimation (HASE). It is known that the window is applied twice in this method—first as a time domain weighting function on a frame of samples, and then again as a weighting function on the calculated autocorrelation, to emphasize coefficients with higher lags

As explained in Ref. [49], all autocorrelation coefficients contain useful information about the signal, its envelope and formants. However, in the presence of various noises and degradations, coefficients associated with lower lags become contaminated more than those with higher lags. In this case, it is more robust to give a greater importance to higher-lag coefficients. Moreover, Morales et al. [25] have empirically confirmed that asymmetric windows with the center point close to a lag which approximately corresponds to the pitch of voiced speech, give the strongest contribution to recognition robustness. This statement is supported by empirical results gained by measuring the spectral distance between clean and noise contaminated AMFCC feature vectors [25].

Various windows were evaluated on more complex speech recognition tasks on widely known databases Aurora2 and Aurora3 [25]. As we have said, Aurora2 database is artificially corrupted with real recordings of noises [52], whereas Aurora3 contains recordings from real and noisy environments such as in-car recordings or different microphones [53]. The results confirm better robustness of the proposed windows in comparison to the symmetric Hamming window. The windows used in the presented practical evaluations are shown in Fig. 24. We need to point out the fact that they were applied as weights to the autocorrelation of the speech signal frame. Therefore, their spectral properties are probably not so important here.

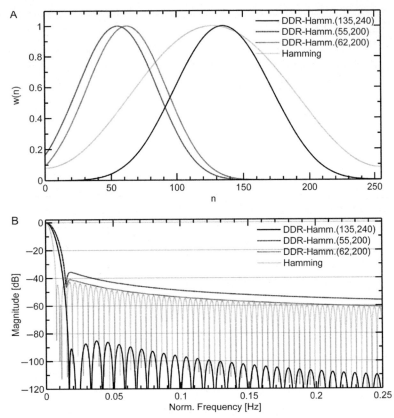

Fig. 24 Time domain presentations (A) and magnitude responses (B) of asymmetric DDR windows and symmetric Hamming window.

4.4.3.3 Asymmetric FIR filter windows

The third approach to using asymmetric windows in speech recognition originates from the FIR filter design field. Windows are finite sequences and their properties can generally be analyzed quite similarly to those of low-pass FIR filters. The only difference is that for the window we usually want the main lobe to be as narrow as possible, whereas this is not common for FIR filters. Other parameters (e.g., stopband attenuation, transition band) have similar meanings in both cases.

However, when designing FIR filters, we frequently specify constant attenuation across the whole stopband. When designing windows, besides the attenuation, the non-zero rate of fall is often desirable for side lobes level. Therefore, Rozman et al. [2,8] used FIR filter design methods to construct asymmetric windows by solving a similar optimization problem with the following specific details:

- The symmetry constraint in case of the linear phase FIR filter is relaxed—windows can be asymmetric.
- Similar to filters, higher relative weight (e.g., 10, 100, 1000) is used for error in the stopband.
- Two possible criteria can be used for error assessment: "phase & magnitude" and "magnitude-only" error functions—Eqs. (13) and (14).
- A window design task using "magnitude-only" error criterion or similar arbitrary criterion represents a complex optimization problem.

When an FIR filter is designed, a substantial set of desired parameters has to be specified. The major question is which main-lobe and transition bandwidth will produce the best windows according to the specification. In addition, a much harder question is how window properties will influence speech recognition performance. The latter two questions are not necessarily related. Therefore, the design of windows as FIR filters is an incremental procedure with many theoretical and practical experiments on several levels of application.

The question of optimal main-lobe width has been partially answered by the limited robustness test of windows with different main-lobe widths while other properties were as similar as possible. The test showed that windows with approximately three times wider main-lobe than the rectangular window of the same length give the most robust speech recognition performance [2]. Therefore, those windows were also used for practical evaluation in comparison to a standard symmetric Hamming window; windows are shown in Fig. 25.

There are two typical examples of windows designed with FIR filter design methods in Fig. 25. The symmetric FIR window was designed with the well-known Parks–McClellan method for design of optimal linear phase FIR filters according to Chebyshev MiniMax error criterion ([14], Eq. (12) in Section 2.3). The asymmetric FIR windows were designed with the general optimization method SOLVOPT [34], according to the magnitude-only error function without the symmetry constraint. It can be seen that the relaxation of symmetry allows up to 8 dB lower side lobes level at the fixed main-lobe width (the difference in attenuation of symmetric and asymmetric FIR windows can be seen in Fig. 25B). This difference seems reasonable because, without the symmetry constraint, all samples can be optimized. On the other hand, in symmetric windows only one half of samples can be exposed to optimization—the other half is fixed by the symmetry constraint.

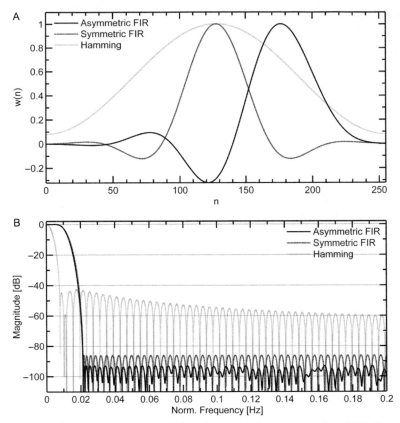

Fig. 25 Time domain presentations (A) and magnitude responses (B) of FIR filter windows and the symmetric Hamming window.

The presented asymmetric windows were evaluated in various speech recognition tasks. Two speech databases and two conceptually different SRS (statistical and neural based) were combined in a practical evaluation of inherent robustness. The clean training mode was used with the highly mismatched testing mode, where common additive noises were added in addition to other common degradations (e.g., convolutive, reverberation) which SRS did not encounter in the training set. The evaluation is presented in Ref. [8]; the results confirm better inherent robustness of speech recognition when asymmetric windows are used.

4.5 Asymmetric windows in FIR filter design

There is a variety of options how asymmetric windows could be utilized in the FIR filter design field. We must be aware that asymmetric impulse responses of FIR filters actually mean the loss of linearity of the phase response.

This actually means that such filters are not suitable for a large group of applications where nonlinear phase distortions are not acceptable. On the other hand, there are applications where symmetry is not needed. Therefore, we can produce better solutions which take advantage of asymmetry. A few typical examples are presented in the rest of this subsection.

If we relax the symmetry constraint, we can design asymmetric FIR filters with a shorter time delay. However, to achieve consistent time delay reduction, an optimization approach is needed. Apaydin tackled this problem with a quadratic programming technique and achieved 12–22% reduction of time delay in passband compared to a symmetric linear phase FIR filter [54].

Sometimes we can also find easier ways to take advantage of asymmetry. Laakso et al. approached the design of an asymmetric window by carefully shifting the symmetric window in the filter design procedure [55]. They showed that the shift operation from the symmetric to the asymmetric window is possible without any substantial loss in spectral properties.

As we have already mentioned, the Windowing method for design of FIR filters is generally suboptimal but easy to implement, particularly for real-time applications. Sac et al. wanted to use this advantage to design Variable Fractional Delay filters by using the window which was extracted from the previously designed optimal filter [16]. In this way they combined the simplicity of the Windowing method with the performance of the more complex optimal method. The resulting solution is much more suitable for real-time operation.

5. Discussion and conclusion

The main purpose of the presented article was to give a comprehensive overview of applications of asymmetric windows in various fields of the digital signal processing area. In addition, we aimed to build a solid base for further research and development steps in this field. We presented the most interesting ideas and approaches related to asymmetric windows. We believe that this article will encourage more researchers to tackle the promising paradigm of asymmetry, which also shows its potentials in the field of digital signal processing.

The presented general features of asymmetric windows are fairly well known. The two most prominent ones are better spectral properties and a shorter processing time delay. Both come as a consequence of the relaxation of the symmetry constraint present in the much more popular and widely used family of symmetric windows.

Asymmetry allows the central point to move more toward both ends of the window, which is how the shorter processing time delay can be achieved. From the time domain viewpoint, the processing delay can be shortened by some simple design methods like concatenation. This method results in asymmetric windows which are nowadays widely used in shorter delay speech coding systems and applications (e.g., ITU Hamming-Cosine and AAC asymmetric windows).

The asymmetry is also accompanied by the possibility of better spectral properties. From the optimization viewpoint, the relaxation of symmetry means wider optimization search space and theoretically better results according to various criteria. We are practically able to achieve this in several different ways. However, the optimization task is usually not the principal challenge. A much more prominent problem is the definition of optimality criteria—establishing exactly which properties of windows produce not just better signal presentation, but also better overall system performance. This is particularly important for more complex applications (e.g., speech recognition, speaker recognition, and verification). In the article, various related examples were presented and evaluated. Hopefully, the work will be continued in the future.

Nonetheless, time delay and spectral properties carry the most prominent general potential of asymmetric windows. However, there are also some other more subtle properties which can obviously help a lot in specific application contexts. We presented several examples of quite surprising effects of asymmetric windows which would be not so obvious to expect or predict. The ability of windows to discriminate closely spaced components even under one DFT bin of a spectral distance, or the ability to achieve better spectral estimations using phase–difference based methods are surely remarkable achievements.

At the same time, these achievements confirm that there are also other specific and subtle, yet unexplored, properties contained in asymmetric windows. Although windows are considered to be a passive concept, the presented results show greater influence of windows than initially expected. We therefore strongly believe that further research will take place in this field, with a lot of related interesting results appearing in the near future.

In the article, a comprehensive overview of related work in the field of asymmetric windows was presented. Since the number of exposed ideas, approaches, methods, concepts, and practical cases is quite large, we have summarized the most important highlights of the presented content in Table 1. Hopefully, Table 1 will be a useful tool for the overview of the field of asymmetric windows in digital signal processing.

Table 1 Summary of presented work on asymmetric windows in Digital Signal Processing.

Authors	Design methods	Summary
Section 4.1 Pitch modification in time domain		
Jung et al. [6]	Concatenation	• Asymmetric time domain data window for PSOLA Overlap-Synthesis • Shows higher SNR after pitch modification
Section 4.2 Shorter time delay processing in time domain		
Section 4.2.1 Asymmetric windows with minimum phase transform		
Su et al. [1]	MP Transform	• Minimum phase transform on existing symmetric windows • Applicable on a subset of symmetric windows • Shorter time delay for the same magnitude response
Section 4.2.2 ITU Hamming–Cosine window in speech coding		
ITU G.729 [21]	Concatenation	• Asymmetric shorter processing time delay window • Simple concatenation of Hamming and Cosine windows
Section 4.2.3 Parabolic exponential windows in speech coding		
Florencio [18,19]	Modulation Concatenation Convolution	• Several related propositions for simple asymmetric windows • Parabolic exponential (PE window—better spectral properties • PE Confirmed shorter time delay in speech coding (FS1016 CELP)
Section 4.3 Asymmetric windows in frequency analysis		
Section 4.3.1 Asymmetric windows in frequency estimation		

Luo et al. [7,41]	Modulation
	Truncation
	Convolution
	Stretch
		• Several proposed asymmetric windows
		• Improvements in frequency analysis:
		  o Better single component detection (truncated, modulated wind)
		  o Better estimations of Phase difference method (truncated wind)
		• Presented figures of merit and magnitude response comparison

Continued

Table 1 Summary of presented work on asymmetric windows in Digital Signal Processing.—cont'd

Authors	Design methods	Summary
Section 4.3.2 Detection of closely spaced components		
Zivanovic et al. [22,23]	Truncation	• Pair of asymmetric windows gained with symmetrical window truncation • Spectrally inferior, but great discrimination property for components less than DFT bin apart
Section 4.4 Asymmetric windows in speech processing		
Section 4.4.1 Speaker recognition		

Sahidullah et al. [28]
Bakshi et al. [46]
	Differentiation	• Asymmetric window containing broader information:
		  o Power spectrum, its derivative and phase response
		• Two empirical evaluations
		  o Confirmed better speaker recognition on two platforms
		  o Confirmed better speaker identification performance

Section 4.4.2 Speaker verification		

Asbai [27,32,47,48]	Phase Modified	• Asymmetrical windows in two subsystems:
		  o Asym-MFCC features improve Voice Activity Detector (VAD)
		  o Asym-MFCC features as acoustic presentation (23 features)
		• Empirical evaluation on speaker verification task:
		  o Asym-MFCC features improve performance and robustness

Alam et al. [24]	Asym. DDR	• Asymmetrical shift of symmetric DDR windows • Confirmed better inherent robustness
Section 4.4.3 Speech recognition		
Alam et al. [26]	Phase Modified	• Consistently better on Aurora2 clean training task, particularly at 0 dB and 5 dB SNR

Morales et al. [25]	Asym. DDR	• Speech recognition task • Pitch-centered asymmetric shift modification of DDR windows • Consistently better robustness on Aurora2, 3 speech databases: o Clean training mode used on Aurora2
Rozman et al. [2,8]	Asym. FIR	• Designed as FIR filters with "arbitrary phase" criterion based optimization (≈ 8 dB lower side lobes than symmetric) • Consistently better than symmetric windows on clean training mode and highly mismatched test conditions
Section 4.5 Asymmetric windows in FIR filter design		
Apaydin [54]	Optimization	• Design of FIR filters with reduced time delay • Quadratic programming optimization with 12–22% time delay improvement
Laakso et al. [55]	Shift	• Fractionally shifted asymmetric window • Fine shift symmetric window while maintaining magnitude resp.
Sac et al. [16]	VFD window	• Extract asymmetric window from the optimal solution • Use it in the Window FIR filter design method • Suboptimal VFD design, but suitable for real-time operation

In addition, a lot of interesting efforts were highlighted and many promising research topics for the future were revealed. Based on the presented content, we are certain that researchers will continue their work in this field and that new researchers will join these efforts.

As the final conclusion, we strongly believe that the idea of asymmetric windows in the field of signal processing deserves proper attention and continuation of research work.

References

[1] L. Su, H. Wu, Minimum-Latency Time-Frequency Analysis Using Asymmetric Window Functions, arXiv, 2016, preprint arXiv 1606.09047.

[2] R. Rozman, Asymmetric Window Functions in Speech Recognition Systems, PhD Thesis. University of Ljubljana, Slovenia, 2005.

[3] V. Blagojević, D. Bojić, M. Bojović, M. Cvetanović, J. Đorđević, Đ. Đurđević, B. Furlan, S. Gajin, Z. Jovanović, D. Milićev, et al., Chapter one-a systematic approach to generation of new ideas for PhD research in computing, Adv. Comput. 104 (2017) 1–31.

[4] B. Nikolic, Z. Radivojevic, J. Djordjevic, V. Milutinovic, A survey and evaluation of simulators suitable for teaching courses in computer architecture and organization, IEEE Trans. Educ. 52 (4) (2009) 449–458.

[5] K. Prabhu, Window Functions and their Applications in Signal Processing, CRC Press, 2013.

[6] C.-J. Jung, M.-K. Ham, M.-J. Bae, On a pitch alteration technique of speech using the asymmetry weighted window, in: Military Communications Conference Proceedings, 1999. MILCOM 1999. IEEE, vol. 2, 1999, pp. 1439–1443.

[7] J. Luo, Z. Xie, X. Li, Asymmetric windows and their application in frequency estimation, J. Algorithms Comput. Technol 9 (4) (2015) 389–412.

[8] R. Rozman, D.M. Kodek, Using asymmetric windows in automatic speech recognition, Speech Comm. 49 (4) (2007) 268–276.

[9] R. Rozman, D.M. Kodek, Improving Speech Recognition Robustness Using Non-Standard Windows, in: EUROCON 2003. Computer as a Tool. The IEEE Region 8, EUROCON, vol. 2, 2003, Url: https://ieeexplore.ieee.org/xpl/conhome/8828/.

[10] F.F. Kuo, J.F. Kaiser, System Analysis by Digital Computer, Wiley, 1966.

[11] L. Rabiner, Techniques for designing finite-duration impulse-response digital filters, IEEE Trans. Commun. Technol. 19 (2) (1971) 188–195.

[12] B. Gold, K. Jordan, A direct search procedure for designing finite duration impulse response filters, IEEE Trans. Audio Electroacoust. 17 (1) (1969) 33–36.

[13] L. Rabiner, B. Gold, C. McGonegal, An approach to the approximation problem for nonrecursive digital filters, IEEE Trans. Audio Electroacoust. 18 (2) (1970) 83–106.

[14] T. Parks, J. McClellan, Chebyshev approximation for nonrecursive digital filters with linear phase, IEEE Trans. Circuit Theory 19 (2) (1972) 189–194.

[15] T. Parks, J. McClellan, A program for the design of linear phase finite impulse response digital filters, IEEE Trans. Audio Electroacoust. 20 (3) (1972) 195–199.

[16] M. Sac, M. Blok, A nearly optimal fractional delay filter design using an asymmetric window, Int. J. Electron. Telecommun. 57 (4) (2011) 465–472.

[17] T. Barnwell, Recursive windowing for generating autocorrelation coefficients for LPC analysis, IEEE Trans. Acoust. Speech Signal Process. 29 (5) (1981) 1062–1066.

[18] D.A. Florencio, Investigating the use of asymmetric windows in CELP vocoders, in: Acoustics, Speech, and Signal Processing, 1993. ICASSP-93., 1993 IEEE International Conference on, vol. 2, 1993, pp. 427–430.

[19] D.A. Floréncio, On the use of asymmetric windows for reducing the time delay in real-time spectral analysis, in: Acoustics, Speech, and Signal Processing, 1991. ICASSP-91., 1991 International Conference on, 1991, pp. 3261–3264.

[20] R. Rozman, Enostavnejša zasnova sistema za razpoznavanje govora/simplified design of the speech recognition system, Elektrotehniški Vestnik/J. Electr. Eng. Comput. Sci. 80 (4) (2013) 171.

[21] I.-T. S. Group and others, Coding of speech at 8 kbits/s using conjugate-structure algebraic-code-excited linear-prediction (CS-ACELP), in: International Telecommunication Union Telecommunication Standardization Sector, Draft Recommendation, Version, vol. 6, 1995.

[22] M. Zivanovic, A. Carlosena, On asymmetric analysis windows for detection of closely spaced signal components, Mech. Syst. Signal Process. 20 (3) (2006) 702–717.

[23] M. Zivanovic, A. Carlosena, Extending the limits of resolution for narrow-band harmonic and modal analysis: a non-parametric approach, Meas. Sci. Technol. 13 (12) (2002) 2082.

[24] M.J. Alam, P. Kenny, D. O'Shaughnessy, On the use of asymmetric-shaped tapers for speaker verification using I-vectors, in: Odyssey, 2012: The Speaker and Language Recognition Workshop, 2012, pp. 256–262.

[25] J.A. Morales-Cordovilla, V. Sánchez, A.M. Gómez, A.M. Peinado, On the use of asymmetric windows for robust speech recognition, Circuits Systems Signal Process. 31 (2) (2012) 727–736.

[26] M.J. Alam, P. Kenny, D. O'Shaughnessy, Robust speech recognition under noisy environments using asymmetric tapers, in: Signal Processing Conference (EUSIPCO), 2012 Proceedings of the 20th European, 2012, pp. 1638–1642.

[27] N. Asbai, M. Bengherabi, A. Amrouche, Y. Aklouf, Improving the self-adaptive voice activity detector for speaker verification using map adaptation and asymmetric tapers, Int. J. Speech Technol. 18 (2) (2015) 195–203.

[28] M. Sahidullah, G. Saha, A novel windowing technique for efficient computation of MFCC for speaker recognition, IEEE Signal Process Lett. 20 (2) (2013) 149–152.

[29] H. Fletcher, Speech and Hearing in Communication, D. Van Nostrand, Princeton, NJ, 1953.

[30] K. Paliwal, K. Wójcicki, B. Shannon, The importance of phase in speech enhancement, Speech Comm. 53 (4) (2011) 465–494.

[31] K.K. Paliwal, L.D. Alsteris, On the usefulness of STFT phase spectrum in human listening tests, Speech Comm. 45 (2) (2005) 153–170.

[32] N. Asbai, M. Bengherabi, F. Harizi, A. Amrouche, Effect of the front-end processing on speaker verification performance using PCA and scores level fusion, in: International Conference on E-Business and Telecommunications, 2013, pp. 359–368.

[33] J.W. Adams, J.L. Sullivan, Peak-constrained least-squares optimization, IEEE Trans. Signal Process. 46 (2) (1998) 306–321.

[34] A. Kuntsevich, F. Kappel, SolvOpt: The Solver for Local Nonlinear Optimization Problems. Matlab, C and Fortran Source Codes, Technical Rep, Institute for Mathematics University of Graz, 1997.

[35] D. Burnside, T.W. Parks, Optimal design of FIR filters with the complex Chebyshev error criteria, IEEE Trans. Signal Process. 43 (3) (1995) 605–616.

[36] L.J. Karam, J.H. McClellan, Design of optimal digital FIR filters with arbitrary magnitude and phase responses, in: Circuits and Systems, 1996. ISCAS'96., Connecting the World., 1996 IEEE International Symposium on, vol. 2, 1996, pp. 385–388.

[37] S. Nordebo, I. Claesson, Z. Zang, Optimum window design by semi-infinite quadratic programming, IEEE Signal Process Lett. 6 (10) (1999) 262–265.

[38] N.Z. Shor, Minimization Methods for Non-differentiable Functions, vol. 3, Springer Science & Business Media, 2012.

[39] U. Čibej, J. Mihelič, Adaptation and evaluation of the simplex algorithm for a data-flow architecture, Adv. Comput. 106 (2017) 63–105.

[40] M. Schnell, M. Schmidt, M. Jander, T. Albert, R. Geiger, V. Ruoppila, P. Ekstrand, G. Bernhard, MPEG-4 enhanced low delay AAC-a new standard for high quality communication, in: Audio Engineering Society Convention 125, 2008.

[41] J. Luo, M. Xie, Phase difference methods based on asymmetric windows, Mech. Syst. Signal Process. 54 (2015) 52–67.

[42] D.P.W. Ellis, A Perceptual Representation of Audio, Massachusetts Institute of Technology, Dept. of Electrical Engineering and Computer Science, 1992.

[43] M. Sahidullah, G. Saha, Design, analysis and experimental evaluation of block based transformation in MFCC computation for speaker recognition, Speech Comm. 54 (4) (2012) 543–565.

[44] R.M. Hegde, H.A. Murthy, V.R.R. Gadde, Significance of the modified group delay feature in speech recognition, IEEE Trans. Audio Speech Lang. Process. 15 (1) (2007) 190–202.

[45] S. Nakagawa, L. Wang, S. Ohtsuka, Speaker identification and verification by combining MFCC and phase information, IEEE Trans. Audio Speech Lang. Process. 20 (4) (2012) 1085–1095.

[46] A. Bakshi, S.K. Kopparapu, S. Pawar, S. Nema, Novel windowing technique of MFCC for speaker identification with modified polynomial classifiers, in: Confluence the Next Generation Information Technology Summit (Confluence), 2014 5th International Conference, 2014, pp. 292–297.

[47] N. Asbai, M. Bengherabi, A. Amrouche, F. Harizi, Improving speaker verification robustness by front-end diversity and score level fusion, in: Signal-Image Technology & Internet-Based Systems (SITIS), 2013 International Conference on, 2013, pp. 136–142.

[48] N. Asbai, M. Bengherabi, F. Harizi, A. Amrouche, Improving the performance of speaker verification systems under noisy conditions using low level features and score level fusion, in: Signal Processing and Multimedia Applications (SIGMAP), 2013 International Conference on, 2013, pp. 33–38.

[49] B.J. Shannon, K.K. Paliwal, Feature extraction from higher-lag autocorrelation coefficients for robust speech recognition, Speech Comm. 48 (11) (2006) 1458–1485.

[50] B.J. Shannon, K.K. Paliwal, Noise robust speech recognition using higher-lag autocorrelation coefficients, in: Proceedings of the Microelectronic Engineering Research Conference, School of Microelectronic Engineering, Griffith University, 2005.

[51] N. Morgan, J. Cohen, S.H. Krishnan, S. Chang, S. Wegmann, Final Report: OUCH Project (Outing Unfortunate Characteristics of HMMs), vol. 10, *CiteSeerX*, 2013, p. 7249. no. 1.395.

[52] H.-G. Hirsch, D. Pearce, The Aurora experimental framework for the performance evaluation of speech recognition systems under noisy conditions, in: ASR2000-Automatic Speech Recognition: Challenges for the New Millenium ISCA Tutorial and Research Workshop (ITRW), 2000.

[53] L. Netsch, Description and Baseline Results for the Subset of the Speechdat–Car German Database Used for ETSI STQ Aurora WI008 Advanced Front–End Evaluation, *Documentation on the German Aurora Project Database CD–ROMs*, 2001.

[54] G. Apaydin, Realization of reduced-delay finite impulse response filters for audio applications, Digital Signal Process. 20 (3) (2010) 620–629.

[55] T.I. Laakso, T. Saramaki, G.D. Cain, Asymmetric Dolph-Chebyshev, Saramaki, and transitional windows for fractional delay FIR filter design, in: Circuits and Systems, 1995., Proceedings., Proceedings of the 38th Midwest Symposium on, vol. 1, 1995, pp. 580–583.

About the author

Robert Rozman received his Doctoral degree in Computer Science from the University of Ljubljana in 2005. He currently works as a senior lecturer at the Laboratory of Algorithmics, Faculty of Computer and Information Science, University of Ljubljana, Slovenia. His research interests include digital signal processing, asymmetric windows, speech recognition, ambient intelligence, sensor networks, IoT, smart cities and smart home or building automation.

## CHAPTER FIVE

Intelligent agents in games: Review with an open-source tool

Matej Vitek, Peter Peer
University of Ljubljana, Faculty of Computer and Information Science, Ljubljana, Slovenia

Contents

Advances in Computers, Volume 116
ISSN 0065-2458
https://doi.org/10.1016/bs.adcom.2019.07.005

Abstract

The field of artificial intelligence has come a long way in the last 50 years, and studies of its methods soon expanded to a field in which they are of great practical value—computer games. The concept of intelligent agents provides a much needed theoretical background for the comparison of various different approaches to intelligent, rational behavior of computer-controlled characters in games. By combining rationality with certain limitations to the capabilities of our agents, we can achieve behavior resembling that of a human player, which is also desirable in games. The goal of this article is to introduce various types of agents that are used in games, show how to implement meaningful, reasonable limitations to agent capabilities into the game world, and provide a freely available, open-source application for the comparison of such agents. Additionally, in this article we show that even the simplest agents can succeed in their tasks in certain task environments, whereas more difficult task environments often require a more sophisticated agent architecture. Our application consists of two task environments with nine agents in total but could easily be extended with additional task environments and agent implementations. In the end, we find the addition of goals into the agent architecture has the biggest impact on the agent's behavior and performance, whereas the state-based approach helps keep our implementation simple and compact.

1. Introduction

The field of artificial intelligence is very broad. Its methods are used in economics, mathematical problems and proofs, medical diagnosis, chess, etc. In this chapter we focus on its use in computer games, where its main use is control of nonplayer characters (NPCs). Our goal is to introduce and compare various types of agents used in the game world. For the purpose of comparison of agents we developed two scenarios, and the entire implementation is available as an open-source tool. The application and this accompanying chapter are intended to provide an educational review, comparison and evaluation, similar to [1].

We first define the task environment (Section 2) and present an overview of the different types of intelligent agents in games, based primarily on the categorization popularized by Russel and Norvig [2] (Section 3). Then we define the test scenarios (Section 4) and implement (Section 5) and compare (Section 6) various intelligent agents that will have to perform the tasks specified by the scenarios.

We use the freely available development platform Unity3D [3] to first create the desired environments and then implement the agents. Our application serves as a foundation for future implementation of new agents, enabling their comparison in already given scenarios, or in newly designed ones. The source code is freely available under the GPL [4] on the GitHub repository [5]. It may be used freely for any future research, educational purposes, or other use, and the repository also gives even the users without programming knowledge the ability to suggest improvements and additional features. The program has been tested in Windows, but, due to Unity's platform independence, should be transferable without major problems to other operating systems.

1.1 What is an intelligent agent?

An agent is a program that acts autonomously in a given environment (which can be virtual or physical). It receives information about the state of the world around it in the form of percepts through *sensors*, which can again be either virtual or physical. It then responds to the information by choosing an appropriate action and executing it via its *actuators*. This process is illustrated in Fig. 1. Formally, the choice of action can be defined as a mathematical function, called the *agent function*, which maps every possible percept sequence to an action.

This definition of an agent is only one of many. Franklin and Graesser [6] demonstrate the plethora of different formal definitions of a software agent and seek to unite their essence. Nwana [7] delves deeper into the subject. Likewise, there are various ways to define intelligence; a more in-depth discussion on the topic from the AI community can be found in [2, 8, 9].

Throughout this article, we adopt Russel and Norvig's definition. Informally, *intelligence* (also known as *rationality*) refers to the agent's ability to always choose the action it considers to be the best in the current situation.

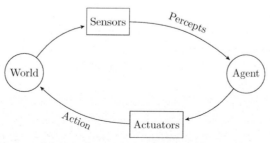

Fig. 1 An agent's decision making and acting in a world.

For a formal definition we first introduce the *performance measure*, which rates the agent's success in performing the task it was given in its task environment. Given its dependence on the specific task, it is generally best if we define the performance measure function ourselves when we define the tasks (thereby defining what we consider to be good, desirable behavior), rather than using a predefined fixed one (see [10]). By convention, the higher the agent's performance measure, the better it performed its task.

Formally, then, an intelligent agent always chooses the action, which will maximize its expected performance measure, given its current percept sequence and its built-in knowledge of the task environment. Note that the performance measure need not be explicitly known to the agent for it to be able to make intelligent decisions, and in fact most of the agents we mention have no knowledge of it. They are, however, still capable of making fairly intelligent (or in some cases even perfectly intelligent) decisions in various task environments.

Note that, at least following this definition, our measure of intelligence depends on the given task environment. Therefore, many agent types may indeed achieve intelligence in certain (simpler) task environments but not in more complex ones.

1.2 Game world

The game world is a virtual environment with rules and laws, affecting the characters and objects in it. These rules need not apply equally to every character, though at least the laws of physics normally do. The characters can be player-controlled (player characters) or computer-controlled (NPCs). Note that the characters are entities with visible bodies and require only that their behavior appear rational and autonomous to the human player. Thus, the simpler ones can often be controlled by programs that do not in fact exhibit true intelligence as we defined it. The intelligent agent approach is merely one of the ways to control the characters (for further discussion on this subject, see [11]). It is, however, a reliable approach, since it guarantees that the characters will indeed appear both intelligent and autonomous, as true intelligence and autonomy are the very requirements we set in our definition of an intelligent agent.

Besides intelligence we usually require one more property of our agents in the game world: fairness. The computer used to control characters will generally at all times have complete knowledge of the current world state. Ideally, however, we would like it to control every character using only information that particular character can currently perceive, according to the rules we have set in our game world. Therefore, we restrict the information available to each of the agents, and in doing so, we in fact programmatically implement the agent's sensors (virtual sensors). Examples of the agent's actions, in turn, are moving around, communicating with other agents, assisting or attacking (hindering) them, etc.

With reasonable restrictions our agents then cannot see through walls, cannot hear sounds miles away, and ideally do not have direct access to the percepts of other agents.

2. Task environment

A task environment encompasses the world in which the agent will act, the task(s) it will have to perform, and the agent itself. To define a task environment we use the PEAS description: we specify what we consider to be good *performance* of our agent (we can do that by defining a performance measure function), what constitutes the physical (or virtual) *environment* (or *world*) and its various states, what *actuators* the agent has available (or equivalently, what actions are available to it under what circumstances), and the *sensors* at its disposal (equivalently, how it receives information about the world state). The latter two are of course necessary as per our definition of an agent, while the performance measure pertains more to the rationality (intelligence) of the agent.

In the game world we often include other desired properties into our performance measure, such as the agent being beatable, imperfect, unpredictable, random, etc. While some of these properties may not appear rational (a rational person would generally not intentionally attempt to be beatable), we often wish our agents to have certain such properties in order to provide a more satisfying gameplay experience to the player. Including them into our performance measure then allows us to still use our definition of an intelligent agent with no adjustments to solve the desired tasks.

When we later evaluate our agents, we can use the performance measure from the PEAS description, as defined by Russel and Norvig [2]. We must be careful to use a large enough sample, depending on how random our agent's performance is. Another example of agent evaluation, useful particularly when we desire human-like behavior from our agent, is the Turing test. It asks whether a human could, after 5 min of conversation, correctly determine whether it was communicating with another human or a machine. A very subjective test in its raw form, however, using a large enough sample size of evaluators, we can achieve fairly reliable results. An extensive Turing-test-based evaluation of intelligent agents was performed in [9]. A slightly more complex approach was used for Turing test evaluation in [12].

2.1 Task environment properties

Task environments can be described with a range of additional properties that can later help us select the appropriate agent type. We distinguish between:

- fully observable, partially observable, or unobservable
- single agent or multiagent
- deterministic or stochastic
- episodic or sequential
- static or dynamic
- discrete or continuous
- known or unknown

task environments (as defined by [2]).

However, it is often not enough to simply list the properties of a task environment. We must consider them more in depth, as we will see in our implementation.

A task environment is *fully observable* if the agent has access to full information of the world state (or at least the parts relevant to its decision making) at all times. If it only has partial information, the environment is *partially observable*. Sometimes we must consider theoretically fully observable environments as partially observable due to processing power and memory limitations. In partially observable environments agents will often require the ability to memorize previous percepts of the world state. The task environment can also be *unobservable* if the agent never receives any information about the environment state (i.e., it has no usable sensors). Even tasks in unobservable task environments can sometimes be reliably solvable.

A task environment is *multiagent* if it contains multiple interacting agents and *single agent* if it contains only one agent. If it contains multiple agents, but the agents' performance measures are independent of each other, the task environment can be considered single agent from the point of view of each of the agents (each of them can consider other agents as simply parts of the world, rather than entities to interact with). In a multiagent task environment relationships between agents can be *competitive* (if their performance measures are negatively correlated—increasing one decreases the other) or *cooperative* (if their performance measures are positively correlated—increasing one increases the other). Competitive environments will often see random behavior come up as rational, while in cooperative environments communication is often a vital rational action choice.

A task environment is *deterministic* if the next world state is completely defined by the current world state and the action(s) executed by the agent(s). If, however, there are multiple possible outcomes to certain actions in a certain state (usually these outcomes occur with certain probabilities), the task environment is *stochastic*. There is never any uncertainty in a fully observable, deterministic task environment.

A task environment is *episodic* if every action selection process is independent of previous actions performed and has no effect on future action selections. In such an environment the agent does not require planning. If action selections are connected, the task environment is *sequential*.

A task environment is *static* if the world state does not change during the agent's action selection process. Otherwise, the task environment is *dynamic*. In a static environment, we do not need to worry about the time the agent takes to choose an action (other than from the viewpoint of user experience), and the agent does not need to keep track of environment changes during its action selection process.

Time, as well as the sets of world states, percepts and available actions, can be either *discrete* or *continuous*. Time, for example, is handled discretely in move-based games, such as chess and tic-tac-toe, as well as turn-based strategy games.

Informally, the distinction between known and unknown task environments is based on whether we know in advance the rules and laws that apply in our world. More formally, if an agent can always predict how its actions will affect the environment, the task environment is *known*. Otherwise, it is *unknown*. Note that stochastic environments can still be known, as long as the agent always knows in advance all the possible outcomes of its available actions. An unknown task environment will generally require the agent to possess the ability to learn and adapt. Fortunately, unknown scenarios are rare in games, as we generally know the rules of the game we are designing agents for.

3. Agent overview

In this article we loosely follow the categorization of general artificial intelligence agents from [2], which is used in most literature as a base.

The categorization is in fact an array of gradual improvements and upgrades to an agent's architecture, although some of them do not in fact require their predecessors, as we show in our implementation. Following the classification from [13], through the majority of this section we are *specializing (S)* the categorization from [2] to the world of games, appropriately adapting the more general agent architectures to our specific needs, as well as introducing the state-based agent—a particularly useful agent architecture in the game world.

In the final section we present some alternative views of agent categorization and agent architecture. These approaches, while different, often achieve very similar agent behavior. We can therefore say that in the final section we are *generalizing (G)* [13] various architectures to a common factor—the behavior of the agents, rather than the concrete architectures and their respective implementations.

For each agent type, we note what types of environments it is appropriate for and what types it may struggle with. We also give a few examples of games utilizing each agent type. Note that the actual implementation of the agents in these games may not exactly follow the architecture we give here for that specific agent type, but the behavior of the agents reflects the given architecture.

3.1 Table-driven agent

An agent can be defined as a mathematical function (the agent function) that maps every possible percept sequence to an appropriate choice of action.

The table-driven agent is a direct implementation of this function, using an array (table) of each possible percept sequence and its respective action choice. The agent simply looks up the appropriate entry in the table at every step and executes the appropriate action (which requires $\mathcal{O}(1)$ time using a hash table).

This type of agent will of course not be usable in practice (even in the simplest environments), as the number of different possible percept sequences is simply too large (often even infinite). Even in tic-tac-toe, an exceedingly simple game, the size of the table would be roughly 300,000.

Still, it is a good theoretical concept that shows how an agent should ideally act.

3.2 Simple reflex agent

The simple reflex agent follows a number of predefined if-then rules set by the programmer. It only uses the current percept when choosing an action, as it has no memory of previous world states. Therefore, it is generally less reliable than more complex agents in partially observable environments and will often fail completely. Infinite loops are a common occurrence, though they can sometimes be prevented by introducing some randomness into our action selection process.

This agent also has no planning ability or memory of its actions so it will usually perform suboptimally in sequential task environments. This does not, however, mean that it cannot function in them—only that it functions less effectively than better, more advanced agents.

It also cannot adapt well, so it will generally fail in unknown environments.

The simple reflex agent is quite limited in its functionality, but it is very easy to implement and undemanding in terms of computer resources.

Simple reflex agents are common even in modern games. Examples of such include, e.g., bystanders in various role-playing games (RPGs), such as the Dragon Age series (2009, 2011, 2014) and the Witcher series (2007, 2011, 2015). These characters move around an area randomly and, if interacted with, respond with a random one of a few predefined pieces of dialogue They are used in modern RPGs mainly to help the ambience and allow the players to immerse themselves into the world with greater ease. They generally have little impact on the gameplay, however. For that purpose, more advanced architectures are generally used (indeed we can find most of the agent types introduced in this section in RPGs).

Simple reflex agents are also found often in (particularly older) racing games, such as various Need for Speed games, F1, etc., as race car driving in the worlds of these games is simple enough that a reflex agent suffices.

3.3 State-based agent

The state-based agent is in fact a subtype of the simple reflex agent, as every state-based agent can be reduced to a number of if-then rules. The resulting simple reflex agent will, however, often use deeply nested if-statements, and it will also often check conditions it would not need to check at every step, so this is not done often.

Since it is technically not a distinct agent type, it is often left out in categorizations in the literature and is not mentioned explicitly. We nevertheless include it as part of our categorization, as it is a very useful implementation paradigm, particularly in games. In fact, the vast majority of intelligent agents in games (beyond the very simplest ones) use the state-based agent as a base for their implementation (see [14]).

The state-based agent is, in essence, a finite-state machine, with each of the states representing a separate, independent simple reflex agent. The agent keeps track of its internal state, and chooses the appropriate action selection function depending on the state it is currently in. It can also transition between states if the appropriate conditions are met. One could say a state-based agent is really a simple reflex agent consisting of simple reflex subagents.

At a high level, the state-based agent could be considered a data flow agent with distinct states exchanging information (data), although the states are executed exclusively, rather than in parallel as is the usual paradigm in data flow architectures.

Similarly to the simple reflex agent, it lacks memory or any planning and adapting abilities. However, due to its higher complexity, it can generally outperform the simple reflex agent in partially observable and sequential task environments.

The state-based architecture is massively used in games, whether in the form described above, or as a base for the more advanced agent architectures.

It was used in one of the oldest 3D games, Tomb Raider (1996), for controlling the enemies Lara (the game's protagonist) faces. These enemies have a peaceful state, in which they simply move around or stand still (e.g., sleeping). They also have an aggressive state to which they transition when Lara approaches them or crosses a certain threshold. In this state they constantly attempt to move in the direction of Lara. When they are close enough, they transition to the attacking state, in which they attempt to bite, hit, or shoot Lara, depending on the type of enemy.

The state-based agent is also common in sport games, such as the EA Sports games FIFA, NBA, NHL, etc. In these games the agents' state transitions may for example happen when a ball is transferred from one player to another or from one team to the other.

Bystanders in older Grand Theft Auto games (up to GTA: San Andreas) are also state-based agents. These characters wander around an area, then when attacked (or in certain cases, when they spot the player), they transition to the attacking or running-away state, depending on the enemy type, their health, etc.

An example from the RPG genre would be the bystanders in The Elder Scrolls V: Skyrim (2011). These characters behave similarly to the reflex agents described above in the Dragon Age and Witcher games, but have the added state of attacking (or running away), to which they transition when they are attacked or witness a crime.

3.4 Memory-based agent

The memory-based agent improves on the previous two with the ability to memorize relevant information about the world state and use it in future action selections.

This is particularly helpful (or even required) in most partially observable task environments, and it is also of help in sequential task environments, where previous actions affect our current action selection process.

We can also use world state and action memory in unknown task environments, namely to figure out the effects our actions have on the world state. Such an agent is called a model-based agent, and the model of the world it uses contains information about the past world states, as well as the effects of our actions on said states. Such an approach is often used alongside the learning agent approach, which we will describe later.

Memory-based agents will, however, still struggle in stochastic task environments, as the (currently unobservable) parts of the world the agent has memorized can change independently of the agent's actions.

The memory-based approach is used often, but generally alongside others (particularly the goal-oriented approach).

It can often be found in guards pursuing the player, as well as enemies in shooting games. Both of these have to remember the player's last location when the player is lost from their view (as otherwise they would immediately abort their pursuit, as they cannot at that time perceive the player).

In the RPG Skyrim certain NPCs remember their previous interactions with a player and use that memory in deciding how to treat the player when interacted with again.

3.5 Goal-based agent

The previous agents followed a fixed set of rules that were completely defined by the programmer. If we attempted to use the same agent to solve a similar task with slightly altered parameters, the agent would generally fail. It will also often fail if it encounters a world state the programmer did not foresee.

With the addition of goals and goal orientation we look to solve these problems and give our agent a degree of autonomy and adaptability. A goal-based agent will no longer simply follow predefined directions. Rather, it will have the ability to select its actions based on which one will help it achieve a goal.

Note that an agent will always have goals—at least one is already given by the task itself. Goal orientation refers to the direct use of goals in the agent's decision making. A goal-based agent therefore needs to be capable of planning—selecting a series of actions that lead to a goal.

As an example, consider an agent whose goal is to reach the end of a level in an adventure game, where the end is blocked by a locked door (here we assume that the agent knows the layout of the world). This agent will need to find and execute a series of actions that will result in its picking up the key, unlocking the door and proceeding to the end.

Many planning algorithms exist—means-ends planning with goal regression, graph planning, partial order planning, etc. These algorithms all require prior knowledge of the requirements and effects of each of the agent's possible actions, and use this knowledge to create a plan for the agent to reach its desired goal state. Note that such a plan will often contain implicit (or in certain algorithms explicit) definitions of new goals, such as picking up the key in the above example. A more detailed description of each of these algorithms is available in [2].

When the action requirements and effects are simple and do not interact with each other in complex ways, state space search is often used. As an example, we consider one of the most common problems in state space search—pathfinding. In pathfinding, the agent must decide at each step in which direction to move in order to reach its destination in a known world layout. Our example agent could, for instance, use one of the above planning algorithms to define the higher-level plan *Pick up key* → *Unlock door* → *Exit*, and then be left with the lower-level pathfinding problem for each of these intermediate goals.

If we attempted to tackle the pathfinding problem without goal orientation, we would need to give our agent precise, predefined directions to reach a destination, such as "go straight ahead for 50 meters," "turn right," etc. If we then tried to send this same agent to a different location, we would need to manually tune its direction choices or it would fail completely.

The most widely used algorithm in pathfinding (as well as many other state space search problems) is A* [15], which is briefly described in Section 5.5. We can attribute its popularity to the fact that it finds the shortest path to a destination quickly and reliably. More recently, improved versions of A*, such as D* variants [16–18], are also often employed, particularly in dynamic environments.

Goal orientation also allows the agent to adapt to changes in the world, as it can always change its current plan if a better alternative presents itself (e.g., a door opens, a bridge is lowered, a guard falls asleep). Goal-based agents therefore work well even in stochastic environments (though they still perform better in deterministic ones), but the biggest improvement is seen in sequential environments.

In certain partially observable environments goal-based agents will still struggle, if the unseen parts of the world are vital to the state space search algorithm. For example, if we wish to use a goal-based pathfinding agent in a world whose layout it does not know, the agent will fail.

Since the pathfinding problem is so common in games, goal orientation is likely the most important upgrade. It can be found for instance in modern versions of memory-based agents in shooting games described above, as agents can now not only recall the location the player was last sighted, but can also then create and execute a plan to reach that location. As a concrete example, we look at The Last of Us (2013), where the enemies (various types of zombie-like creatures, as well as regular humans) remember the player's location after seeing or hearing them, and will path toward it when they can no longer perceive the player.

Because of how fundamental the pathfinding problem is, methods for solving it are often already given in the game engine itself and therefore we could in fact consider the above agents to be simply memory-based agents (rather than goal-based). However, goal orientation can be found in other aspects of agent decision making as well. For an example, we look again at Skyrim, where followers (characters following and assisting the player) exhibit goal-based behavior. They have a set of predefined goals they may choose (for instance, defeating an enemy that has engaged the player or protecting the player from the attack) and will use different methods to achieve the goal, depending on the current situation (if they are too far away, they will first move closer, if there are corpses around, they may use a reanimation spell, etc.).

3.6 Utility-based agent

As we mentioned previously, the A* algorithm will (under certain assumptions, as described in Section 5.5) always return the shortest path to a goal. However, the goal-based agent does not particularly care about that—to the goal-based agent the important property of A* is that it finds the path quickly and reliably.

The utility-based agent, on the other hand, attempts not only to reach a goal, but to do so in the best possible way. We can consider the task environment as a set of states the agent can reach from the current state through its actions (this is called the state space). The agent's task is then to find the set of actions in this space that will lead it from the current state to a goal state. Several goal states can often represent the same goal (e.g., a won game of football, a destination in our world) but with different properties. For example, a 1:0 victory, a 2:0 victory, and a 6:1 victory may all represent a won game of football (which was our goal), but they are not entirely the same—some may be more desirable than others. Likewise, the states we go through may also not all be equally desirable—for example, in a pathfinding problem we may want to avoid dangerous areas, so solutions containing states in which our character passes through a dangerous area are less desirable.

To the goal-based agent all goal states are equivalent, and any solution that reaches a goal state is good enough. The utility-based agent on the other hand can rate goal states and solutions based on their desirability and perform actions leading toward the most desirable goal state (or best combination of goal states).

The utility function represents the agent's approximation of the expected performance measure in a given goal state. The utility-based agent attempts to maximize this function through its actions. Note that this agent is the first one in our categorization to actually be aware of its performance measure function—the performance measure is not only a tool for our evaluation of the agent but is used directly in the agent implementation.

The biggest improvement over a goal-based agent is seen in stochastic and partially observable environments. In the aforementioned pathfinding problem in an unknown world, the utility-based agent can choose between the currently visible locations, based on which one is expected to maximize its odds of reaching the goal destination.

Additionally, when there are several goals the agent wishes to achieve, the utility-based agent can find the best path between them, and when there are mutually exclusive goals, it can select the best possible combination of the goals to achieve. Note that goals are not necessarily physical locations. They can be more abstract concepts, such as helping an ally and attacking an enemy. The two are often mutually exclusive goals, but the utility-based agent will have the ability to properly balance the two, based on their desirability.

An example from the RPG genre would be the members of a player's party in Dragon Age: Inquisition (2014). These characters need to constantly make decisions between mutually exclusive goals during battle. Should they be attacking the enemy or protecting an ally? Should they attempt to block a spell, ignore it, or run away from its area of effect? The player can set up some basic pointers to behavior they want from the character, but most of the decisions will have to be made by the character's AI, and depend on a multitude of factors the agent has to consider.

Utility-based behavior can also be found in modern real-time strategy games in the AI opponents. In these games the agent has to constantly react to the players' and other agents' actions, weighing how critical these responses are when several responses need to be made at once (if it is attacked from two sides at once, for instance). The world is also partially observable, so it may require certain probabilistic guesses by the agent. Last, the pathfinding problem is not quite as simple, since the agent also has to take into account the enemies faced along the way—a safer path may be better than a shorter one, though not necessarily. An example of such pathfinding is given in [19].

As above, other games that require the agent to replace actual human players also often feature utility-based agents. The bots in Counter-Strike: Global Offensive (2012), for instance, need to constantly make tactical decisions: Should I hold position and defend or look for the enemy? Should I obey the radio command given by a player or continue with my own plan? Should I help a teammate or continue with the mission? If engaged in a fight, should I be moving left and right, standing still or running away? All of these depend on a multitude of factors, which should all be weighted with how they affect the agent's chance of winning the round—their respective utilities.

3.7 Learning agent

The learning agent is the most complex of the agents listed here and is the only one that functions well in unknown task environments. It is rarely used in games in the industry, as in the game world we require reliable, tested agents, while learning agents are based on machine learning and data mining and are thus unreliable and difficult to thoroughly test. It also requires a long lifetime to improve its behavior enough, and we generally want our agents in games to behave intelligently from the start. Due to the unreliability of results and the complexity of the task environment that would arise from attempting to compare this agent to the previous ones, we leave it out of our implementation, but we nevertheless present a brief description, as it is an important agent type in general.

This type of agent is, however, often found in academic research in the field of video game AI, as seen in [20–23]. In addition, recent advances in machine vision and deep learning have been combined with established reinforcement learning methods to achieve high quality performance in simple games, even when virtually no information about the game is given to the agent in advance (most prominently in [24], which spurred many similar experiments, such as [25–28]).

The learning agent consists of two elements: the learning element and the performance element. The performance element is what in the previous introduced agent types we considered to be the entire agent—it selects the actions based on the percepts it receives. The learning element rates the behavior of the performance element and proposes improvements. It also at times suggests actions that would otherwise be discarded as suboptimal, if they have the potential to give more information about the world and thereby help improve the agent's behavior in the future.

An example of a learning agent in games would be the creature AI in Black & White (2001), which employs a kind of player-supervised reinforcement learning to build decision trees in combination with neural networks to continuously improve the creatures' behavior.

Neural networks were also used in Creatures (1996), another game where player-supervised learning is used to teach the AI how to behave.

Note that, despite this architecture being the most complex in our categorization, it was used even in older games, in the cases where creatures learning from the player was an integral part of the game.

Last, we cannot forget chess playing agents. Many modern chess agents use various machine learning techniques to improve their gameplay, learning from the vast databases of past chess games, as well as from their own games. Additionally, learning agents can also be used to teach chess, as in [29].

3.8 Alternative agent architectures

As mentioned in the beginning of this section, the categorization we have presented is not the only one. There are various different approaches to agent architecture categorization, some general like this one and some designed for more specific purposes.

The BDI (belief-desire-intention) architecture is designed for logical programming languages and follows Bratman's human practical reasoning model [30]. Described in our terms, belief represents the agent's model of the world, desire represents the agent's goal(s) and intention represents the action choice. While this architecture resembles our definition of a goal-based agent, it focuses more on the representation of the above three concepts, rather than focusing on finding the path to the goal (as we did in goal-based agents). The search part is unimportant to the architecture and is performed by the language itself (which is why logical programming languages are used). The aforementioned game Black & White used this architecture for its AI.

A widely used architecture in modern games is the behavior tree (BT), first defined in [31]. A BT is a mathematical model representing the agent's behaviors. Each node of the tree represents a certain task (behavior), with each of its children being a subtask, down to the leaves, which represent very simple single tasks. More precisely, each internal node is a control flow node, which, depending on its type, will execute its children until one succeeds, until one fails, or all in parallel. Each leaf is either an action or a condition. This architecture is flexible and allows us to model a wide array of behaviors. It is used in many modern games, such as the Halo series from Halo 2 (2004) onward, the Bioshock series (2007, 2010, 2013), and Spore (2008).

Conte [32] uses a different approach to agent categorization and distinguishes between rational and intelligent agents—two terms we consider equivalent. In this categorization rational agents, along with goal-directed agents, are a subset of intelligent agents, which are in turn a subset of goal-oriented agents. Therefore, all agents are goal-oriented (i.e., are designed to reach a certain goal, even if they are not aware of it). The definition of intelligent agents is similar to ours, while the rational and goal-directed agents are defined similarly to our utility-based and goal-based agents, respectively.

Two examples of architectures designed for more specific purposes are the hybrid architectures InteRRaP [33] and TouringMachines [34], described and compared in a game world in [35].

TouringMachines is a three-layer architecture with a control subsystem that assigns control to the appropriate layer at each point in time. The reactive layer functions similarly to a simple reflex agent. The planning layer is similar to a goal-based agent, but, rather than searching a state space, uses predefined schemas of solutions. The modeling layer maintains a world model and determines goals to be used by the planning layer.

InteRRaP also consists of three layers. At the bottom is the behavior layer, which again behaves like a simple reflex agent. In the middle is the planning layer, functioning similarly to the one in TouringMachines. At the top is the cooperation layer, which functions as a utility-based agent, determining goals that are the best not only for this agent but also for others cooperating with it. At each step the action choice process is executed bottom-to-top. First the behavior layer attempts to choose an appropriate action. If it is unsuccessful, the planning layer attempts to do the same, and if it fails as well, the cooperation layer chooses the action. The action is then executed in a top-to-bottom fashion. The layer that selected the action executes it, using lower layers to accomplish subtasks (an architecture that divides tasks into subtasks like this is called a subsumption architecture).

3.9 Summary

Below is a summary of the main agent types introduced in this section:

- The *simple reflex agent* (SRA) performs simple tasks by following predefined if-then rules.
- The *state-based agent* (SBA) is a more complex form of the SRA, which includes states and state transitions.
- The *memory-based agent* (MBA) possesses the ability to memorize certain key elements of past observations and therefore makes use of the entire percept history, rather than just the current percept.
- The *goal-based agent* (GBA) is aware of the goals it is trying to achieve (and can even define its own goals). It uses planning and state space search to choose the actions that lead to these goals.
- The *utility-based agent* (UBA) improves on the GBA by implementing the ability to measure the quality of its goals and the paths to them.
- The *learning agent* (LA) uses methods from the field of machine learning to alter its behavior, based on the effect its past actions have achieved.

4. Testing scenarios

We have described the various agent types we will be comparing, now we need an appropriate task environment to compare them in. While in practice we would normally design agents to fit the task environment, in this case we reverse the process and instead design the task environment to fit our agents. Since the goal of this work is the evaluation and comparison of different agent types, we design the scenarios to illustrate the agents' strengths and weaknesses as clearly as possible.

One problem we face in our scenario design is the fact that goal orientation is such an impactful addition to the agent's architecture, that tasks that require it usually cannot be solved by agents that do not possess the abilities of goal-based agents (state space search and planning). Rather than using an overly convoluted scenario or having nongoal-based agents appear entirely unintelligent and useless, we implement two testing scenarios—one for agents with goal orientation and one for agents without it.

In this section we define the two scenarios, describe their properties and describe the user controls available in our application.

4.1 Main menu

The main menu, which can be seen in Fig. 2, allows us to select the scenario and agent type. To select an agent type, simply click on it with the mouse cursor. Certain agents in the second scenario will on selection present us with the choice to load the normal or alternative (moved) version of the scenario. We discuss in Section 5.5 why this is necessary.

We are also given the option of using a first person controller in either scenario to control the character ourselves (to get acquainted with the environment and the task or compare our own behavior to that of the agents).

4.2 First scenario

The first scenario is a simple target practice scenario from the world of first person shooter (FPS) games. We describe it using the PEAS description, starting with the environment and the task, as well as the agent's built-in knowledge of the task environment.

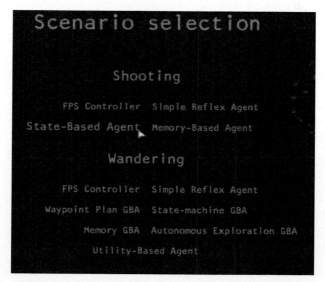

Fig. 2 Main menu.

The world consists of a small cubical room, in which the agent resides. A target is located on one of the room's walls. When the agent hits the target or 30 s have passed since its appearance, the target disappears and a different one appears on a different wall. The targets will, throughout a single scenario run, keep appearing in the same order (after the last one the first one appears again), randomly determined at the start of the run. The agent's task is to hit as many targets as it can in the given time. The scenario from the agent's point of view can be seen in Fig. 3.

The agent does not know in advance where the targets will be located, how many targets there are, or the order they will appear in. It does know that both the locations and the order are fixed (after the start of the run). This allows us to demonstrate the value of memory of past percepts, while still allowing agents without it to accomplish the task (though obviously not as well).

In our test the scenario ends after 2 min and we use four targets, one on each wall. In the bottom right corner of the screen (Fig. 3) is the time remaining, and in the bottom left is the score (number of targets hit). We can use this score as the performance measure for our agents.

The actions available to our agent are camera rotation (simulating mouse movement) and shooting (in the direction of the camera, as in FPS games). Shots are not perfectly accurate and can go a few degrees off target. They also cause the camera to jerk upwards slightly, due to weapon recoil. Also, the rate of fire is limited appropriately. Camera rotation allows the agent to aim, while shooting allows it to hit the targets.

The primary sensor of our agent is sight. As already described, we use a programmatic approximation of the sensor, as the computer in truth always has complete knowledge of the world state. We will describe

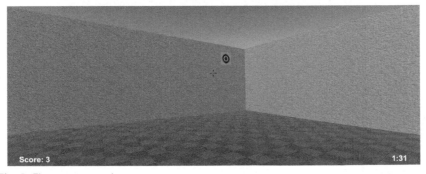

Fig. 3 First test scenario.

Table 1 PEAS description of the first scenario.

Performance	Environment	Actions	Sensors
Number of targets hit	Room with targets on walls	Rotate camera, shoot	Sight, hearing, clock

Table 2 Properties of the first scenario.

Observ.	No. of agents	Determ./stoch.	Epis./seq.	Stat./dynam.	Disc./cont.
Partially observable	Single agent	Stochastic	Sequential	Dynamic	Continuous

our implementation of sight later. As a secondary sensor, the agent hears whether it hit the target or not when shooting. It also has an internal clock to keep track of the time passed.

We provide a concise PEAS description of our task environment in table form in Table 1.

Now let us take a look at the properties of our task environment. First we simply list them in Table 2.

The properties are not so black-and-white, however. For instance, while the task environment is technically stochastic, as a target will disappear after 30 s regardless of the agent's then-selected action, this has little effect on the agent's behavior, as it will happen very rarely (and even when it does, it does not affect the agent too much). Therefore, we can consider the scenario deterministic, while accounting for spontaneous target switches in our agent implementations.

Similarly, we can in practice consider the scenario static. Dynamic changes (changes during the action selection process) will appear at most once every 30 s, while the action selection happens several times every second, and even if a dynamic change were to cause a wrong action selection, the effect this would have on our agent's behavior would be insignificant (as no single action massively impacts our agent's performance).

Additionally, while the scenario is sequential, it does not require any advanced planning algorithm. Sequentiality is obvious, as the camera rotation direction determines whether (and how fast) our agent will find the target, and aiming affects the agent's ability to hit the target. Executed actions thus affect future action selections. The correlation between past and future actions, however, is simple enough that we can solve it in our implementation through simple built-in rules, rather than using more complex planning techniques.

The environment is indeed partially observable, as we can only see a portion of the world at any point in time. The world states are continuous (as target locations and camera looking direction are both represented by real numbers), as are actions (camera rotation direction is also represented by real numbers) and percepts. All of these will therefore have to be approximated with floating point numbers. Time is technically continuous, but we can discretize it to steps corresponding to the frame rate.

We see that the only problematic property in this scenario is partial observability. The other undesired properties can be ignored or easily dealt with. This scenario will show the usefulness of simpler agents with built-in, deterministic rules, as well as illustrate the effect memory has on agent behavior in partially observable environments.

The user can exit a scenario and return to the main menu by pressing the Esc button. At the end of the scenario the final score is displayed, and the user can return to the main menu by pressing any key. In the first person controller the user can move around with the WASD keys, turn with mouse movement and fire with the left mouse button or the Ctrl key.

4.3 Second scenario

The second scenario is more complex and includes the pathfinding problem we described as a basic problem for goal-based agents.

The environment makes use of the freely available Viking Village project, as well as tools (axes) and tool racks, all of which are available on the Unity Asset Store [36]. It is comprised of the village and three tool racks, the locations of which are known to the agent at the start of the scenario. The agent also knows the layout of the world. Each of the racks has a neighborhood, defined by a set of cuboids (boxes). An example of such a neighborhood can be seen in Fig. 4, where each of the boxes is

Fig. 4 Neighborhood of a tool rack.

highlighted as a yellow wireframe. The boxes are all fully accessible to the agent and do not contain any obstacles that would block the agent's movement or sight. When the agent reaches a tool rack, a number of axes (5 in our testing) spawn at random positions in its neighborhood. The agent is then tasked with finding the axes and returning them to the rack. The agent's initial (spawn) location is random (though for simplicity of implementation, we simply randomly choose one from a predefined set). The user sees yellow spotlights above currently relevant racks and axes, but these spotlights are not visible to the agent (Fig. 5).

The scenario ends after a certain time (7 min in our test) and the agent can also end it prematurely if it decides it has completed its task as well as it could. As a performance measure we use the total number of axes successfully returned to tool racks (tracked in the top left corner of the screen—see Fig. 5). Since this score alone is not enough to distinguish certain agents (most notably the ones that will always reliably return all axes), we also evaluate the agents by their time taken. We use the time in minutes, multiplied by -1, in our performance measure—higher performance measures implies better performance, while higher time taken means worse performance. Since we are more interested in the agent's score than the time taken, we weight it with a coefficient high enough to dominate the time component of our function. Our performance measure can then be defined as $f = 10 \times score - time$ in minutes, for example. Due to the weight coefficient of 10, the score will always have a bigger impact on the performance measure than the time (as the maximum time taken is 7 min).

The agent's actions are rotation (turning around) and walking straight ahead. The primary sensor is again sight. The agent also always knows its current position and, as in the first scenario, has an internal clock to keep track of time.

Fig. 5 Second test scenario.

Table 3 PEAS description of the second scenario.

Performance	Environment	Actions	Sensors
No. of retrieved axes and time needed	Viking village: houses, boats, barrels, etc.	Turning, walking straight ahead	Sight, GPS, clock

Table 4 Properties of the second scenario.

Observ.	No. of agents	Determ./stoch.	Epis./seq.	Stat./dynam.	Disc./cont.
Partially observable	Single agent	Stochastic	Sequential	Dynamic	Continuous

We again provide a table representing a concise PEAS description in Table 3.

In Table 4 we list the properties of the second scenario. Note that the table of properties is exactly the same as the one we saw for the first scenario (Table 2), despite this one being, as we noted in the description, significantly more complex.

The scenario is partially observable, but the agent now sees far less of the world than it did in the first scenario (since the world is larger and contains a lot of obstacles that obstruct the agent's vision).

Stochasticity is far more impactful this time and is not negligible, mainly due to the axe spawn positions being determined randomly (with probabilities evenly distributed over the given rack's neighborhood). We will need to tackle this problem by exploring the racks' surrounding areas in order to find the axes.

As in the previous scenario, we can again ignore dynamic changes, which only appear in the form of axe positions changing under the effect of gravity right after they are spawned.

Sequentiality poses a bigger challenge this time, particularly in the form of the pathfinding problem—this problem always implies a sequential task environment, as every movement affects future pathing choices. We will therefore require at least goal-based behavior from our agents, at least to find paths between positions in the world. Since time is also a criterion in our evaluation, sequentiality can also be seen in the higher-level problem of selecting the correct positions in the world to visit and the order in which to visit them.

The world states, actions, percepts, and time are all continuous. Since the computer's representation of data is discrete (e.g., floating point approximations for real number values), the agent's perceptions (e.g., perceived axe locations) and action selection (e.g., rotation amount) will be approximate.

However, the approximations are close enough to the real values that this will not noticeably impact our agent's performance. In other words, our agent does not need any special mechanisms to deal with approximation errors and this property is therefore immaterial to our agent selection.

In this scenario we have several undesirable properties we cannot ignore: partial observability, stochasticity, and sequentiality. Due to the last one in particular we will implement goal orientation into our agents from the start.

The same controls as were available in the first scenario are available in this one, and there are some additional functionalities as well:

- Clicking the left mouse button will change the camera. The user can cycle through three cameras: first-person (used also in the sight implementation), third-person OTS (over-the-shoulder) and isometric.
- Moving the mouse left and right rotates the camera in isometric view.
- The scroll wheel controls the zoom in the third person and isometric views.
- The N key toggles navigation display mode (Fig. 6), in which the NavMesh structure (described later), the waypoints, and the current path (computed with the A* algorithm) are drawn. The NavMesh appears in light blue, the waypoints appear as red spheres, and the path is a broken line, drawn in red.

Unfortunately Unity's NavMesh triangulation calculation is not exact, so its position had to be manually readjusted, and it is still not drawn properly in certain places. Additionally, since drawing lines is a complex, resource-intensive process, we draw the NavMesh without

Fig. 6 Navigation display mode—the NavMesh in *light blue*, current planned path as a *red line*, and waypoint goals as *red globes*.

A

B

Fig. 7 Difference in NavMesh displays. (A) NavMesh in navigation display mode.
(B) Actual NavMesh in Unity Editor.

edges, so it appears as a flat shape, rather than a set of visible triangles.
A better approach is to view the NavMesh in the Unity Editor directly
(in the Window menu choose Navigation and then in the scene view
check Show NavMesh and optionally Show HeightMesh). The difference can be seen in Fig. 7.
In the first person controller, the user can again move using the WASD keys and
turn using the mouse, while the left mouse button in this scenario does not
have a function.

5. Agent implementation

In both scenarios we begin by implementing simple agents and then
gradually upgrade them toward more complex ones. Before we implement
the agents, however, we must first consider their primary sensor in both
scenarios—sight.

Note that all the algorithms are available in pseudocode in Appendix.

In the older games sight was usually replaced with a simple detection zone, which would trigger a percept in the agent when the player (or another agent) entered it. More recent games generally use simplified field of view (FOV) approximations, which is the approach we will employ as well. More advanced algorithms, as well as their optimizations can be found in [37].

Our implementation of sight is based on frustum culling and tests three conditions:

1. *Distance*: We could calculate the distance cutoff relative to the size of the object (larger objects can be seen at longer distances), but in our scenarios we simply use a constant distance cutoff for the far clipping plane, since the relevant objects are all of the same size (targets in the first and axes in the second scenario). We can also cut off items that are too close (near clipping plane), but we ignore this in our implementation.

2. *FOV bounds*: We approximate the agent's field of view with a pyramid (some use a cone representation instead). Objects are only seen if they intersect this vision pyramid. Testing whether an object intersects the pyramid is as simple as checking the horizontal and vertical angles between the agent's looking direction and the directions of the object's vertices against the agent's FOV angle bounds. The cone representation only requires one angle check per vertex rather than two, but the pyramid is more accurate in terms of what the player would see on their screen if they were in the agent's position.

3. *Occlusion*: The object is not visible if the paths to all its vertices from the agent's location are obstructed. We test this using raycasting, which is a feature provided by Unity's physics engine. We cast rays from the agent's eye position to the object's center and the object's vertices within the FOV bounds (see the previous step). If any of these rays reach the object without collisions, we consider the object to be visible.

Since objects can be quite complex, we use their bounding box vertices in steps 2 and 3, rather than their actual vertices. An object's bounding box is a rectangular cuboid, just large enough to completely enclose the object. Additionally, the world objects' colliders are approximations of the actual objects as we see them to simplify collision detection. This can lead to a degree of inaccuracy, namely, the agent can detect objects that should not be visible to it, but these errors are generally unnoticeable. The agent also does not distinguish between seeing only a small corner and seeing the full object.

Having implemented sight, we can now move on to the implementation of our agents.

5.1 First scenario: Simple reflex agent

The simple reflex agent moves the camera around randomly. If the target is in sight, it slows down the camera rotation and starts shooting. It changes the rotation direction every 3 s (changing it too often would likely cause the agent to get stuck). It obviously only uses the current percept in its action selection process.

This is a very simple implementation, and we could of course improve the performance of this agent by adding additional rules for it to follow (as stated previously, we could implement the entire state-based agent as a large simple reflex agent). However, simplicity is the one point in favor of simple reflex agents, and thus we keep the number of rules to the minimum.

5.2 First scenario: State-based agent

We now upgrade our agent to a state machine, which allows us to do precisely what we wanted to do in the previous section—add additional rules—while still keeping the implementation as simple and compact as possible.

This agent has three distinct internal states (Fig. 8):

- *Searching*: The agent moves the camera to the right constantly, while randomly moving it up or down. If it spots a target, it transitions to the Aiming state.
- *Aiming*: The agent moves the camera toward the target. We use spherically interpolated rotation steps to achieve more player-like camera movement. When the agent is looking directly at the target, it transitions

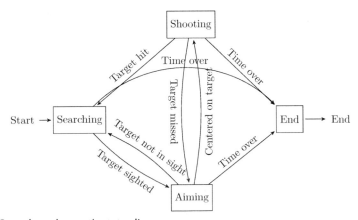

Fig. 8 State-based agent's state diagram.

to the Shooting state. If the target is not visible any more (which happens when 30 s have passed and a target switch occurs), the agent transitions back to the Searching state.

- *Shooting*: The agent fires a single shot. If the target was hit, it transitions to the Searching state, otherwise it transitions back to the Aiming state.

The agent still only uses the current percept to choose an action.

5.3 First scenario: Memory-based agent

As the final step in the first scenario we add memory to our agent. In the scenario description we noted that the agent knows in advance that targets will always appear in the same order, but the previous two implementations did not make use of that fact.

When this agent spots a target, it remembers it as the successor of the previous target if it has information of the previous one. Note that it will always have this information, except in two cases: when it finds its very first target or when 30 s have passed since the last target switch and the agent therefore has to delete the predecessor information, as it is no longer valid—the predecessor has changed.

The agent keeps the states from Fig. 8, with a few changes in transitions (Fig. 9). The transition from Searching to Aiming now occurs if the agent has knowledge of the successor to the one that appeared previously, even if it cannot currently see that successor. Likewise, if the agent has knowledge of the successor, the transition from Aiming back to Searching only occurs after 30 s and not when the target is not in sight.

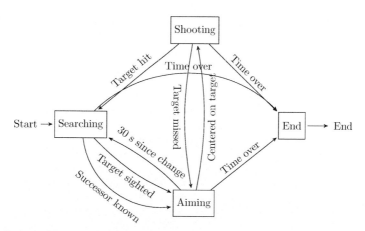

Fig. 9 Memory-based agent's state diagram.

This agent uses both the built-in knowledge and the entire percept history to maximize its performance measure and thus, by our definition, behaves intelligently.

As we will see in the next section, we do not need more complex agents in a scenario like this.

5.4 Second scenario: Simple reflex agent

We start with the simple reflex agent in the second scenario also, simply to demonstrate how the lack of goal-based behavior impacts our agent's performance in this scenario. The simple reflex agent moves straight ahead constantly. If it hits an obstacle, it performs a random turn (90–270 degrees) and continues. Since the agent does not have a bumper sensor, it will infer whether it hit an obstacle from its current speed (calculated from the last position, current position and time passed between the two). If the current speed is too low, it will assume it has hit an obstacle.

This agent is of course an extreme case of simplicity—it exhibits no intelligence whatsoever. It is implemented merely as a clear example of the poor performance brought on by the lack of goal orientation. It is not meant to be representative of the best possible behavior we can achieve without goals, as we could clearly improve its performance in the same way we did in the first scenario. Instead, however, we begin with the far more impactful improvement—goal-based behavior.

5.5 Second scenario: Basic goal-based agent

All the agents' architectures (aside from the SRA from Section 5.4) in this scenario consist of two layers (Fig. 10). The bottom (low-level) layer is tasked with solving the pathfinding problem between points passed to it by the top (high-level) layer. It implements goal-based behavior and mainly utilizes methods already provided by the Unity libraries, such as the NavMesh data structure and the A* algorithm. This layer remains more or less the same throughout the agent upgrades in Sections 5.6–5.9.

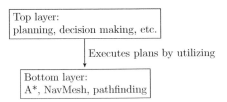

Fig. 10 Second scenario agent architecture.

The top layer is in charge of more general decision making and planning. It is implemented by us in full and represents the upgrades of our agents.

The basic goal-based agent's top layer is still only a simple reflex agent. It follows a predefined plan of waypoints, specified by the programmer. Such a plan obviously requires that the programmer know the locations of the tool racks and their neighborhoods at the time of implementation. However, this is not what was not specified in the scenario description—we only know that this data will be available to the agent at the start of the scenario run. The lack of data about the rack locations can easily be dealt with (simply have the agent insert their locations into the plan when those locations are known). Adding an appropriate selection of points from a neighborhood, however, poses a more difficult problem for a simple agent to solve.

To demonstrate the shortcomings of our approach, we use two scenario cases for our agents (barring the simple reflex agent) in our testing—in the first case we assume the programmer knows the rack locations at implementation time, while in the second we move the racks to simulate the lack of that knowledge. This agent's plan remains the same in both cases and it will therefore perform poorly in the second case. It does, however, at least use the programmer's knowledge along with the lower level's goal orientation to behave somewhat successfully in the first case.

The lower level is implemented with the use of the A* algorithm. This is currently still the most popular algorithm in pathfinding and in many more general graph traversal problems [38]. It is a heuristic algorithm—i.e., it uses a quality estimation function—that finds a shortest path between two nodes in a graph $G = (\mathcal{V}, \mathcal{E})$ (where $\mathcal{V}$ is a set of vertices, and $\mathcal{E} = \{(u,v); u,v \in \mathcal{V}\}$ is a set of edges between those vertices). In pathfinding the nodes represent locations in the world, while the edges between them represent physical path segments, which can be trivially traversed by walking in a straight line.

The A* algorithm requires a graph representation of the pathfinding space. We provide this representation in the form of a navigation mesh (NavMesh) (see [39] for implementation details). The NavMesh is a data structure that represents all the walkable surfaces of the world with convex polygons (triangles are commonly used). Either the polygons or their vertices are then used as nodes in the graph. In the first case graph edges connect nodes that represent adjacent polygons. In the second case the graph edges represent the polygon edges (or in some implementations all straight paths between vertices). The creation of a NavMesh is generally done offline (before the agent starts its run), but certain approaches allow online corrections as well (see [40]). On the other hand, building a full mesh online (required in worlds whose layout is not known in advance) is a far more difficult problem.

5.6 Second scenario: State-based goal-based agent

Having implemented the bottom layer, we can now focus on upgrading the top layer to achieve better performance. We start by adding states, just as we did in the first scenario (Fig. 11).

- *Rack Approach*: The agent moves toward a tool rack, using the lower layer to plan the path to it. When it reaches a rack for the first time, it reads the predefined plan of the neighborhood exploration (specified by the programmer as in the previous agent) and transitions to the Waypoint Approach state. When it returns to a rack after traversing the neighborhood, it continues in this state to the next rack (or concludes the run if it has returned to the last rack).
- *Waypoint Approach*: The agent moves toward the current waypoint, again using the lower layer to plan the path. If it reaches the waypoint,

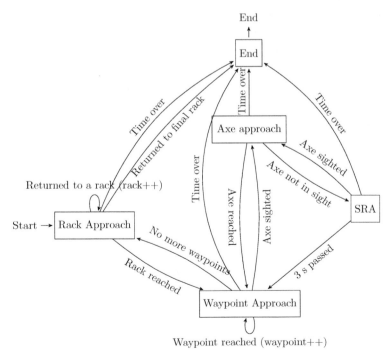

Fig. 11 Goal-based state-based agent's state diagram.

it continues to the next one in this state. If the reached waypoint is the last one in the current neighborhood exploration plan, it transitions back to the Rack Approach state to return to the current rack. If at any point it spots an axe, it transitions to the Axe Approach state.

- *Axe Approach*: The agent moves toward a visible axe. If it reaches the goal (thereby picking up the axe), it transitions back to the Waypoint Approach state. Note that the agent has no memory of the world state, which means that if it loses the axe from view, it will not know its location anymore and will lose the information about its current goal. This is why if the agent loses the axe from view, it transitions temporarily to the SRA state, which can help it either pick up the axe (if it is very close already) or at least spot it again.
- *SRA*: The agent (similarly to the SRA from Section 5.4) moves straight ahead constantly and turns if it hits an obstacle. After 3 s it transitions to the Waypoint Approach state. If an axe is sighted before that, it transitions to the Axe Approach state.

This agent now uses the current percept in its decision making and does not rely on the programmer's knowledge alone. This will make it perform better in both scenario cases, though its behavior will still be unreliable in the second one.

5.7 Second scenario: Memory-based goal-based agent

As we did in the first scenario, we continue by adding memory of the world state to our agent. When the agent spots an axe, it now remembers its location and uses it as an additional waypoint with higher priority. We can now remove the SRA state, and our agent will now more easily deal with cases where it spots multiple axes at once. It also remembers how many axes it has picked up in a neighborhood and immediately concludes the exploration if it has picked up all 5.

This agent uses the entire percept history in its decision making, but still in large part relies on the programmer's knowledge. It will perform slightly better than the previous agent in both scenario cases, but this upgrade is more of a stepping stone toward better subsequent upgrades.

5.8 Second scenario: Autonomous exploration agent

The previous agents possessed goal-based behavior in the bottom layer (pathfinding), but the top layer still followed rigid, predefined rules, which is why these agents would fail in a task with slightly altered parameters (as we will see in the second scenario case with moved racks).

In order to achieve better autonomy and adaptation in our agent, we add goal orientation to the top layer as well. The main change in this upgrade is in the neighborhood exploration step. Previously, the agent would read a predefined plan of waypoints to follow after it reached a rack. This upgrade will allow the agent to choose an appropriate set of points in the neighborhood autonomously rather than rely on predefined ones. The agent's goal is to select a set of points that will allow it to at least once see every point in the neighborhood (recall that, as per our scenario specification, the agent knows these neighborhoods at run time). That is to say, the goal is full coverage of the entire neighborhood.

Since the boxes that define a neighborhood do not contain vision-blocking obstacles, the agent will be able to see each box in its entirety from any point in that same box. That is why it is enough to take one random point from each box and use it as a waypoint. The agent should then turn around to see the entire box and proceed to the next one. The reason we take a random point from each box rather than a fixed one (such as the center or a corner) is because, due to the slight inaccuracy of our sight implementation, specific pairs of points inside boxes can appear, where one is not visible from the other. By using random points and additional traversals if the first one does not find all five axes, we can deal with even these (very rare) occurrences.

Recall that the goal-based agent does not distinguish between different solutions to the goal (full coverage of the neighborhood) based on their desirability, which is why this solution is perfectly fine for it. The agent will also visit boxes in whatever order the engine returns information about them.

Aside from this neighborhood exploration waypoints algorithm, the agent will behave similarly to the previous ones (using the points from the algorithm as it did the predefined waypoints previously).

This agent will be able to adapt to changes in the task parameters and is the first to guarantee a successful scenario run (i.e., returning all axes) regardless of the rack locations and their neighborhoods, even if it may not be the fastest. It again uses the entire percept history, and it also uses the built-in knowledge to maximize the number of tools picked up. It does not yet minimize the time taken, so it does not fully maximize its expected performance measure, but it still achieves decent intelligence.

5.9 Second scenario: Utility-based agent

First, let us consider what the goals of the previous agent were at each step: reaching all three racks, exploring their respective neighborhoods and picking up all the found axes in these neighborhoods. The previous agent already guaranteed successful achievement of all three goals, but it did not guarantee the best possible solutions to these goals. In order to achieve the best possible performance, we define a utility function for each of the goals.

When approaching the racks the only thing that impacts the performance measure is time taken. In this stage we can define the performance measure as $f(t) = -d(t)$, where t is the rack traversal and $d(t)$ is its distance (since the agent moves with constant speed, the distance determines the time taken). When picking up the tools, the performance measure depends on time taken and number of tools picked up. Since we already defined the goal as getting *all* the tools, we again only distinguish between solutions based on their required time, so we can define the utility function the same way we did previously. In the stage of planning a neighborhood exploration, the performance measure impacting factors are the coverage (which affects the number of axes found) and, once again, the time taken. The number of found tools is far more important to the performance measure, so rather than balancing the coverage against the time, we want to guarantee full coverage, so we define the utility function as

$$f(p, t) = \begin{cases} -d(t); & p \text{ covers the entire neighborhood} \\ -\infty; & \text{otherwise,} \end{cases}$$

where p is the set of selected points in the neighborhood and t is, again, a traversal between them. These utility functions are now what the agent will be attempting to maximize at each stage.

Each stage requires that the agent determine the fastest possible traversal of a set of points. This is known as the travelling salesman problem (TSP). Before we attempt to solve it, however, let us first take a look at the neighborhood exploration waypoint selection process. We have to use an estimate of the best solution, as there are too many possible choices of points. As a simple measure of the quality of a solution, we can use the number of waypoints in it. This measure works fairly well in practice. We then wish to find a solution with the lowest number of waypoints that still guarantees full coverage of a neighborhood. The algorithm we follow then is

1. Initialize a list of boxes to all the boxes in the neighborhood.
2. Select a random box from the list and remove it from the list.

3. Select a random point from the selected box and add it to the list of exploration waypoints.
4. Remove from the list all boxes that are fully visible from the selected point. We use an approximation of full visibility, testing the visibility of the 8 vertices of each box from the selected point (raising the point first to the eye level). If all the 8 vertices of a box are visible, the box is fully visible, otherwise not.
5. If the list still contains any boxes, go back to step 2. Otherwise the algorithm concludes.

We repeat this algorithm several times (20 in our implementation), and keep the best solution (the one with the fewest points). Note that, since our full coverage test is approximate, we need to consider the possibility that all axes will not be found on the first traversal (even though this happens very rarely in practice). Much like the autonomous exploration agent, the utility-based agent simply repeats the entire process if it returns to a rack without all the axes.

Another improvement we make to the agent is that when it spots an axe, it no longer goes to pick it up directly. Instead it adds it to the list of exploration waypoints, immediately after the waypoint closest to it. If it has already spotted all of the remaining axes, it removes all the waypoints from the list, except for the axe locations, computes the shortest traversal between them (TSP again), and picks them up in the order given by the computation.

Now that we have improved the agent's waypoint selection and axe collection processes, we need to solve the travelling salesman problem in order to finish maximizing the utility functions. There are many approaches to solving this problem, as it is one of the most studied NP-hard [41] problems in computer science and graph theory. In our implementation we use a combination of an exact TSP algorithm, a nearest-neighbor algorithm and 2-opt local optimization (see Algorithms 11 and 12 in Appendix for the respective pseudocodes of the latter two).

The original version of the problem can be stated as follows: In a weighted graph, find the least-weight Hamiltonian cycle. In other words, find the path through the graph that visits every vertex exactly once and then returns to the original point, with the least sum of edge weights. Our variation is slightly different. First of all, we require a Hamiltonian path rather than a cycle (we do not need to return to the starting position). Second, we have a fixed starting point (the agent's current location). The latter would not matter in the original version of the problem (as it does not matter where in a cycle we start), but it does in our variation. For edge weights we could use the Euclidean distances between points, but those would lead to inaccurate results. Instead, we use the distances of paths computed by the A* algorithm.

On small input sizes (in our implementation, inputs with number of waypoints less than 8) we use an exact naive algorithm. It simply checks every possible permutation of the graph vertices to find which one has the least total weight. Since there are $(n - 1)!$ permutations on an input of size n (the starting point is fixed), this algorithm would fail quickly on larger inputs.

That is why on larger inputs we instead use an inexact approach. We first use a nearest-neighbor algorithm [42] to get an initial solution. This algorithm starts off with a list containing only our starting position, then simply repeatedly appends the waypoint nearest to the last waypoint in the list.

After we get an initial solution, we use the 2-opt local optimization algorithm [43] to see if we can improve it. It is a popular algorithm, designed specifically for the travelling salesman problem to remove intersecting path segments, which lead to suboptimal solutions. It does require that the triangle inequality hold for any triangle of graph edges. This is true in our case due to the guarantee of A* to always return a shortest path between two points (no path computed with A* can be longer than a combination of two paths beginning in the first point and ending in the second). The 2-opt algorithm inverts every possible subpath to see if it can find a better overall solution. If it manages to improve the solution, it takes it as the current solution and repeats the process on it. The algorithm halts when no subpath inversion improves the solution. We also limit the number of steps (to 30 in our implementation) to limit the computation complexity of this procedure (this does not in practice noticeably affect our agent's performance). Note that the algorithm's time complexity is cubic in the size of the input, and nearest-neighbor's is linear. Therefore this approach greatly improves the performance on larger input sizes (at the cost of exactness), compared to the naive algorithm (whose time complexity was factorial).

Having solved the travelling salesman problem, the agent now fully maximizes the utility function at every stage. In the traversal of the racks, it computes the fastest possible one (since each rack is visited before and after the neighborhood exploration, we can indeed treat this stage as independent). In the neighborhood exploration, it finds the shortest traversal between the smallest set of points that offer full coverage of the neighborhood. Lastly, it picks up the axes in the shortest time possible.

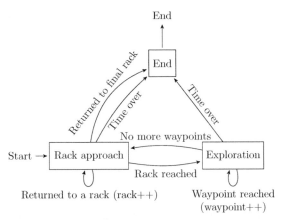

Fig. 12 Utility-based agent's state diagram.

We have additionally reduced the number of required states (from Fig. 11) to two—the SRA state is no longer required and the agent now picks up axes in the Exploration state rather than in a separate state. The resulting state diagram can be seen in Fig. 12.

Note that our implementation does not contain the utility function explicitly, but the agent still at every stage attempts to maximize the appropriate utility function. Also note that all intense computation is done in the planning phase, which only happens a few times in a run, rather than at every action selection step, so these computation times do not critically impact our agent's performance.

This agent utilizes the entire percept history as well as all the built-in knowledge well. It chooses actions that are expected to both maximize the number of axes returned and minimize the time taken. Thus, it fully maximizes its expected performance measure using its percept history and built-in knowledge and therefore fits our definition of intelligence perfectly.

6. Results

In this section we measure the performance of our agents and compare them. For more reliable and accurate results we run each agent several (10) times and use the average measurements for our performance measure for comparison. We also provide the standard deviation (SD), which represents the stability of the agent's performance.

Table 5 Average scores of the agents in the first scenario.

	SRA	SBA	MBA
Score	0.6	14	57
SD	0.52	1.22	6.8

6.1 First scenario

In Table 5 we see the results of the agents in the first scenario. The score denotes the average number of targets hit per scenario run. Recall that a single scenario run lasts 2 min.

The simple reflex agent (SRA) managed to at least hit a target in more than half of the scenario runs. Given its simplicity and randomness this is perhaps even slightly above expectations. It certainly did not exhibit any real intelligence, but it was not a complete failure either. Its actions still somewhat increase its expected performance measure compared to an agent that would choose them completely at random and as such it is at least a good first step toward intelligence. The fairly high standard deviation can be attributed to the randomness of the agent's implementation as well as the binarity of the score (it either got 1 or 0, both of which are relatively far from the average).

The state-based agent (SBA) performed fairly well in the scenario. It already maximized the expected performance measure based on only the current percept. It did not yet use its given knowledge of the task environment, so it is a sort of a half-way milestone on the way to intelligence. The low standard deviation is due to the fact that it always behaves similarly, regardless of the target positions.

The memory-based agent (MBA) used its built-in knowledge by remembering the order of the targets. As we can see it outperformed the other two by far. Note that, while the performance measures (scores) of the other two increase linearly over time, this one does not (or rather increases with two distinct linear speeds—one while it still has to search for targets, the other once they start repeating). This agent is an example of more or less perfect intelligence in this scenario, as it uses both the built-in knowledge and the entire percept history to maximize its performance measure. Its performance is slightly more dependent on the target positions, as well as more affected by the randomness of missed shots, to which we can attribute the slightly higher standard deviation.

Even if we analyse the agents more subjectively, we can say the MBA exhibited a convincing approximation of intelligence. If we ignore the unnaturally precise movements and inhuman reflexes (both of which could be corrected in the code, but are fairly irrelevant from the viewpoint of intelligence), the only thing that spoils the feeling of human-like intelligence is the camera movement while searching for targets, which still appears some-what too random and unnatural.

6.2 Second scenario

In Tables 6 and 7 we can see the results of the agents' performances in the second scenario. In Table 6 are the results of the first scenario case (where the programmer's knowledge of the rack locations and neighborhoods is valid), while Table 7 contains the results from the second scenario case (with moved racks and neighborhoods). We leave out the simple reflex agent in the second scenario case as (due to its randomness) it is unlikely to perform any better or worse than in the first case.

Table 6 Average scores and times of agents in the first case of the second scenario.

	SRA	GBA	SBA	MBA	AEA	UBA
Score	0.2	9.2	15	15	15	15
SD	0.45	1.92	0	0	0	0
Time (m:s)	7:00	4:31	5:43	4:21	4:30	3:13
SD (s)	0	13	49	61	34	15
Performance measure	−5	87.5	144.2	145.7	145.5	146.8

Table 7 Average scores and times of agents in the second case of the second scenario.

	GBA	SBA	MBA	AEA	UBA
Score	1	8.7	10.7	15	15
SD	1	2.5	1.5	0	0
Time (m:s)	5:24	6:18	5:12	3:09	2:38
SD (s)	8.7	18	42	31	6.9
Performance measure	4.6	80.7	101.8	146.9	147.4

The neighborhoods in the second case are somewhat smaller and more open (in terms of visibility) in order to fit into the accessible world, which is why we must be careful when comparing agents' performances between scenario cases.

The maximum time of a single run is 7 min and there are 15 axes available in total. The performance measure is calculated as $10 \times$ avg. score $-$ avg. time in minutes (see Section 4.3 for the explanation). We provide the standard deviation for both score and time.

The simple reflex agent (SRA) returned two axes in one of the runs randomly, but the rest of the runs were complete failures. It utilizes its location and clock but does not use them to increase its performance measure. Rather, it chooses its actions completely randomly, relative to the task. It therefore does not fit our notion of intelligence in any way.

The basic goal-based agent (GBA) performed decently in the first case, as it on average returned two thirds of the axes. It only uses its percepts on the bottom layer to help its pathing. It uses its given knowledge somewhat, but without the ability to adapt to changes. That is why we can only describe it as somewhat intelligent in the first case of the scenario, while in the second it is no more intelligent than the simple reflex agent.

The state-based goal-based agent (SBA) performed well in the first case, always returning all the axes. To increase the number of returned axes it utilized the programmer's knowledge as well as the current percept. Therefore, in the first case it already possessed a degree of intelligence. In the second case it only used the current percept, not using any background knowledge. It therefore still possessed marginal intelligence even in the second case (returning slightly over half the axes on average), but not nearly comparable to the first case.

The memory-based goal-based agent (MBA) performed similarly to the SBA. It improved slightly on the time, as well as (in the second case) the score. Since it used the entire percept history in both cases, it was slightly more intelligent than the SBA in both, however, it was still noticeably less effective in the second case.

The autonomous exploration agent (AEA) was the first to perform reliably well in both scenario cases, which implies autonomous behavior and good adaptation—two properties the previous agents lacked. It always returned all the axes, though it took more time in the first scenario case than the MBA. It used both the built-in knowledge and the percept history to maximize the score. The percepts also decreased the time taken (through premature exploration termination, much like the MBA). Therefore, it behaved with decent intelligence in both scenario cases.

The utility-based agent (UBA) achieved the best performance in both scenario cases. Note that the difference in performance measure to the AEA (and the SBA/MBA in the first scenario case) is low due to the relatively small impact of time on the performance measure. However, if we look at the time score, the UBA's superiority is apparent. Since it used both the built-in knowledge and the percept history to maximize its expected performance measure, it was a good approximation of intelligence in both scenario cases.

Even with subjective examination the UBA behaved with quite human-like intelligence. The exact 360 degree rotations in the exploration stage are slightly unnatural, but this is again irrelevant from the viewpoint of intelligence. Aside from that, what bothers a human observer the most is the over-coverage of the neighborhoods when exploring. Humans can remember with very little effort (at least approximately), which parts of the world they have searched already, but to an agent building such a model of the already explored world is a difficult problem. That is why we used the simpler approach with calculated points and 360 degree rotations, which appears somewhat unnatural.

7. Conclusion

We compared the behavior of various artificial intelligence agents in two scenarios that represented situations from different game worlds. We demonstrated the usefulness of simpler agents in certain situations, as well as the necessity of upgrades to more complex agents in others.

The biggest impact was introduced by goal-based behavior, which gave our agents autonomy and the ability to adapt to task parameter changes, as well as dynamic changes in the environment. We noted the difference between knowledge given to the programmer at the time of implementation and knowledge given to the agent at the start of its run—the latter requires a certain amount of autonomy to be used properly.

The categorization of agents we used differs from the one proposed in [2] as we added the state-based agent. This is in fact a subclass of the simple reflex agent, but is important enough in games to warrant a category all to itself. This approach allows a compact, easy to read implementation of behavior that would require redundant condition checks as well as deeply nested and complex statements in a basic simple reflex agent implementation—defeating the purpose of such an agent. This is why this approach is so often used in games (and more generally) and it is the approach we used in most of our agents as well. Additionally, in the second scenario, we deviated slightly from the categorization to show that the agent functionalities need not necessarily be implemented in the order described by Russel and Norvig (who introduce them as upgrades in a fixed, rigidly defined order).

In both scenarios we managed to achieve a good approximation of intelligent behavior in our agents. We could improve on the subjective appearance of human-like behavior in our agents, but aside from that they performed their tasks intelligently.

Unity's NavMesh and navigation algorithms posed some problems as, even with the addition of the HeightMesh, they do not function as well on rugged, uneven terrain. The navigation agent also at times struggles when given a goal it is very close to and occasionally gets stuck. As these are problems with the development tool and not our implementation it is difficult to deal with these issues in an elegant and robust way.

Our tool could be expanded to unknown and multiagent scenarios in which the learning ability, which we did not implement, is often desired or even required. An appropriate scenario to include both properties would be a scenario from a real-time strategy game. Such a scenario would be unknown and multiagent, but with a limited number of agents. Another appropriate tool to help with this implementation would be SEPIA [44], designed precisely for the implementation and academic study of intelligent agents in such games.

Lastly, we could focus on agent optimization and balancing intelligence against resource usage and computational intensity. An example of research into such optimization can be seen in [45], which provides a detailed view into some of the intricate issues arising in existing AI Planning methods. Such optimization is vital in multiagent environments with large numbers of agents, particularly in games where these agents are all controlled by one computer. In this work, the only benefit of the simpler agents was simplicity and compactness of code. However, in modern games, due to the complexity of their environments, as well as real-life limitations (budget, deadlines, etc.), such simple agents are still utilized, despite more advanced, more intelligent architectures being available.

Appendix. Pseudocode

The source code is freely available under the GNU GPL v3 on the GitHub repository: https://github.com/MatejVitek/Intelligent-Agents.

Algorithm 1 Sight implementation.

Input: camera, item, max_distance
Output: item_visible: *boolean*

---

if |camera.position − item.position| > max_distance **then**
    return *false*
for vertex ∈ ({item.position} ∪ item.bounding_box.vertices) **do**
    forward_vector = camera.looking_direction_vector
    item_vector = vertex − camera.position
    if (
        HorizontalAngle(forward_vector, item_vector) >
            camera.FOV.horizontal_angle / 2 **and**
        VerticalAngle(forward_vector, item_vector) >
            camera.FOV.vertical_angle / 2 **and**
        RaycastSuccessful(camera.position, vertex)
) **then**
        return *true*
return *false*

Algorithm 2 First scenario: simple reflex agent.

Internal: direction, last_change_time
Percept: *sight*, current_time

---

if current_time > last_change_time + 3 **then**
    direction = GetRandomDirection()
    last_change_time = current_time
if TargetVisible() **then**
    Shoot()
    UseLowSpeed()
else
    UseHighSpeed()
MoveCamera(direction)

Algorithm 3 First scenario: state–based agent.

Internal : state, vertical_direction, last_change_time
Percept : *sight*, *hearing*
Initialize: state = Searching

---

switch state **do**
case Searching **do**
   MoveCameraRight()
   MoveCameraUpDown(vertical_direction)
   ChangeDirectionIfNeeded(vertical_direction, last_change_time)
   if TargetVisible() **then**
      state = Aiming
case Aiming **do**
   MoveCameraTowardsTarget()
   if LookingDirectlyAtTarget() **then**
      state = Firing
   else if not TargetVisible() **then**
      state = Searching
case Firing **do**
   Shoot()
   if TargetHit() **then**
      state = Searching
   else
      state = Aiming

Algorithm 4 First scenario: memory-based agent.

Internal : state, vertical_direction, last_change_time,
            next_target_change_time
Memory : successors, previous, current
Percept : current_time, *sight*, *hearing*
Initialize: state = Searching

---

switch state **do**
case Searching **do**
   HandleCameraMovement(vertical_direction, last_change_time)
   if previous.exists **and** successors.HasSuccessorOf(previous) **then**
      current = successors.SuccessorOf(previous)
      state = Aiming
   else if TargetVisible() **then**
      current = *visible target*
      if previous.exists **then**
         successors.Add(current as successor to previous)
      state = Aiming

(Continued)

Algorithm 4 First scenario: memory–based agent—Cont'd

---

case Aiming **do**
    MoveCameraTowardsTarget()
    if current_time > next_target_change_time **then**
        previous = current
        next_target_change_time += 30
        state = Searching
    else if LookingDirectlyAtTarget() **then**
        state = Firing
case Firing **do**
    Shoot()
    if TargetHit() **then**
        previous = current
        next_target_change_time = current_time + 30
        state = Searching
    else
        state = Aiming

Algorithm 5 Second scenario: simple reflex agent.

Internal: last_position
Percept: current_position, time_from_last_update

MoveForward()
if |current_position − last_position| < time_from_last_update **then**
    Rotate(Random(90,270))
last_position = current_position

Algorithm 6 Second scenario: basic goal-based agent.

Knowledge: waypoints
Internal   : current_waypoint
Initialize   : waypoints.Add(FindRackLocations() to correct indices),
           current_waypoint = waypoints.first

MoveToward(current_waypoint)
if GoalReached(current_waypoint) **then**
    if waypoints.next.exists **then**
        current_waypoint = waypoints.next
    else
        EndScenario()

Algorithm 7 Second scenario: state-based goal-based agent.

Knowledge: given_waypoints
Internal : state, waypoints, current_waypoint, racks, current_rack
Percept : *sight*
Initialize : racks = FindRackLocations(), current_rack = racks.first

---

switch state **do**
case ApproachingRack **do**
   MoveToward(current_rack)
   if GoalReached(current_rack) for the first time **then**
      waypoints = GetWaypoints(current_rack, given_waypoints)
      current_waypoint = waypoints.first
      state = ApproachingWaypoint
   else if GoalReached(current_rack) again **then**
      if racks.next.exists **then**
         current_rack = racks.next
      else
         EndScenario()
case ApproachingWaypoint **do**
   MoveToward(current_waypoint)
   if AxeVisible() **then**
      state = ApproachingTool
   else if GoalReached(current_waypoint) **then**
      if waypoints.next.exists **then**
         current_waypoint = waypoints.next
      else
         state = ApproachingRack
case ApproachingTool **do**
   MoveToward(*visible axe*)
   if not AxeVisible() **then**
      state = SRA
   clsc if GoalReached(*visible axe*) **then**
      state = ApproachingWaypoint
case SRA **do**
   MoveForwardAndTurnIfBlocked()
   if AxeVisible() **then**
      state = ApproachingTool
   else if *3 seconds have passed* **then**
      state = ApproachingWaypoint

Algorithm 8 Second scenario: memory-based agent.

Knowledge: given_waypoints
Internal : state, waypoints, current_waypoint, racks, current_rack
Memory : priority_waypoints, tools_picked_up
Percept : *sight*
Initialize : racks = FindRackLocations(), current_rack = racks.first

---

switch state **do**
case ApproachingRack **do**
   `// Same as ApproachingRack in Alg. 7`

(Continued)

Algorithm 8 Second scenario: memory-based agent—Cont'd

---

case ApproachingWaypoint **do**
    MoveToward(current_waypoint)
    if GoalReached(current_waypoint) **then**
        if waypoints.next.exists **then**
            current_waypoint = waypoints.next
        else
            state = ApproachingRack
    if AxeVisible() **then**
        priority_waypoints.Add(*visible axe*)
        state = ApproachingTool
case ApproachingTool **do**
    if AxeVisible() **then**
        priority_waypoints.Add(*visible axe*)
    MoveToward(priority_waypoints.first)
    if GoalReached(priority_waypoints.first) **then**
        tools_picked_up += 1
        priority_waypoints.RemoveFirst()
    if tools_picked_up == 5 **then**
        state = ApproachingRack
    else if priority_waypoints == ∅ **then**
        state = ApproachingWaypoint

Algorithm 9 Second scenario: autonomous exploration agent.

Internal : state, boxes, current_waypoint, racks, current_rack
Memory : priority_waypoints, tools_picked_up
Percept : *sight*
Initialize: racks = FindRackLocations(), current_rack = racks.first

---

switch state **do**
case ApproachingRack **do**
    MoveToward(current_rack)
    if GoalReached(current_rack) for the first time **then**
        boxes = GetBoxes(current_rack)
        current_waypoint = GetRandomPointInBox(boxes.first)
        state = Exploring
    else if GoalReached(current_rack) again **then**
        if racks.next.exists **then**
            current_rack = racks.next
        else
            EndScenario()
case Exploring **do**
    MoveToward(current_waypoint)
    if GoalReached(current_waypoint) **then**
        if not boxes.next.exists **or** tools_picked_up == 5 **then**
            state = ApproachingRack
        else
            current_waypoint = GetRandomPointInBox (boxes.next)
    if AxeVisible() **then**
        priority_waypoints.Add(*visible axe*)
        state = ApproachingTool
case ApproachingTool **do**
    // Same as ApproachingTool in Alg. 8

Algorithm 10 Second scenario: utility-based agent.

Internal : state, waypoints, current_waypoint, racks, current_rack
Memory : waypoints, priority_waypoints, tools_picked_up
Percept : *sight*
Initialize: racks = FindRackLocations(), SolveTSP(racks), current_rack
    = racks.first

---

switch state **do**
case ApproachingRack **do**
   MoveToward(current_rack)
   if GoalReached(current_rack) for the first time **then**
      waypoints = GetBestExplorationWaypoints(current_rack)
      SolveTSP(waypoints)
      current_waypoint = waypoints.first
      state = Exploring
   else if GoalReached(current_rack) again **then**
      if racks.next.exists **then**
         current_rack = racks.next
      else
         EndScenario()
case Exploring **do**
   MoveToward(current_waypoint)
   if GoalReached(current_waypoint) **then**
      TurnAround()
      if not waypoints.next.exists **or** tools_picked_up == 5 **then**
         state = ApproachingRack
      else
         current_waypoint = waypoints.next
   if AxeVisible() **then**
      priority_waypoints.Add(*visible axe*)
      closest = FindClosest(waypoints, *visible axe*)
      waypoints.Add(*visible axe* after closest)
      if priority_waypoints.count == 5 − tools_picked_up **then**
         waypoints = priority_waypoints
         SolveTSP(waypoints)
         current_waypoint = waypoints.first

Algorithm 11 Nearest-neighbor algorithm.

Input: points
Output: sorted_points

---

sorted_points = [points.first]
points.RemoveFirst()
while points $\neq \emptyset$ **do**
   nearest = *null*
   for p $\in$ points **do**
      if $|p - \text{sorted\_points.last}| < |\text{nearest} - \text{sorted\_points.last}|$ **then**
         nearest = p
   sorted_points.Add(nearest)
   points.Remove(nearest)

Algorithm 12 2-Opt optimization algorithm.

Input: points

---

points.RemoveFirst()
steps = 0
while steps++ < 30 **do**
   if not 2-Opt(points) **then**
      break

boolean function 2-Opt(points)
   for i ∈ range [1..points.size-1] **do**
      for j ∈ range [i+1..points.size] **do**
         current = Invert($points, i, j$)
         if TotalDistance(current) < TotalDistance(points) **then**
            points = current
            return *true*
   return *false*

/* Invert subpath between indices i and j, inclusive   */
Point[] function Invert(points, i, j)
   result=[]
   for k ∈ range [1..i-1] **do**
      result.Add(points[k])
   for k ∈ range [j **downto** i] **do**
      result.Add(points[k])
   for k ∈ range [j+1..points.size] **do**
      result.Add(points[k])
   return *result*

References

[1] B. Nikolic, Z. Radivojevic, J. Djordjevic, V. Milutinovic, A survey and evaluation of simulators suitable for teaching courses in computer architecture and organization, IEEE Trans. Educ. 52 (4) (2009) 449–458.
[2] S. Russel, P. Norvig, Artificial Intelligence: A Modern Approach, third ed., Upper Saddle River: Prentice-Hall, 2010.
[3] Unity3D, 2017. Online at https://unity3d.com/.
[4] GNU General Public License v3, 2017. Online at http://www.gnu.org/licenses/gpl-3.0.html.
[5] GitHub Repository, 2017. Online at https://github.com/MatejVitek/Intelligent-Agents.
[6] S. Franklin, A. Graesser, Is it an agent, or just a program? A taxonomy for autonomous agents, in: Third International Workshop on Agent Theories, Architectures, and Languages, Springer-Verlag New York, Inc., New York, NY, USA, 1996.
[7] H.S. Nwana, Software agents: an overview, Knowl. Eng. Rev. 11 (03) (1996) 205–244.

[8] S. Legg, M. Hutter, Universal intelligence: a definition of machine intelligence, Minds Mach. 17 (4) (2007) 391–444. ISSN: 1572-8641, https://doi.org/10.1007/s11023-007-9079-x.

[9] P. Hingston, A turing test for computer game bots, IEEE Trans. Comput. Intell. AI Games 1 (3) (2009) 169–186. ISSN: 1943-068X, https://doi.org/10.1109/TCIAIG.2009.2032534.

[10] E. Norling, Agents for Games and Simulations, Springer, Berlin, Heidelberg, 2009.

[11] J. Gemrot, C. Brom, T. Plch, Agents for Games and Simulations II, Springer, Berlin, Heidelberg, 2011.

[12] C. Brandao, L.P. Reis, A.P. Rocha, Evaluation of embodied conversational agents, in: 8th Iberian Conference on Information Systems and Technologies (CISTI), IEEE, 2013, pp. 1–6.

[13] V. Blagojević, D. Bojić, M. Bojović, M. Cvetanović, J. Đordević, . Đurdević, B. Furlan, S. Gajin, Z. Jovanović, D. Milićev, V. Milutinović, B. Nikolić, J. Protić, M. Punt, Z. Radivojević, ž. Stanisavljević, S. Stojanović, I. Tartalja, M. Tomašević, P. Vuletić, Chapter one—A systematic approach to generation of new ideas for PhD research in computing. in: A.R. Hurson, V. Milutinović (Eds.), Creativity in Computing and DataFlow SuperComputing, Advances in Computers, vol. 104 Elsevier, 2017, pp. 1–31, https://doi.org/10.1016/bs.adcom.2016.09.001.

[14] M. Buckland, Programming Game AI by Example, first ed., Wordware Publishing, Inc., 2005.

[15] P.E. Hart, N.J. Nilsson, B. Raphael, A formal basis for the heuristic determination of minimum cost paths, IEEE Trans. Syst. Sci. Cybernet. 1 (2) (1968) 100 107. ISSN. 0536-1567, https://doi.org/10.1109/TSSC.1968.300136.

[16] A. Stentz, Optimal and efficient path planning for partially-known environments. in: IEEE International Conference on Robotics and Automation, vol. 4 IEEE, 1994, pp. 3310–3317, https://doi.org/10.1109/ROBOT.1994.351061.

[17] S. Koenig, M. Likhachev, Fast replanning for navigation in unknown terrain, IEEE Trans. Robot. 21 (3) (2005) 354–363. ISSN: 1552-3098, https://doi.org/10.1109/TRO.2004.838026.

[18] A. Stentz, The focussed D* algorithm for real-time replanning, in: Proceedings of the International Joint Conference on Artificial Intelligence, Morgan Kaufmann Publishers Inc., San Francisco, CA, USA, 1995.

[19] J. Hagelbäck, Hybrid pathfinding in starcraft, IEEE Trans. Comput. Intell. AI Games 8 (2016) 319–324. ISSN: 1943-068X, https://doi.org/10.1109/TCIAIG.2015.2414447.

[20] N.C. Hou, N.S. Hong, C.K. On, J. Teo, Infinite Mario Bross AI using genetic algorithm, in: IEEE Conference on Sustainable Utilization and Development in Engineering and Technology (STUDENT), IEEE, 2011, pp. 85–89.

[21] D. Wang, A.H. Tan, Creating autonomous adaptive agents in a real-time first-person shooter computer game, IEEE Trans. Comput. Intell. AI Games 7 (2) (2015) 123–138. ISSN: 1943-068X, https://doi.org/10.1109/TCIAIG.2014.2336702.

[22] P.G. Patel, N. Carver, S. Rahimi, Tuning computer gaming agents using Q-learning, in: IEEE Federated Conference on Computer Science and Information Systems (FedCSIS), IEEE, 2011, pp. 581–588.

[23] P. Bailis, A. Fachantidis, I. Vlahavas, Learning to play monopoly: a reinforcement learning approach, in: Proceedings of the 50th Anniversary Convention of The Society for the Study of Artificial Intelligence and Simulation of Behaviour, AISB, 2014, pp. 399–401.

[24] V. Mnih, K. Kavukcuoglu, D. Silver, A.A. Rusu, J. Veness, M.G. Bellemare, A. Graves, M. Riedmiller, A.K. Fidjeland, G. Ostrovski, Human-level control through Deep Reinforcement Learning, Nature 518 (7540) (2015) 529–533.

[25] Z. Chen, D. Yi, The Game Imitation: A Portable Deep Learning Model for Modern Gaming AI, Stanford University, Stanford, CA, 2016.

[26] K. Ranjan, A. Christensen, B. Ramos, Recurrent Deep Q-Learning for PAC-MAN, Stanford University, Stanford, CA, 2016.

[27] M. Stevens, S. Pradhan, Playing Tetris with Deep Reinforcement Learning, Stanford, CA, 2016.

[28] C. Tessler, S. Givony, T. Zahavy, D.J. Mankowitz, S. Mannor, A deep hierarchical approach to lifelong learning in minecraft, arXiv:1604.07255 (2016).

[29] M. Guid, M. Možina, C. Bohak, A. Sadikov, I. Bratko, Building an intelligent tutoring system for chess endgames, in: CSEDU 2013, 2013, pp. 263–266.

[30] M.E. Bratman, Intention, Plans, and Practical Reason, Harvard University Press, 1987.

[31] R.G. Dromey, From requirements to design: formalizing the key steps. in: First International Conference on Software Engineering and Formal Methods, IEEE, 2003, pp. 2–11, https://doi.org/10.1109/SEFM.2003.1236202.

[32] R. Conte, Rational, Goal-Oriented Agents, Springer, New York, 2012.

[33] J.P. Müller, M. Pischel, The agent architecture InteRRaP: concept and application, DFKI, 1993, (DFKI Research Reports) Tech. rep.

[34] I.A. Ferguson, Touring machines: autonomous agents with attitudes, Computer 25 (5) (1992) 51–55. ISSN: 0018-9162, https://doi.org/10.1109/2.144395.

[35] K.S. Løland, Intelligent Agents in Computer Games, Ph.D. thesis, Norwegian University of Science and Technology, Norway, 2008.

[36] Unity3D Asset Store, 2017. Online at https://www.assetstore.unity3d.com/en/.

[37] D.M. Kuiper, R.Z. Wenkstern, Agent vision in multi-agent based simulation systems, Autonomous agents and multi-agent systems 29 (2) (2015) 161–191.

[38] W. Zeng, R.L. Church, Finding shortest paths on real road networks: the case for A*, Int. J. Geogr. Inf. Sci. 23 (4) (2009) 531–543.

[39] Unity3D Navigation Manual, 2017. Online at http://docs.unity3d.com/Manual/Navigation.html.

[40] W. van Toll, A.F. Cook, R. Geraerts, A navigation mesh for dynamic environments, Comput. Anim. Virtual Worlds 23 (2012) 535–546, https://doi.org/10.1002/cav.1468.

[41] E. Bonomi, J.L. Lutton, The N-city travelling salesman problem: statistical mechanics and the metropolis algorithm, SIAM Rev. 26 (4) (1984) 551–568.

[42] S. Dhakal, R. Chiong, A hybrid nearest neighbour and progressive improvement approach for Travelling Salesman Problem. in: International Symposium on Information Technology (ITSim), vol. 1 IEEE, 2008, pp. 1–4, https://doi.org/10.1109/ITSIM.2008.4631549.

[43] I. Mavroidis, I. Papaefstathiou, D. Pnevmatikatos, Hardware implementation of 2-opt local search algorithm for the traveling salesman problem, in: 18th IEEE/IFIP International Workshop on Rapid System Prototyping (RSP), ISSN 1074-6005, IEEE, 2007, pp. 41–47, https://doi.org/10.1109/RSP.2007.24.

[44] S. Sosnowski, T. Ernsberger, F. Cao, S. Ray, SEPIA: a scalable game environment for artificial intelligence teaching and research, in: Proceedings of the Twenty-Seventh AAAI Conference on Artificial Intelligence, AAAI, 2013, pp. 1592–1597.

[45] E. Jacopin, Game AI planning analytics: the case of three first-person shooters, in: Proceedings of the Tenth Annual AAAI Conference on Artificial Intelligence and Interactive Digital Entertainment, AAAI, 2014, pp. 119–124.

About the authors

Matej Vitek is a Junior Researcher and Ph. D. candidate at the Faculty of Computer and Information Science, University of Ljubljana, Slovenia. He received his Bachelor's Degree from the Faculty of Computer and Information Science and the Faculty of Mathematics and Physics in 2015 and his Master's Degree in 2018 from the same two institutions. His fields of interest are artificial intelligence, machine learning, computer game development, deep learning, and biometry.

Peter Peer is an Associate Professor at the Faculty of Computer and Information Science, University of Ljubljana, Slovenia. He received his Ph.D. degree in Computer Science from the same institution in 2003. Within his post-doctorate he was an invited researcher at CEIT, Donostia—San Sebastian, Spain. He teaches courses on Operating systems, Game technology and virtual reality, and Image-based biometry. His research interests include artificial intelligence, computer vision, biometry, and computer games. He participated in several national and EU funded R&D projects and published more than 75 research papers in leading international peer reviewed journals and conferences. He is a member of the IAPR and IEEE, where he also served as chairman of the Slovenian IEEE Computer chapter for 4 years.

CHAPTER SIX

Using clickstream data to enhance reverse engineering of Web applications

Marko Poženel, Boštjan Slivnik
University of Ljubljana, Faculty of Computer and Information Science, Ljubljana, Slovenia

Contents

Abstract

Due to advances in Web technologies, existing Web applications are rewritten or replaced by new ones. As a result of either ad hoc or agile development, many of them lack proper technical documentation. Nevertheless, the domain knowledge built into these applications is valuable, which is why the reverse engineering, an activity aimed at detecting software components and their interrelationships to provide multiple views of software systems at a higher level of abstraction, of existing Web applications is becoming an important issue.

Advances in Computers, Volume 116
ISSN 0065-2458
https://doi.org/10.1016/bs.adcom.2019.07.006

Apart from the static reverse engineering based on examining the system's source code, analyzing the dynamic aspect of Web applications often proves worthwhile. One important source of data the dynamic analysis of a Web application can be based on are HTTP server log files. User sessions, results of clickstream analysis, and session reconstruction in particular, can be used as the basis for the first automatic step of reverse engineering, employed in order to gain a quick insight into Web application's source code.

It is shown how clickstream data can be used to reveal not only the intensity of connections between individual Web application's source code artifacts but also the overall structure of a Web application. The extracted structure, based either on code artifacts names or their usage, is presented visually as an ATG, with code artifacts belonging to the same application module grouped together. Because session reconstruction is an inherently probabilistic process and thus in general produces noisy data, clustering the code artifacts becomes a challenging task. It is shown that multidimensional scaling and even a simple graph drawing approaches yield better representation of the application transition graph than hierarchical clustering.

The method was tested against the results obtained by an expert (the author of the Web application used as a test case). Additionally, the method can also be used for verifying the structure obtained by manual reverse engineering of the application's source code.

Abbreviations

AJAX	asynchronous JavaScript and XML
ASM	atomic section model
ATG	application transition graph
CIM	component interaction model
MDS	multidimensional scaling
RIA	rich Internet application

1. Introduction

With the advance of Web technologies, many existing Web applications are rewritten or replaced by new ones, which means that they are totally reengineered. In many cases, the source code is available, but reliable and comprehensive documentation is missing.

There are various reasons why the documentation might be missing:

1. Some Web applications were written in a hurry without using any proper software development methodology which would ensure proper documentation.
2. The existing documentation might be insufficient as the application was produced using agile software development methodology [1] which gives priority to *working software over comprehensive documentation* [2].

3. In an application's lifetime the source code was modified either during maintenance or when new functionalities were added, but quite often many modifications are insufficiently documented or not documented at all.

It happens all too often that in such cases, reverse engineering must be performed to extract business logic from the application's source code.

Reverse engineering is a software reengineering activity aimed at *detecting software components and their interrelationships to provide multiple views of software systems at a higher level of abstraction* [3]. It could be considered a postcomprehension method if compared to simulation, which is a pre comprehension method [4]. The other two major activities are comprehension and enhancement of software systems. The former is about finding out what the software components detected during reverse engineering actually do and the latter is about modifying and improving software systems.

Reverse engineering of a Web application seeks to understand its structure and functionality, and in the end it should produce (visual) representation of the application's structure. Visual representation is preferred as it has been shown that graph-based comprehension tools offer a more efficient basis for comprehension than, for example, relational database-based tools [5].

In many cases the number of components, i.e., individual Web pages and/or behind the scene codes for generating Web pages, might be relatively large. Therefore, simply displaying connections between components is not enough. One should, if at all possible, find which components form a certain part of the Web application, e.g., modules implementing individual functionalities, so that later comprehension is easier.

The structure of the article is as follows. After the description of reverse engineering topics specific to Web applications in Sections 2 and 3 gives an overview of different approaches to reverse engineering. As this article aims at extracting the structure of a Web application using clickstreams, Section 4 introduces application transition graphs (ATG), while Section 5 describes preprocessing of clickstreams and user sessions. The metrics used for clustering and visualization of ATGs are defined in Section 6, and Section 7 presents methods used for clustering and final visualization of ATGs.

2. On reverse engineering of Web applications

When the World Wide Web appeared in 1989/90, it consisted of a number of interconnected Web pages. Soon after Web applications appeared, their development and deployment only accelerated in time.

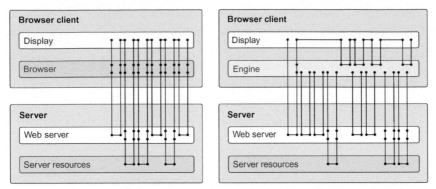

Fig. 1 The difference between traditional and rich Internet applications regarding the internal components and the communication model (https://blog.heliossolutions.in).

Regarding the way a Web application interacts with a user on one side and with the server on the other side, Web applications can be divided into two main groups:

1. *Traditional Web applications*

   As shown in Fig. 1 (left), traditional Web applications consist of (i) a Web browser on the client side and of (ii) a Web server, e.g., Apache HTTP Server by Apache Software Foundation of Internet Information Services by Microsoft. The application in the browser communicates with the server in a synchronous way: whenever a user asks for something, an HTTP request sent by the browser and the reply is obtained from the server.

2. *Rich Internet applications*

   As shown in Fig. 1 (right), a rich Internet application (RIA) consists of (i) an engine running within a browser on the client side and of (ii) a Web browser. However, the engine communicates with the server in an asynchronous way: along with the requests resulting from explicit user actions, many requests are generated by an application behind the scene, i.e., without user intervention or even without their knowledge. This is achieved by employing AJAX (Asynchronous JavaScript And XML) or similar technologies supported by numerous modern Web development frameworks.

Virtually all new Web applications nowadays are RIAs, but many older traditional Web applications are still in use. If not abandoned, they usually need to be replaced by new ones using new Web technologies. Although technologies change, many business processes stay more or less the same and for the reasons described above many existing Web applications must often be reverse engineered.

Regardless of the Web application type, every request sent to a Web server, whether it resulted from a user's click or not, can be logged by the Web server. Initially, logging on the server side was implemented for Web server debugging purposes but it turned into an indispensable feature that is rarely turned off during configuration. Hence, there are two resources which can be the basis for reverse engineering of a Web application:

1. *Source code*

   The first resource is the application's source code. It includes the source code of all application's components and provides a complete picture of the connections among these components. However, the automatic discovery of components vital for understanding application structure might prove to be difficult.

2. *Clickstream data*

   The second resource are HTTP server access log files as they contain (server side) clickstream data. Clickstream data is a recorded sequence of HTTP requests resulting from clicks made by users while using a Web application and browsing its Web pages [6]. Not all components of the application can be clicked and such components are therefore not visible in log files. However, a lot of important information about the application can still be extracted, primarily the intensity of connections among individual components.

Although clickstream data has not been used often in reverse engineering, it is worth investigating this approach as well [7].

At this point, one must realize that this investigation focuses on old and therefore predominantly traditional Web applications which must be reverse engineered in order to be rewritten, and available HTTP server log files which contain simple events recorded by the server. However, if an application keeps a log of its user sessions by itself, independently from the HTTP server, these log files are extremely useful and should of course be used. But in the vast majority of cases, old Web applications do not produce these logs so the reverse engineer must cope with simple HTTP log files instead.

In general, clickstream data found in HTTP server log files can be used for various analyses of a Web application and its users' behavior [6, 8]. These analyses are most often performed to provide the basis for different personalization strategies aimed at better services for individual users, to help with marketing approaches and perception of business opportunities which would be missed otherwise, to improve system performance or increase system security [9, 10]. In this study, however, clickstream data is transformed into individual user sessions and used to discover the structure of a Web application as a part of the reverse engineering process.

Provided that clickstream data is collected for a sufficiently long period of time, it may provide a reliable basis for automatic clustering of Web application's artifacts. As it is shown in this article, user sessions obtained from clickstream analysis can significantly enhance the identification of application's central components and the understanding of its overall structure. Furthermore, we investigate the methods for estimating the reliability of reverse engineering based on clickstreams.

2.1 Case study: eStudent, a student information system

Let us first introduce eStudent, a traditional pre-AJAX non-RIA Web application which will be used as a test case for the methods presented later on.

A student record information system called eStudent, deployed at the University of Ljubljana between 2003 and 2012, has been used as a case study before [11] and is used here as well. Other similar Web applications, especially Web server log files (even if anonymized) are notoriously hard to get as companies are afraid of releasing too much information about their business processes or clients due to privacy concerns in the EU and elsewhere.

There are two kinds of eStudent artifacts used in this study:

1. *Source code of individual Web pages*

    The source code of eStudent consists of 354 Web pages. The pages are parameterized and each individual page represents either a functionality or a part of a functionality (if a page is used as a part or in a sequence with another page).

    The names of Web pages are made up of words separated by underscores. There are 326 distinct words altogether but a typical Web page name consists of 3–5 words. For instance, an eStudent Web page implementing a student's registration for a written exam is named (in Slovene original) VZ63_STUDENT_PRIJAVA_NA_PISNI_IZPIT_STUD.

2. *Log files containing clickstream data*

    Clickstream data for eStudent has been stored in a data Webhouse for several years. User sessions over the span of almost 2 years were acquired from data Webhouse 5.471.097. Of all these user sessions only the sessions with length between 10 and 15,000 were used. Shorter and longer sessions are mostly created by Web robots or represent poorly reconstructed sessions which have been left out.

    The majority of user sessions have between 20 and 30 Web page requests. Longer sessions are not very common and belong to advanced users.

3. Related work

Due to the technological importance of reverse engineering, it is no surprise that it is a well investigated field in software engineering and that many tools for reverse engineering have been produced so far. The list (by no means exhaustive) includes Moose [12], GUPRO [3], Columbus [13], CodeCrawler [14], SolidFX [15], and SQuAVisiT [16]. However, these tools are made for reverse engineering of traditional, i.e., no-Web applications written in languages like C, C++, Java, and Cobol (although some of them, like SQuAVisiT, offer support for JavaScript and PL/SQL, for instance). Tools for reverse engineering of Web applications, e.g., Rigi [17] (with Web support based on Web2Rfs [18]) or ReWeb [19], are far less common.

Most of these tools provide support for visual software analytics, code attributes and code metrics, program comprehension based on extraction and visualization of source code artifacts, etc. However, the vast majority of all these tools use only the static analysis of the program code for reverse engineering. In other words, only very few of these tools perform reverse engineering on the basis of usage data. Furthermore, a systematic literature review [20] of IEEE, ACM, Science Direct, Springer and SEI (Software Engineering Institute) databases and digital libraries yielded 175 papers relating to reverse architecture, extraction, mining, recovery or architecture dis covery. Based on this review, it has been established [21] that:

- only 16 out of 175 papers, i.e., 9.14%, report on recovering the dynamic aspects of architecture, and
- only 3 out of 175 papers, i.e., 1.71% report on recovering both static and dynamic aspects of architecture.

A lot has happened in the field of reverse engineering since 2010. To provide a more complete summary we performed an additional extensive search for papers on reverse engineering in software published after that date. We found 22 new papers. Four of them focus on reverse engineering of Web applications, eight papers cover static aspects of reverse engineering, and seven use a dynamic approach, while three combine static and dynamic approaches. The rest of the papers are either surveys [22] or show the historical evolution of reverse engineering approaches [23]. Some of these papers present reverse engineering approaches in more detail [24, 25].

Riva et al. [26] combine static and dynamic information about the system to acquire its architecture and use directed graphs to show system structure. Since this paper was published in 2002, it was included in the aforementioned review, but many papers combining static and dynamic approaches have appeared since. For instance, Bruneliere et al. [27] presented MoDisco, a model-driven reverse engineering (MDRE) framework for facilitating reverse engineering to acquire legacy systems' APIs. MDRE applies model-driven techniques and principles to reverse engineering. Bergmayr et al. [28] proposed a framework for reverse engineering behavioral aspects from existing Java code bases named fREX. fREX enables the discovery of behavior models from a Java code and improves the comprehension of acquired models using fUML diagrams. Bernardi et al. [29] proposed an approach which integrates reverse engineering techniques with model-driven Web engineering methods in order to reduce evolution effort while improving the quality of the modified Web application.

Sections 3.1–3.4 present the latest work related to the fundamental reverse engineering approach.

3.1 Static approaches to reverse engineering

Legacy systems age over time and need to be replaced by newer ones while preserving the embedded business knowledge. Rabelo et al. [30] state that organizational information systems often suffer from poor maintenance over time and become obsolete. The business knowledge that is located in the source code has to be obtained for a reengineering process. They propose an approach for a business process recovery from the source code. The approach uses static analysis and is based on the knowledge discovery meta-model (KDM) [31], standard and heuristic rules. The recovered models are presented in an intuitive graphic notation, so they are easily understandable and compliant with the business process model and notation (BPMN). A standard for modernizing a legacy system using KDM is presented in Ref. [31].

In their paper, Peréz-Castillo et al. [32] also propose and validate a method for recovering and rebuilding business processes from legacy information systems. Their approach uses static analysis as a reverse engineering technique with a source code as the key software artifact, following model-driven development principles. The results are presented in the form of KDM models and business process models. They state that the proposed approach offers possible extraction of business knowledge needed for the system to evolve and is less time-consuming than process redesign by experts from scratch. In Ref. [33] the authors performed a series of case studies to empirically validate the presented business process mining methods using analysis and meta-analysis techniques. The obtained result shows that the presented business process mining methods are suitable for recovering business processes in an effective and efficient manner.

Trias et al. [34] present an approach for migration of Web applications to content management systems (CMS) using architecture-driven modernization. The reengineering process is composed of three classic stages: (i) the reverse engineering stage, (ii) the restructuring stage, and (iii) the forward engineering stage. The paper focuses on the reverse engineering stage, where KDM models are generated from the source code using static analysis.

Garces et al. [35] present a white-box transformation approach which changes application architecture and the technological stack without losing business value and quality attributes. The approach fully automates the migration of graphical interface components and CRUD logic, while the migration of the PL/SQL code is done manually. In the first step, the technology specific model is obtained from the legacy source code, which is then used in the second step to generate the target model.

A significant part of recent legacy applications are Java Enterprise Edition (JEE) applications. JEE are multilanguage systems which often rely on JEE container services that abstract the complexity of the runtime environment, but can also hide useful component dependencies. Shatnawi et al. [36] presented a novel static code analysis approach to analyze JEE applications. The authors presented JEE RE challenges and proposed strategies for addressing them. They also developed a Modisco based tool called DcJEE for identifying a program dependency call graph.

On the other hand, Bozkir et al. [37] proposed a dynamic-based approach for getting visual similarities among Web pages by using structure and vision-based features. The approach consists of a visual inspection of DOM trees and a computer-vision-based method for defining page structure.

Salihu et al. [38] compared GUI Reverse Engineering Techniques focusing on mobile applications. They studied how GUI reverse engineering techniques are useful for mobile applications. They found out that the dynamic approach is widely used for RE of GUI applications while the static approach is rarely used. The static approach enables extracting more exact and complete information from the system but it fails to acquire the behavior data of GUI applications. The dynamic approach results in more incomplete data but is better in acquiring the behavior of GUI applications.

3.2 Dynamic approaches to reverse engineering

Joorbachi et al. [39] presented a reverse engineering approach for acquiring a
state model of an iPhone application capturing user interface states and transitions among them. Their dynamic approach was implemented and tested
using the tool ICRAWLER, which is capable of exercising a given iPhone
application, automatically navigating it, and analyzing UI changes.

Legacy Web applications are replaced with RIAs with developers rapidly adopting RIA tools and the modernization of legacy Web applications
being a popular topic [40]. However, the transition process often lacks a
systematic approach. Conejero et al. [40] presented a systematic process
for modernizing legacy Web applications into RIAs.

An AJAX-based Web applications analysis can be challenging since
they possess a number of properties different from non-AJAX Web applications [41]. AJAX applications are more difficult to comprehend, maintain
and thus reverse-engineer. In this context, Matthijssen and Zaidman [42]
proposed a tool called FireDetective, which facilitates the understanding
of AJAX-based Web applications. The tool uses dynamic analysis to retrieve
execution data from the client and server side. For each page, FireDetective
visualizes a JavaScript execution trace with a corresponding trace view,
server side code view, and page resource list.

Mesbah et al. [41] described a novel dynamic analysis approach for
crawling AJAX-based applications through automatic dynamic analysis of
user-interface-state changes that creates an application model. The resulting
model can be used for program analysis, comprehension, or testing. They
implemented the approach in an open source tool CRAWLJAX, which uses
the Selenium [43] framework which programmatically interacts with the
user interface.

An insight into application behaviors through explicit models can improve
development, testing, validation, and maintenance. Such behavioral models
can be used to implement model-based testing and can be valuable where
the source code is not accessible or amenable to static analysis. However, such
models are challenging to mine. Schur et al. [44] used a dynamic analysis
approach to mine behavior models from Web applications which support
multiuser workflows. They implemented a tool PROCRAWL which applies
dynamic analysis, i.e., generalizes upon observed application executions
(traces).

As Schur et al. [44], Walkinshaw et al. [45] proposed an approach for
reverse engineering of software behavior models. It uses extended finite state

machines (EFSMs), which combine control-specific and data-specific behavior, and addresses the problems of inflexibility and nondeterminism of EFSMs. The approach does not rely on source code analysis and combines the established state-merging techniques with arbitrary data classifier inference algorithms.

3.3 Static and dynamic approaches to reverse engineering

Gimblett et al. [46] present a lightweight formal method for user interface model discovery, whereby a model of an interactive system is discovered by simulating user actions. The models created are directed graphs where nodes represent system states and edges correspond to user actions. Their approach yields reusable and abstract API for user interface discovery.

Silva et al. [47] propose a hybrid approach for reverse engineering of Web application user interfaces, which combines positive aspects of static and dynamic analysis. Their approach uses the dynamic analyses of the application at runtime, while the static analyses of the source code of event handlers are done during interaction, which yields a more complete model of user interface. In Ref. [47] the authors applied the approach on top thousand globally most popular Web sites and presented the analysis of those sites in terms of events and variables used in control flow constructs.

To avoid pitfalls of static and dynamic analysis, Silva and Campos [48] proposed a hybrid approach for analyzing Web applications to support the process of engineering and reengineering through handlers' reverse engineering. They focused on the user interface layer of Web applications. The static part includes event handlers' source code analysis during interaction, while the dynamic analysis is applied to identify those event handlers at runtime.

3.4 Static and dynamic reverse engineering of Web applications

In 2000, Ricca and Tonella introduced ReWeb where they applied a static-only approach to reverse engineering of Web applications even though such approach disregards an important part of information regarding application architecture [49]. Since then a lot has happened in the field of reverse engineering (of Web applications).

In one of the more recent approaches, Shatnawi et al. [50] claim that understanding and reusing large APIs which use OOP is a complex and tedious task. They proposed an approach for identifying components with corresponding classes and methods from object-oriented APIs using reverse engineering. Their approach, however, uses static code analysis only. Likewise, Cloutier et al. [51] presented a reverse engineering tool for Web applications called WAVI. It uses static analysis and is able to extract elements coming from essential Web languages such as HTML, CSS, and JavaScript. The tool uses heuristics to acquire connections between application elements and enables the visualization of application structure using intuitive force-directed and class diagrams. Furthermore, Pan et al. [52] used static-only reverse engineering for removing advertisements from Android applications. They were less interested in the application model and more in identification and successful isolation of the advertisement code. In [53] the authors evaluated three commercial Web tools for reverse engineering capabilities for content and structure using the framework REEF based on static Web site analysis.

Unlike Pan et al., Le et al. [54] recently successfully used a dynamic approach to reverse engineering for testing access control and resolving security issues in Web applications. Similarly, two tools for reverse engineering of AJAX-based Web applications apply a dynamic approach as well. The first tool is the aforementioned FireDetective [42]. It records execution traces of both the JavaScript code executed in the browser and the server side code. The other tool is DynaRIA [55]. It implements an approach for comprehension of RIA implemented in AJAX.

Similarly, the tool WANDA produces accurate UML documentation of Web application architecture and its dynamic behavior [56]. However, the application must be instrumented and executed to collect dynamic information.

Most importantly, several groups used an approach which uses dynamic as well as static information in order to acquire Web application architecture. For instance, Lucca et al. [57] defined the WARE approach for reverse engineering of Web applications. It combines static and dynamic approaches to acquire presentation with UML diagrams which enable a better understanding of existing undocumented Web applications. They also propose reverse engineering as a means of supporting effective Web application maintenance. Swearngin et al. [58] use reverse engineering with static and dynamic analysis for acquiring a model of Web application interaction. They implemented a tool called Genie which models application's commands and proposes other interaction models, e.g., a voice command model. Genie provides a solution for enhancing interaction with existing websites.

In Ref. [59] it is claimed that *a static analysis of (the) source code may be extremely difficult (and in general unfeasible) because of the presence of dynamic generation of the HTML code that is part of the application under analysis.* The recovery of a Web application architecture model is presented using dynamic analysis performed by a ReWeb module Spider which is used for automatic extraction of an explicit-state model of a Web application. Spider requires that user input is simulated in order to compute probabilities of transitions between individual Web pages and to produce the model of a Web application formulated as a Markov chain. The paper includes an example with six pages and eight transitions, but it does not give information about applying the method to a real-world application. However, a real-world application would contain hundreds of Web pages and even more transitions; no analysis and potential clustering of identified Web pages is mentioned.

Instead of the statistical data obtained by executing a Web application (on real or simulated input data) for the purpose of reverse engineering, in order to obtain the dynamic component of an application's behavior, our approach uses existing real-world clickstream data containing a large number of real user sessions. The resulting application model is meant to improve the comprehension of the targeted application as it resembles the application more accurately. Furthermore, our method includes clustering a set of identified Web pages as well as the analysis of the resulting clustering. To the best of our knowledge, no work on using clickstreams for reverse engineering of Web applications has been published so far apart from the work of Rožanc and Poženel [60].

4. Application transition graph of a Web application

In reverse engineering, the structure of a Web application should be uncovered and its (visual) representation should be produced. To represent the structure of a Web application, the *atomic section model* (ASM) was selected in this study [61]. Initially, the ASM was defined for testing Web applications, but it has been shown that it is useful in reverse engineering as well [11].

The ASM consists of two parts. At the higher level of abstraction, it consists of the ATG which describes how individual components of a Web application are integrated. A component is an HTML page or a separate piece of code. A piece of code can be many different things depending on Web technology, e.g., a Java applet or servlet, a dynamic page (in the Oracle Portal context), etc. Formally, the ATG of a Web application is a quadruple $\mathcal{A} = \langle \Gamma, \Theta, \Sigma, \alpha \rangle$ where

- Γ denotes the set of all components (CIMs, see the next paragraph),
- Θ denotes the set of all transitions among components,
- Σ denotes variables defining possible states of the presentation layer, and
- α denotes a set of all diverse starting pages (usually one).

Note that $\langle \Gamma, \Theta \rangle$ denote a directed graph of all components of a Web application.

At the lower level, the structure of each individual component is described by the *component interaction model* (CIM). In other words, many CIMs, each representing an individual HTML page or another supporting piece of code, are combined into a single ATG of a Web application. Formally, the CIM of a Web application component is a quadruple $\mathcal{C} = \langle S, A, CE, T \rangle$ where

- S represents the set of start components that call $\mathcal{C}$,
- A represents the set of atomic sections of $\mathcal{C}$,
- CE stands for a regular expression describing the structure of $\mathcal{C}$, and
- T represents a set of transitions from and to (other) components.

Note that according to CIM's original definition [61], S is a component (and not a set of components as specified above) because "it is assumed that each software component has a unique start page" [61]. However, in many applications most components can be accessed from many other components.

Once the ATG is constructed in the process of reverse engineering, it represents the global structure of an application. Hence, if ATG's nodes are successfully clustered, the clusters can help the reverse engineer to quickly identify different parts of the application and/or which tasks are being performed by the application. To understand how each task is performed, the CIMs of the corresponding components must then be constructed and understood.

In forward engineering it takes time to design and implement an application. However, producing all CIMs and the ATG needed for testing an application takes only a limited amount of time because the internal structure of the application is known to developers. However, the situation is different in reverse engineering. The construction of the ATG (without the internal structure of CIMs) based on the source code is reduced to a relatively simple parsing of each component's source code and then aggregating the results.

As an example, the ATGs obtained from the source code of eStudent are shown in Fig. 2, where each rectangle represents a single code

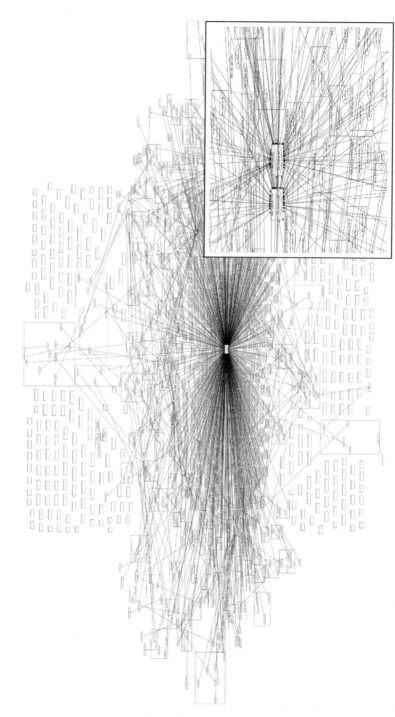

Fig. 2 Automatically generated ATG for eStudent application: even a fraction of a zoomed-in view (shown in the bottom right corner) remains hard to comprehend.

artifact, i.e., an application Web page. Free floating rectangles represent unused code artifacts; they are either remnants of older versions or they are used for debugging and development purposes only. It is obvious that such graphs are very hard to comprehend, which makes further processing absolutely necessary.

5. From raw clickstream data to user sessions

If an ATG of a Web application is automatically generated from the application's source code, i.e., using the static approach, the obtained result as the one shown in Fig. 2 is likely to be of no use. To improve the presentation of the ATG, a dynamic component of all information which can be gathered about the Web application must be taken into account. Web server access log files should therefore be considered as well.

The HTTP server usually serves multiple users at the same time and records their HTTP accesses to a single log file in the same order as they are received. The records of different users are therefore interwoven in that single log file. However, each record in the log file represents a separate HTTP request without an explicit link to other requests in the same user session, which is defined as a chronologically ordered sequence of clicks or resulting HTTP requests from the start till the end of user's one time usage of a Web application [62].

However, as HTTP server log data files lack specific contextual data, i.e., they do not identify sessions or users, critical events, or store Web form information [63], user session identification is a complex task and presents unique challenges. To extract an individual user session from the clickstream data stored in HTTP server log files, a clickstream data conflation and transformation process must be performed. As illustrated in Fig. 3, it typically consists of the following phases:

1. *Data gathering and cleaning*

    Web access log files are analyzed, synchronized, and cleaned. First, the HTTP access log format, i.e., CLF, ECLF, is identified. The size of input data is reduced as much as possible without losing the integrity of the specified data grain. Finally, HTTP log records which are not important for further clickstream processing are discarded. These are, for instance, requests for pictures or CSS specifications, or requests to ad servers.

2. *User identification, user session reconstruction, and verification*

    In the main phase users are first identified. Commonly used mechanisms for user identification are based on IP numbers or cookies.

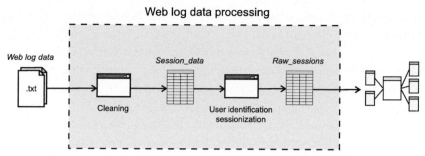

Fig. 3 Architecture of clickstream processing and user session identification.

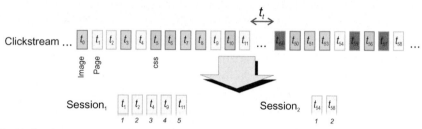

Fig. 4 Session reconstruction process.

Web users' requests are then assigned to identified Web users in correct time order. According to the defined timeout between two consequent user sessions, individual sessions are identified. Web user requests which were used for calculating session attributes and are not used any more, are discarded as illustrated in Fig. 4. Reconstructed individual user sessions are now ready to be consolidated.

However, session reconstruction is based on heuristic methods and is therefore error prone. The two common sources of difficulties with session reconstruction are IP sharing and parallel browsing [64]. In this phase additional steps can be taken in order to further improve the quality of user session reconstruction. One of the possibilities is detection and separation of interleaved sessions [64].

3. *Storing session data in a format suitable for further analyses*

In the last step of clickstream processing, data is consolidated and stored in a database, e.g., a star schema [65] formatted Data Webhouse [66, 67], suitable for various data analyses [10, 68, 69].

Faithful session reconstruction can be a challenging task [63, 70]. An HTTP protocol is stateless and user session pages are therefore not directly linked to each other in log files. Unlike proactive strategies [71, 72] typically based on using cookies or appending client-specific data on a Web page request, a reactive session reconstruction process [73, 74] relies only on the data found in the Web access log file containing loosely structured or even missing data [63]. Reactive strategies apply various heuristic approaches which cope with various session reconstruction problems like user identification type [73], Web robots detection [75], session identification heuristics [73, 74], back button [75, 76], and parallel-browsing behavior [77, 78]. As a result, Web sessions may not be reconstructed absolutely correctly. Despite the deficiencies, reactive strategies are attractive since they do not raise privacy concerns (as Web users show opposition to cookies and tracking browsing activity [79]). With the aim of respecting users' privacy and sometimes strict data privacy protection laws, this study uses a reactive strategy.

In their work, Sen et al. [80] note that creating clickstream data is relatively easy, so the Web analytics market is growing fast. Nevertheless, they think that the use of clickstream has still not exploited all its potential, and they look for possible reasons for that. They believe that emphasis must shift from the data side toward the analysis side and understanding of data.

The final result of session reconstruction is a list of sessions where each session is given a unique identification code. Some session identification codes might be missing since Web robot sessions and purely reconstructed sessions have been omitted (as explained in Section 2.1).

Each session is an ordered sequence of page requests triggered by user's clicks. Each page request is represented with (a) its sequence number within a session and (b) the name of the requested Web page. Some request sequence numbers might be missing as some requests do not reflect user's clicks and are therefore irrelevant.

The initial part of the printout of reconstructed user sessions is shown in Fig. 5. The first 4 sessions were omitted, requests 10 and 11 are omitted from session 5, sessions 6 and 7 are omitted, and so on. The reconstructed user sessions as shown in Fig. 5 represent the input to the reverse engineering methods described below.

6. Metrics for clustering the ATG of a Web application

To cluster Web application pages and thus relevantly visualize the application's ATG, the distance between these pages must obviously be defined. Among many possible definitions of distance, two definitions are used:

1. the one based on page names and
2. the other based on application usage data.

```
# session export file
# session length: [10,1500]
#
# ------------------------------------------------
# attributes: session_id, sequence, page_name
# ------------------------------------------------
#
5       1       /
5       2       PORTAL30_wwsec_app_priv_login
5       3       portal30_sso_wwsso_app_admin_ls_login
5       4       portal30_sso_prijava
5       5       portal30_sso_preveri_geslo
5       6       portal30_wwsec_app_priv_process_signon
5       7       VZ63_STUDENT_ZACETNI_PARAMETRI
5       8       VZ63_STUDENT_TEKOCA_OBVESTILA
5       9       VZ63_STUDENT_PRIJAVA_KOLOKVIJ
5       12      VZ63_STUDENT_PRIJAVA_KOLOKVIJ
5       13      VZ63_STUDENT_PRIJAVA_KOLOKVIJ
5       14      VZ63_LOGOUT
5       16      PORTAL30_wwsec_app_priv_logout
5       17      portal30_sso_wwsso_app_admin_ls_logout
5       18      /logout_fri
8       1       /
8       2       PORTAL30_wwsec_app_priv_login
8       3       portal30_sso_wwsso_app_admin_ls_login
8       4       portal30_sso_prijava
```

Fig. 5 The initial part of the printout of reconstructed user sessions.

The first definition relies on the names found in the source code and thus exposes the static information about the Web application, whereas the second definition uses Web server access log files and exposes the dynamic information about the behavior of a Web application.

6.1 The distance based on the component name

The first approach is based on the assumption that application developers used appropriate and systematic naming of pages depending on the part of the application to which an individual page belongs. This is based on a well-established finding that automatic program analysis and comprehension can be based on identifiers and names of program entities in general [81, 82].

Therefore, one should be able to identify the key parts of an application simply by observing that pages with similar names are likely to belong to the same application part while pages with different names are likely to belong to different parts of the application.

To cluster the set Γ of all Web application components, all components must be mapped into appropriate vector space in order to define distances between any two components.

A component name is usually a multiword identifier which must be split into individual words first. Different naming conventions are used for constructing multiword identifiers, with letter-case separated words (also called medial capitals in Oxford English Dictionary or simply CamelCase) and delimiter separated words (called snake case if the underscore serves as a delimiter) being most widely used.

The let function *split* maps the name of a Web application component into a multiset of all words appearing in its name (a multiset $\mathcal{A}$ is a pair (A, m) where A is a set and $m : A \longrightarrow \mathbb{N}$; furthermore, let $a \in \mathcal{A}$ iff $a \in A$). The actual function *split* depends heavily on the naming convention used in the Web application which is being reverse engineered. A set of all the words which component names consist of can thus be expressed as

$$W = \{w \mid \exists \gamma \in \Gamma : w \in split(\gamma)\}.$$

Let w_i be the ith word in W according to some fixed ordering of W.

All components of Γ are mapped into an n dimensional vector space V over $\mathbb{Z}$ where $n = |W|$. Each word $w_i \in W$ is mapped into a basis vector $\vec{w}_i = (a_1, a_2, \ldots, a_n)$ where $a_j = 1$ if $i = j$ and $a_j = 0$ if $i \neq j$. Hence, if $split(\gamma) = (A_\gamma, m_\gamma)$, a component $\gamma \in \Gamma$ is mapped into a vector

$$\vec{v}_\gamma = \sum_{i=1}^{n} a_i \vec{w}_i \quad \text{where} \quad a_i = \begin{cases} 0 & w_i \notin A_\gamma \\ m_\gamma(w_i) & w_i \in A_\gamma \end{cases}$$

Once all components are mapped into vector space, the cosine similarity between vectors $\vec{v}_1$ and $\vec{v}_2$ defined as

$$sim(\vec{v}_1, \vec{v}_2) = \cos \phi = \frac{\vec{v}_1 \cdot \vec{v}_2}{\|\vec{v}_1\| \quad \|\vec{v}_2\|},$$

is used as the distance between two vectors.

As the component a_i of the vector $\vec{v}_\gamma$ represents the number of occurrences of the word w_i in the name of the Web component γ, it is always nonnegative, i.e., $a_i \geq 0$, and therefore $sim(\vec{v}_1, \vec{v}_2) \in [0, 1]$.

Based on the cosine similarity the distance matrix $D_n \in \mathbb{Z}^{n \times n}$ (index n means *names*) contains elements $d_{i,j}$ for $i, j \in \{1, 2, \ldots, n\}$ where $d_{i,j} = sim(\vec{v}_i, \vec{v}_j)$.

The data about all application pages is also stored in a data Webhouse. We acquired 354 distinct application pages from a star schema page dimension representing application pages. The number of pages obtained from page dimension (354) is greater than the number of pages obtained from clickstream data (318), since some application pages had never been accessed in the period when clickstream data were collected. However, these pages still exist in the system and have to be dealt with. Application page names were tokenized and mapped into vector space using cosine similarity between vectors (Section 6). Application page names were given by developers so that they could be easily broken into page name tokens, e.g., VZ63_STUDENT_EXAM_APPLICATION. The data about cosine similarity between page vectors was stored to a distance matrix D_n (index n denotes names) of size 354 × 354.

6.2 The distance based on Web application usage

After a session is reconstructed, a set of all pages for which at least one request is recorded in the log file(s), and a set of user sessions become available. Let $\mathcal{P}$ and $\mathcal{S}$ denote the set of all pages and the set of all user sessions, respectively.

Each session in $\mathcal{S}$ is a sequence of Web page requests issued one after another. Formally, session $s \in \mathcal{S}$ consisting of $n(s)$ consecutive requests for Web pages $p_1, p_2, \ldots, p_{n(s)}$ is

$$s = \langle p_1, p_2, \ldots, p_{n(s)} \rangle \in \times_{i=1}^{n(s)} \mathcal{P}$$

and thus $\mathcal{S} \subseteq \times_{i=1}^{\infty} \mathcal{P}$.

In order to carry out the clustering process, attributes or measures have to be defined. Web pages in a Web log file have few attributes which can be relied upon in the clustering process. Each Web page has the attributes such as the click rate or the number of average distinct appearances in a session, which can be used for attribute based clustering, but clustering results based on those attributes were not very promising. In order to use the multi-dimensional scaling (MDS) clustering algorithm which can use the distance between two objects for the clustering process, we had to determine the appropriate measure which defines the similarity between two pages which would be used as input to the distance based clustering algorithm. We first defined the similarity between two individual Web pages, which represents a basis for determining the distance between two individual pages.

The similarity between two Web pages p_i and p_j is based on the number of times both Web pages appear in the same Web session s and is defined as

$$sim(p_i, p_j) = \begin{cases} 1 & p_i = p_j \\ \dfrac{|\{s \in S | p_i \in s \wedge p_j \in s\}|}{|S|} & \text{otherwise} \end{cases}$$

Further, the distance between two Web pages is derived from the similarity between these two Web pages and is defined as

$$dist(p_i, p_j) = \begin{cases} 1 - sim(p_i, p_j) & sim(p_i, p_j) > 0 \\ \infty & \text{otherwise} \end{cases}$$

With the defined distance between two pages, we can further define the distance matrix $D_u \in \mathbb{Z}^{|\mathcal{P}| \times |\mathcal{P}|}$ (index u means *usage*) with elements $dist_{i,j}$ for $i, j \in \{1, 2, ..., |\mathcal{P}|\}$, where p_i and p_j are the ith and jth page in $\mathcal{P}$ according to some fixed ordering of pages and $\mathcal{P}$. If two pages appear in same sessions many times, it implies that these two pages should most likely belong to the same cluster, so their distance must be small. To summarize, two pages without direct transitions are at distance ∞, while two pages which both appear in every user Web session are at distance 0. Again, matrix D_u represents the input to the clustering algorithms computed by Orange and the custom implemented graph drawing program proposed by Kamada–Kawai [83].

7. Clustering and visualizing Web application's ATG

Out of many different clustering and visualizing methods, the following three were used to cluster and visualize the Web application's ATG in this study:

- hierarchical clustering,
- the graph drawing algorithm proposed by Kamada and Kawai [83], and
- multidimensional scaling (MDS) [84].

In these methods, the distances as defined earlier were used.

Hierarchical clustering was used first because it combines objects at lower levels into clusters on higher levels. Since what was sought was the multilevel architecture of a Web application, hierarchical clustering was deemed ideal for this purpose. For a reverse engineer, it seemed an attractive method for clustering ATG because of its simplicity, e.g., the number of clusters does not need to be specified and any valid measure of distance can be used. Furthermore, data mining suites such as Orange provide interactive dendrogram visualization of how clusters are merged as one moves up the hierarchy.

The other two methods, namely graph drawing and MDS, were used because they arrange objects according to distances between them even if their locations are not known. It was expected that clusters could be identified by placing objects, i.e., Web applications' artifacts, in a two-dimensional plane using either of these two methods. The results of the experiments which tested these expectations are described below.

7.1 ATG visualization using hierarchical clustering

In order to cluster the application's ATG, hierarchical clustering was tried first. It was performed by means of Orange using the following four different distance metrics:

- distance based on Web application usage (as defined in Section 6.2),
- distance based on component names (as defined in Section 6.1),
- distance based on the combination of Web application usage and component names, and
- random distance.

The last one was used as a reference, measuring the amount of information which could be extracted using the first three.

The basic clusterings, the one based on component names and the other based on component usages, are shown in Fig. 6. Even though the ordering of components, i.e., Web pages, in the column next to each clustering differs, it is obvious that these two clusterings are significantly different. A former eStudent developer manually inspected these two clusterings and concluded that none of them matched his perception of the eStudent application. Moreover, he established that none of these two provided any new insight and he could not envisage how either of them would be of any help to a future reverse engineer.

The resulting clusterings, for different numbers of clusters, were compared to the manual clustering produced by the former eStudent developer with the inside knowledge of eStudent structure. Fig. 7 shows the evaluation of the resulting clusterings using four well-established metrics for comparing clusterings, namely mutual information (MI), normalized mutual information (NMI), Fowlkes–Mallows index $\mathcal{F}$ (F Measure), and Jaccard index.

If the clusterings based on usage or on a combination of names and usage are considered, one cannot determine the number of clusters. Even worse, MI and NMI indicates that more clusters yield better results, which contradicts the opinion of the former eStudent developer. The Jaccard index does not perform significantly better. Even the Fowlkes–Mallows index, which gives the best results at first glance, indicates that the number of clusters might be anything from 4 to 12 (while the correct number given by the

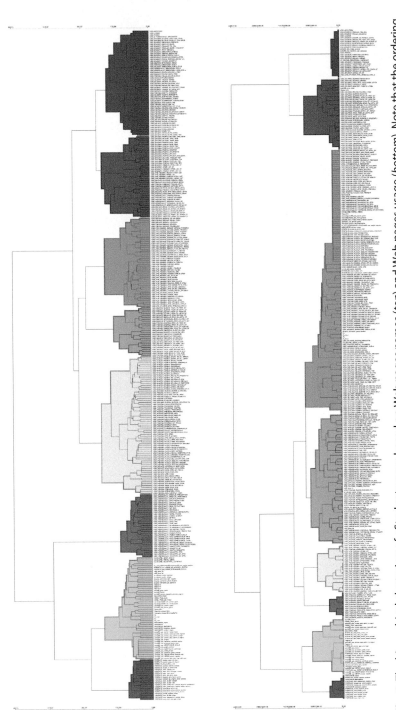

Fig. 6 The hierarchical clustering of eStudent pages based on Web pages names (*top*) and Web pages usage (*bottom*). Note that the ordering of Web pages (*horizontally*) is different in both clusterings.

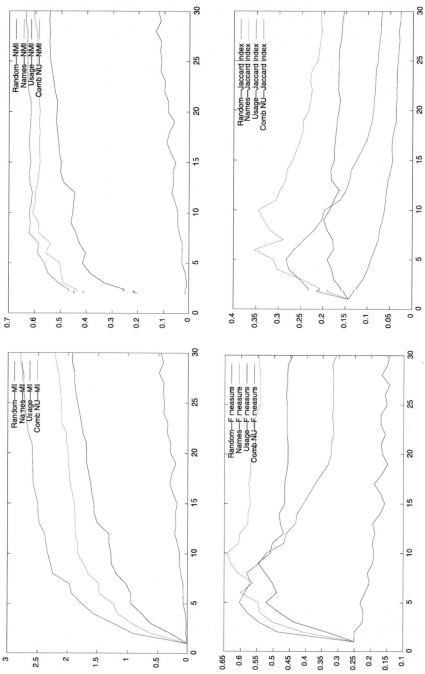

Fig. 7 The evaluation of the results of hierarchical clustering using different metrics (*top*: mutual info., norm. mutual info., *bottom*: Fowlkes–Mallows index $\mathcal{F}$ (F Measure) and Jaccard index) with random distance, distances based on names and usage, and a distance based on combination of names and usage. The number of clusters and the value of each metric are shown along x- and y-axes, respectively.

eStudent developer is 8). Furthermore, once the dendrogram produced by hierarchical clustering was manually compared to the clusters provided by the eStudent developer, the distance between many objects in the dendrogram did not match the distance in clusters.

Note that the produced clusterings based on names yield relatively good results. However, the names in the studied Web application were chosen very carefully and result in eight inherent clusters. If the names had been chosen a bit less systematically, the results of name-based clusterings in Fig. 7 would not bear resemblance with the application's structure.

It was concluded that no distance metrics yields satisfactory results. By far the most probable reason for poor performance of hierarchical clustering is the relatively flat structure of the analyzed Web application. As it turned out, the application's structure and the automatically determined number of clusters, i.e., application's modules, were not appropriate.

It was established that someone who does not know the reverse engineered application would probably be unable to determine correct clusters. In addition, Web pages from rarely used application parts were mostly classified into wrong clusters.

As the eStudent developer confirmed this finding, hierarchical clustering was abandoned.

7.2 ATG visualization using Kamada–Kawai graph drawing algorithm

In the first approximation, the ATG of a Web application can be visualized using any graph drawing algorithm which accounts for distances between pages.

The Kamada–Kawai graph drawing algorithm was selected because of its simplicity [83]. It is a classic force-directed algorithm: a spring-like force attracts connected vertices while a repulsive force repels each pair of vertices in a graph (all vertices are electrically charged with the same charge). The algorithm places all vertices in a plane, applies the spring-forces relating to the lengths of edges, i.e., known distances between vertices, and uses simulated annealing to find the equilibrium.

More precisely, the Kamada–Kawai method assumes that the distortion of the graph's layout is proportional to the energy

$$E = \sum_{i=1}^{n-1}\sum_{j=1}^{n-1}\left(\frac{1}{2}k_{ij}\left(|p_i - p_j| - l_{ij}\right)^2\right)$$

of the underlying spring model where p_1, p_2, ..., p_n denote graph's vertices and hence $|p_i - p_j|$ denotes the distance between vertices p_i and p_j in the two-dimensional layout. Furthermore, let d_{ij} denotes the shortest path between vertices p_i and p_j. Quantities l_{ij}, the desired distance between these two vertices, and k_{ij}, the strength of the spring between them, are defined as

$$l_{ij} = L \cdot d_{ij} \quad \text{and} \quad k_{ij} = \frac{K}{d_{ij}^2},$$

respectively, where K is a constant and

$$L = \frac{L_0}{\max_{i! = j} d_{ij}}$$

for an L_0 which stands for the desired length of a single edge.

The optimal layout is the one with minimal energy E, which is a function of $2n$ parameters x_1, x_2, ..., x_n and y_1, y_2, ..., y_n, where a pair (x_i, y_i) represents the coordinates of the vertex p_i, for $i = 1, 2, ..., n$.

In order to find the optimal layout, one would need to find the global minimum of function $E(x_1, x_2, ..., x_n, y_1, y_2, ..., y_n)$. This is generally very hard to achieve so the Newton–Raphson method is used to find the local minimum (just as in the original paper [83]) and to yield a satisfactory layout. The local minimum of function E (and therefore the final layout) is reached when

$$\frac{\partial E}{\partial x_m} = \frac{\partial E}{\partial y_m} = 0$$

If $|p_i - p_j|$ is substituted with $\sqrt{(x_i - x_j)^2 + (y_i - y_j)^2}$, the partial derivations of E by x_m and y_m are expressed as

$$\frac{\partial E}{\partial x_m} = \sum_{i \neq m} \left(k_{mi} \left((x_m - x_i) - \frac{l_{mi} \, (x_m - x_i)}{\sqrt{(x_m - x_i)^2 + (y_m - y_i)^2}} \right) \right)$$

$$\frac{\partial E}{\partial y_m} = \sum_{i \neq m} \left(k_{mi} \left((y_m - y_i) - \frac{l_{mi} \, (y_m - y_i)}{\sqrt{(x_m - x_i)^2 + (y_m - y_i)^2}} \right) \right)$$

for any $m = 1, 2, ..., n$.

Instead of solving this system of $2n$ inter-dependent nonlinear equations and moving every vertex in every step, only the vertex with the largest value

$$\Delta_m = \sqrt{\left(\frac{\partial E}{\partial x_m}\right)^2 + \left(\frac{\partial E}{\partial y_m}\right)^2}$$

is moved in a single step. Hence, in step t, the coordinates of vertex p_m change according to equations

$$x_m^{(t+1)} = x_m^{(t)} + \delta x \quad \text{and} \quad y_m^{(t+1)} = y_m^{(t)} + \delta y$$

while all coordinates of all other vertices, i.e., x_i and y_i for $i \neq m$, remain unchanged. The distances δx and δy are computed using the following pair of linear equations:

$$\frac{\partial^2 E}{\partial x_m^2}(x_m^{(t)}, y_m^{(t)})\delta x \;+\; \frac{\partial^2 E}{\partial x_m \partial y_m}(x_m^{(t)}, y_m^{(t)})\delta y = -\frac{\partial E}{\partial x_m}(x_m^{(t)}, y_m^{(t)})$$

$$\frac{\partial^2 E}{\partial y_m \partial x_m}(x_m^{(t)}, y_m^{(t)})\delta x \;+\; \frac{\partial^2 E}{\partial^2 y_m}(x_m^{(t)}, y_m^{(t)})\delta y = -\frac{\partial E}{\partial y_m}(x_m^{(t)}, y_m^{(t)})$$

It is assumed that the formulae for the second derivatives of function E can be derived from the first derivatives by the reader.

The overall algorithm for finding the local minimum of function E iterates as long as the largest Δ_m is above the preselected threshold.

As the algorithm organizes Web pages visually into clusters, the relations between pages enable the reverse engineer to spot different groups of Web application's components belonging together. Namely, the closer Web pages are according to the selected distance, the closer they ought to be in ATG visualization.

The resulting visualization of the ATG shows which Web pages of the application are used together in user Web sessions. As a general rule, users are authorized to access only the application's Web pages that belong to the user group which users are enrolled into. Application Web pages usually cannot be accessed from other user groups, e.g., students can only access student Web pages and cannot access Web pages which belong to professors' user group.

It is expected that the resulting clustered ATG shows groups of pages which belong to some user group or are otherwise connected. From the clusterization it can be seen how many distinct user roles an application has. A Web application can also have a group of pages which are common for all user roles (e.g., login pages). The resulting visualization also indicates such pages, which are usually located in the middle among other surrounding clusters.

To visualize eStudent's ATG, five different distance metrics were used within the Kamada–Kawai graph drawing algorithm:

- distance based on Web application usage (as defined in Section 6.2),
- distance based on component names (as defined in Section 6.1),
- distance based on component names with each name impaired by one random mutation (as defined in Section 6.1),
- distance based on component names with each name impaired by two random mutations (as defined in Section 6.1),
- distance based on a combination of distances based on Web application usage and component names, and
- random distance.

The last one was again used as a reference to measure the amount of information which could be extracted using the first four metrics. Furthermore, the distances based on component names impaired by one or two random mutations were introduced because component names in eStudent were carefully given using a very systematic scheme, which is something that cannot always be expected from application developers.

The results are shown in Figs. 8 and 9. Individual components are colored in regard to the manual clustering of the former eStudent developer. In reality, however, no colors are available if they were, no clustering would be needed.

First, it is obvious that the random distances used in Fig. 9 (right) produce no meaningful visualization. Second, if the ATG is visualized using component names as shown in Figs. 8 (left) and 9 (left and center), it is possible to clearly detect one cluster which is shown in cyan and includes "system" pages used by all different kinds of users. However, components belonging to other clusters are best drawn together if application usage is employed as shown in Fig. 8 (right).

All the obtained clusterings were compared to the clustering which the application's developer proposed as the reference clustering. The reference clustering consisted of 8 page clusters where each cluster block consisted of 19–73 application pages. In order to rank the performance of methods according to reference clustering, we need a method to determine the similarity between the reference and the obtained clusterings.

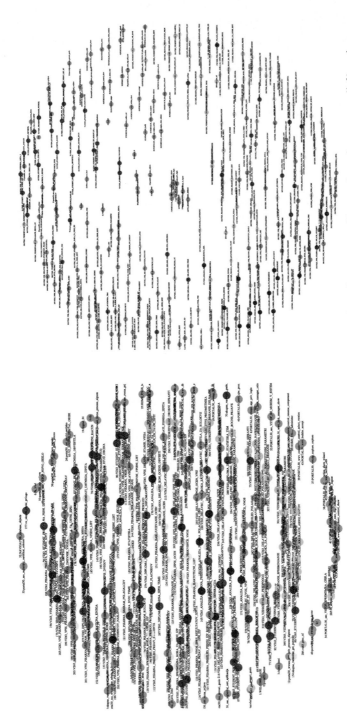

Fig. 8 Visualization of the eStudent ATG using Kamada–Kawai graph drawing algorithm using a distance based on names (*left*) and usage (*right*): *colors of different application's artifacts corresponds to clusters provided by an eStudent developer.*

Fig. 9 Visualization of the eStudent ATG using Kamada–Kawai graph drawing algorithm using a distance based on names impaired with one or two mutations (*left and center*) and using a random distance (*right*): *colors* of different application's artifacts corresponds to clusters provided by an eStudent developer.

To compare the obtained (resulting) clustering with the expert proposal (reference clustering), we colored Web pages in each expert cluster in a different color. The pages in the obtained clusterings were then colored with conterminous colors of reference clustering. The similarity between the reference and obtained clusterings can be estimated visually. Ideally, the number of clusters in both clusterizations should match, while all pages with the same color should appear close to each other and visually separated from other clusters. In reality that will seldom be the case due to numerous reasons:

1. *Viewing angle.* An expert might have a slightly different view of the system than users do. A functional view (functional silo) and a business process view (business processes which cut across a functional silo) produce different results. User sessions are closer to the business process view, since within each user session there are one or more tasks which may contain pages from different functional parts.

2. *The power of user groups.* The resulting clustering might be overfitted to the viewpoint(s) of the user groups which produce the majority of Web sessions. Pages appearing mostly in sessions of perhaps important but small user groups are not shown to be as strongly related as they are.

   For instance, in the eStudent application the vast majority of sessions is produced by students. Support Web pages, which are used in tasks of virtually all user groups, are hence disproportionally attracted to Web pages intended for exclusively student tasks. On the other hand, there are only a few administrator sessions recorded, which is why administration pages are scattered around far more than it is the case with student pages. However, the influence of different and unknown user groups is hard to normalize since the classification of sessions is not known (otherwise clustering would not be needed).

3. *Usage frequency.* Frequently visited Web pages carry more weight and are therefore placed close to their related pages. The placement of rarely visited pages, especially if they are used for performing tasks shared by various user groups, seems more random and error prone.

   Although there is a huge number of student sessions in eStudent, only a few of them contain pages related to thesis submission. Furthermore, as many pages for thesis submission are shared among students, professors and support staff, the thesis related pages are not placed as close together as they should be.

4. *Bad data.* There are inconsistencies in session data due to errors in the preprocessing phase.

One must realize, however, that during reverse engineering one cannot rely on the manual clustering produced by the application's developer. In other words, the circles symbolizing individual components in Figs. 8 and 9 would not be colored, which would make the identification of individual clusters much harder.

Finally, it is worth mentioning that the method can be modified to produce a spatial representation of the graph in more, say N, dimensions instead of just two dimensions. One must just replace each particle, i.e., Web page $p = (x, y)$ with an N-dimensional tuple $p = (x_1, x_2, \ldots, x_n)$ and modify the application of the Newtown-Raphson method accordingly. The visualization becomes harder if not impossible, but if the N-dimensional spatial representation is transformed into 2D presentation, this is already a step toward MDS.

7.3 ATG visualization using multidimensional scaling

In the final attempt, the Orange tool was used for visualizing the application's ATG using MDS [84].

MDS is a method well suited for placing different objects whose locations are not known into multidimensional space while preserving distances among them. It applies an iterative improvement of object locations which minimizes the value of a stress function.

MDS was used with two different initial placements of data points: random and PCA (principal coordinates analysis or Torgesen scaling) initialization. MDS using random initialization positions initial data cases randomly and then rearranges them according to the similarity of individual cases. MDS with PCA initialization uses classical MDS and positions data cases along principal coordinate axes. The results for both initial placements are shown in Fig. 10. Better results were achieved using PCA (Torgesen) initial placement.

Despite the fact that in Fig. 10 we can identify main application clusters which consist of pages with similar application background, we still lack information how each cluster connects to other clusters and how pages are interconnected in the cluster itself. In addition, smaller page clusters close to bigger page clusters can be erroneously assigned to those bigger clusters. In order to further improve ATG cluster visualization, we introduce the most frequently used transitions between pages to the clustered ATG.

Frequently used transitions between pages can be iteratively added to the clustered ATG. The more page transitions between pages are added, the

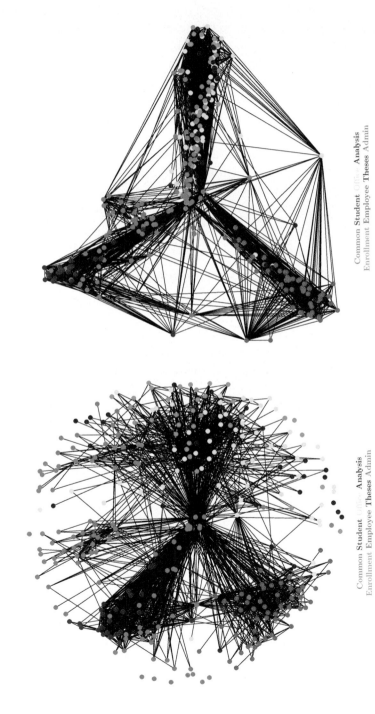

Common **Student** Office **Analysis**
Enrollment **Employee** Theses Admin

Common **Student** Office **Analysis**
Enrollment **Employee** Theses Admin

Fig. 10 MDS with random (*left*) and PCA (*right*) initialization.

more clusters start to appear. The reverse engineer can start with a small volume of top frequently used page transitions, 100 at most, and increase the number iteratively until they are satisfied with the solution. Iterative adding of most frequently used transitions allows the reverse engineer to identify the most frequently used (and therefore most important) parts of the system. Fig. 12 shows the iterative construction of the final clustered ATG solution in four steps, adding more and more frequently used transitions.

Clusters of pages which could be identified earlier can now be further confirmed. Furthermore, less clearly visible smaller clusters which could be previously erroneously assigned to bigger neighboring clusters can now be clearly identified. Transitions between pages in a cluster make it clearly distinct from other clusters. In Fig. 10 such a cluster is the cyan cluster which represents system pages common to all page users.

7.4 The clusterization process in reverse engineering practice

Of all three clustering methods, MDS produced the best results. The result obtained using MDS (Torgesen) is shown in Fig. 11. The positions of individual Web pages are the result of MDS while each different color indicates one of the eight reference clusters, e.g., a student, a professor or a student office, as indicated by the former eStudent developer. In Fig. 11 two Web pages are connected if there is a transition from one page to another in at least one Web session.

The links between application pages are an important part of the presentation of the final solution. The links visually aggregate pages in the result clustering so that clusters are even more prominent. However, the number of links should not be too high, otherwise the transparency can be lost. The reverse engineer can build the final presentation of clustering by gradually adding the most common links between pages to the clusterized ATG:

1. The reverse engineer starts with the clusterization graph with no links at all and gradually adds links between pages, from the most common ones to the least frequent ones.
2. By adding more and more links, individual clusters start to stand out. More added links result in more visible clusters.
3. When additional links do not result in more information on the architecture of the Web application, the reveres engineer gets the final solution.

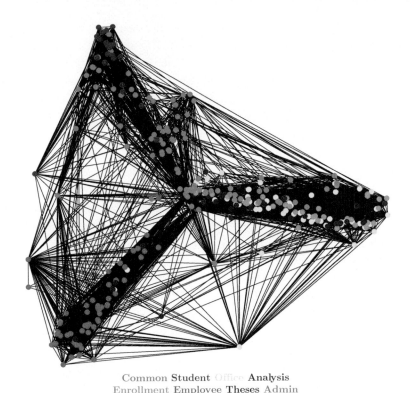

Common Student Office Analysis
Enrollment Employee Theses Admin

Fig. 11 The full ATG of eStudent as clusterized using multidimensional scaling (individual Web pages are colored according to the reference clustering produced by the past eStudent developer).

As an illustration, consider the construction of the final clustering of eStudent's ATG as shown in Fig. 12:

1. The reverse engineer first added 200 most frequently used links to an empty clustering graph (Fig. 12, top left). The upper left arm of the ATG contains the majority of the most frequent transitions. It is hence very likely that this arm represents the user group creating most sessions.

2. With 600 most frequently used page transitions added (Fig. 12, top right), four clusters become more clearly formed. Most Web pages are already interconnected, while there is a number of pages (mostly colored gray but this is not known by the reverse engineer) in the right arm which are still unconnected. It can be suspected that these pages belong to the least frequently used part of the system.

3. With 1100 links (Fig. 12, bottom left) it becomes clear that the pages in the center of the presented ATG form a very important part of the system

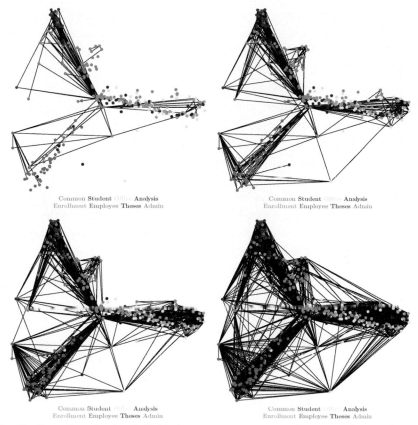

Fig. 12 Clustering eStudent's ATG in four steps: using 200 and 600 most frequently used transitions (*top left* and *right*), and using 1100 and 2300 most frequently used transitions (*bottom left* and *right*).

shared by all user groups. Furthermore, it is very likely that the bulge on the upper left arm forms yet another cluster.

4. Furthermore, with 2300 added links (Fig. 12 (bottom right)) the observations from the previous step are confirmed. However, without previous pictures the bottom right picture alone (or the one shown in Fig. 11) cannot provide the insight provided by all four pictures together.

The number of transitions in each step, i.e., 200, 600, 1100, and 2300, was chosen by the reverse engineer interactively during the reverse engineering process. In principle the visualizations of ATG containing different numbers of transitions are investigated and those considered informative are kept. Finally, after adding some more links, the graph becomes cluttered so the reverse engineer stops the process and obtains the final solution, presented in Fig. 11.

The observations above can be deduced from Figs. 11 and 12 even if the reference clustering is not known, i.e., if all Web pages are shown in the same color. However, the following lessons learnt with the reference clustering applied, must be well noted:

1. Web pages colored in yellow, green, and dark red, i.e., regular student records pages, enrollment pages, and theses, are closely interconnected. However, they cannot be distinguished based on usage data. Even though these are three different activities, they are performed by the same user group, i.e., student office, and thus they appear in the same user sessions.

2. However, a small number of Web pages positioned outside the areas covered by clearly recognizable clusters can still be linked to the correct cluster. Take the two isolated pages on the left side (one red and one blue): with only 200 most frequent transitions displayed, it becomes evident they belong to the upper left arm. This is not possible to see if too many transitions, e.g., 2300, are applied.

3. The administrator pages, shown in gray, are attracted to the right arm. Therefore, it might seem they are just a part of another cluster, but after a detailed examination of pictures with a small number of transitions shown, i.e., 200 and 600, it can be seen that the gray pages are not connected with the rest in the right arm. The reverse engineer might therefore suspect that they form a unique cluster belonging to a small user group which rarely performs its activities.

4. The bulge of the top left arm are basically the pages shared by all user groups. However, the preponderance of student sessions pulls them out of the center toward the student pages aggregated on the tip of the top left arm. The reverse engineer must be aware such anomalies can happen and must therefore not neglect the appearance of the power groups recognizable in pictures with a small number of transitions.

Finally, in case of eStudent the expert who provided the reference clustering took a functional rather than a process view of the system. He aggregated the pages used for performing certain activities. As the student office performs more activities than all other user groups combined, its many activities were split into four different clusters while all activities of any other user group belong to the same cluster.

However, sessions are generally process–oriented, i.e., used for performing a single user task. The only exception are student office sessions. These are generally very long (lasting hours or even entire working days) and comprise many different student office tasks belonging to different functional areas. Therefore usage-based clusterization to some extent differs from the expert's one. Nevertheless, the results are encouraging.

As we indicated in Section 6, we used page names, i.e., static source of data, to perform Web page clustering. Grouping Web application pages based on their page names may yield good results. Unfortunately, this method is highly dependent on (a) whether the naming is defined properly during the analysis phase and is systematically applied afterward, and (b) whether the naming follows a functional view or a process view of the application.

On the other hand, the method based on usage data is independent of the naming and always gives the same results. It is therefore more objective and takes the actual use of the system into account. The application parts which are seldom used or are not used at all during the period of Web log file data generation, might not appear in the final ATG.

8. Conclusion

Reverse engineering of old Web applications with inadequate or non-existent documentation can be challenging. In this text we presented a reverse engineering approach for acquiring a model of a Web application from Web log files. We have proven that Web log files can be a useful source for the reverse engineering process, and that the possibility of reverse engineer's incremental building of the final application model may significantly enhance their understanding. In addition, as seen in our study, usage data analysis yields a more consistent application model than the analysis based on page names.

The clustering of the application's ATG was performed using three techniques: hierarchical clustering, graph drawing, and MDS, with the last one producing the best results. Our approach toward reverse engineering of Web applications uses incremental addition of transitions between Web pages to the clusterized ATG in order to further improve the insight into application's structure. The reason for better performance of a graph drawing algorithm and MDS compared to hierarchical clustering might be the consequence of a relatively flat structure of a Web application which was used as a test case. However, as the architecture of many traditional Web pages is relatively flat, we believe our findings can be used in reverse engineering of traditional Web applications in general. Furthermore, the acquired application model gives the reverse engineer an initial insight into application structure and can be later used as a starting point for further reverse engineering processes.

Alongside the metrics based on usage data retrieved from Web server log files, we also investigated the metrics based on page names, i.e., Web application's component names. However, such metrics is often too dependent on the quality of a particular naming scheme and its consistent implementation. Usage data, on the other hand, is insensitive to these issues.

Finally, it is worth emphasizing that the presented method fits naturally into the extract-abstract-view metaphor, which is the reference architecture for reverse engineering tools [5].

References

[1] R.C. Martin, Agile Software Development, Principles, Patterns, and Practices, Prentice Hall, Englewood Cliffs, NJ, USA, 2002.
[2] K. Beck, M. Beedle, A. Van Bennekum, A. Cockburn, W. Cunningham, M. Fowler, J. Grenning, J. Highsmith, A. Hunt, R. Jeffries, et al., Manifesto for Agile Software Development. Retrieved: June 4, 2013, from http://agilemanifesto.org (accessed: December 29, 2017).
[3] J. Ebert, B. Kullbach, V. Riediger, A. Winter, GUPRO—generic understanding of programs (an overview), Electron. Notes Theor. Comput. Sci. 72 (2) (2002) 10.
[4] B. Nikolic, Z. Radivojevic, J. Djordjevic, V. Milutinovic, A survey and evaluation of simulators suitable for teaching courses in computer architecture and organization, IEEE Trans. Educ. 52 (4) (2009) 449–458. ISSN: 0018-9359, https://doi.org/10.1109/TE.2008.930097.
[5] C. Lange, H.M. Sheed, A. Winter, Comparing graph-based program comprehension tools to relational database-based tools, in: Proceedings of the 9th International Workshop on Program Comprehension (ICPC 2001), IEEE, Toronto, Canada, 2001, pp. 209–218.
[6] I.-H. Ting, C. Kimble, D. Kudenko, A pattern restore method for restoring missing patterns in server side clickstream data, in: Lecture Notes in Computer Science, Proceedings of the 7th Asia-Pacific Web Conference on Web Technologies Research and Development (APWeb'05), vol. 3399. Springer-Verlag, Shanghai, China, 2005, pp. 501–512.
[7] V. Blagojević, D. Bojić, M. Bojović, M. Cvetanović, J. Đorđević, . Đurđević, B. Furlan, S. Gajin, Z. Jovanović, D. Milićev, A systematic approach to generation of new ideas for PhD research in computing, in: A.R. Hurson, V. Milutinović (Eds.), Advances in Computers, vol. 104. Elsevier, 2017, pp. 1–31.
[8] F. Benevenuto, T. Rodrigues, M. Cha, V. Almeida, Characterizing user navigation and interactions in online social networks, Inf. Sci. 195 (2012) 1–24.
[9] L.D. Paulson, Building rich web applications with Ajax, Computer 38 (10) (2005) 14–17. ISSN: 0018-9162, https://doi.org/10.1109/MC.2005.330.
[10] C.T. Lopes, G. David, Higher education web information system usage analysis with a data webhouse, ICCSA (4) 3983 (2006) 78–87.
[11] I. Rožanc, B. Slivnik, Using reverse engineering to construct the platform independent model of a web application for student information systems, Comput. Sci. Inform. Syst. 10 (4) (2013) 1557–1583.

[12] S. Ducasse, M. Lanza, S. Tichelaar, MOOSE: an extensible language-independent environment for reengineering object-oriented systems, in: Proceedings of the Second International Symposium on Constructing Software Engineering Tools (CoSET 2000), 2000, pp. 24–30, Limerick, Ireland.

[13] R. Ferenc, Á. Beszédes, M. Tarkiainen, T. Gyimóthy, Columbus—reverse engineering tool and schema for C++, in: Proceedings of the International Conference on Software Maintenance (ICSM 2002), 2002, pp. 172–181. Montreal, Quebec, Canada.

[14] M. Lanza, S. Ducasse, Polymetric views—a lightweight visual approach to reverse engineering, IEEE Trans. Soft. Eng. 29 (9) (2003) 782–795.

[15] A. Telea, L. Voinea, An interactive reverse engineering environment for large-scale C++ code, in: Proceedings of the 4th ACM symposium on Software visualization (SoftVis'08), Herrsching am Ammersee, Germany, 2008, pp. 67–76.

[16] M. van den Brand, S. Roubtsov, A. Serebrenik, SQuAVisiT: a flexible tool for visual software analytics, in: Proceedings of the 13th European Conference on Software Maintenance and Reengineering (CSMR'09), 2009, pp. 331–332. Genova, Italy.

[17] H.M. Kienle, H.A. Müller, Rigi—an environment for software reverse engineering, exploration, visualization, and redocumentation, Sci. Comput. Program. 75 (4) (2010) 247–263.

[18] J. Martin, L. Martin, Web site maintenance with software-engineering tools, in: Proceedings of the 3rd International Workshop on Web Site Evolution, 2001, pp. 126–131. Florence, Italy.

[19] F. Ricca, P. Tonella, Understanding and restructuring web sites with ReWeb, IEEE MultiMedia 8 (2) (2001) 40 51.

[20] B. Kitchenham, R. Pretorius, D. Budgen, O.P. Brereton, M. Turner, M. Niazi, S. -Linkman, Systematic literature reviews in software engineering—a tertiary study, Inf. Soft. Technol. 52 (8) (2010) 792–805.

[21] M. Monroy, J.L. Arciniegas, R. Julio, An approach to recovery and analysis of architectural behavioral views, in: Proceedings of 2015 Chilean Conference on Electrical, Electronics Engineering, Information and Communication Technologies (CHILECON 2015), IEEE, Santiago, Chile, 2015, pp. 923–930.

[22] M. Torchiano, M.D. Penta, F. Ricca, A.D. Lucia, F. Lanubile, Migration of information systems in the Italian industry: a state of the practice survey, Inf. Softw. Technol. 53 (1) (2011) 71–86. ISSN: 0950-5849, https://doi.org/10.1016/j.infsof.2010.08.002.

[23] P. Tramontana, D. Amalfitano, A.R. Fasolino, Reverse engineering techniques: from web applications to rich internet applications, in: 2013 15th IEEE International Symposium on Web Systems Evolution (WSE), IEEE, 2013, pp. 83–86.

[24] P. Aho, T. Kanstren, T. Räty, J. Röning, Automated extraction of GUI models for testing. in: A. Memon (Ed.), Advances in Computers, vol. 95. Elsevier, 2014, pp. 49–112. https://doi.org/10.1016/B978-0-12-800160-8.00002-4.

[25] N. Walkinshaw, Chapter 1—Reverse-engineering software behavior, in: A. Memon (Ed.), Advances in Computers, Advances in Computers, vol. 91, Elsevier, 2013, pp. 1–58, https://doi.org/10.1016/B978-0-12-408089-8.00001-X.

[26] C. Riva, J.V. Rodriguez, Combining static and dynamic views for architecture reconstruction, in: Proceedings of the Sixth European Conference on Software Maintenance and Reengineering (CSMR 2002), IEEE, 2002, pp. 47–55.

[27] H. Bruneliere, J. Cabot, G. Dupé, F. Madiot, MoDisco: a model driven reverse engineering framework, Inf. Softw. Technol. 56 (8) (2014) 1012–1032.

[28] A. Bergmayr, H. Bruneliere, J. Cabot, J. Garcia, T. Mayerhofer, M. Wimmer, fREX: fUML-based reverse engineering of executable behavior for software dynamic analysis, in: IEEE/ACM 8th International Workshop on Modeling in Software Engineering (MiSE), 2016IEEE, 2016, pp. 20–26.

[29] M.L. Bernardi, G.A. Di Lucca, D. Distante, M. Cimitile, Model driven evolution of web applications, in: 2013 15th IEEE International Symposium on Web Systems Evolution (WSE), IEEE, 2013, pp. 45–50.

[30] L.A.P. Rabelo, A.F. do Prado, W.L. de Souza, L.F. Pires, An approach to business process recovery from source code, in: 2015 12th International Conference on Information Technology-New Generations (ITNG), IEEE, 2015, pp. 361–366.

[31] R. Pérez-Castillo, I.G.-R de Guzmán, M. Piattini, Knowledge discovery metamodel-ISO/IEC 19506: a standard to modernize legacy systems, Comput. Stand. Interfaces 33 (6) (2011) 519–532. ISSN: 0920-5489, https://doi.org/10.1016/j.csi.2011.02.007.

[32] R.P. Pérez-Castillo, I.G.-R de Guzmán, M. Piattini, Business process archeology using MARBLE, Inf. Softw. Technol. 53 (10) (2011) 1023–1044. ISSN: 0950-5849, https://doi.org/10.1016/j.infsof.2011.05.006, (special section on Mutation Testing).

[33] R. Pérez-Castillo, J.A. Cruz-Lemus, I.G.-R de Guzmán, M. Piattini, A family of case studies on business process mining using MARBLE, J. Syst. Softw. 85 (6) (2012) 1370–1385. ISSN: 0164-1212, https://doi.org/10.1016/j.jss.2012.01.022 (special issue: Agile Development).

[34] F. Trias, V de Castro, M. Lopez-Sanz, E. Marcos, Migrating traditional web applications to CMS-based web applications, Electron. Notes Theor. Comput. Sci. 314 (Suppl. C) (2015) 23–44. ISSN: 1571-0661, https://doi.org/10.1016/j.entcs.2015.05.003 (cLEI 2014, the XL Latin American Conference in Informatic).

[35] K. Garcés, R. Casallas, C. Álvarez, E. Sandoval, A. Salamanca, F. Viera, F. Melo, J.M. Soto, White-box modernization of legacy applications: the Oracle forms case study, Comput. Stand. Interfaces 57 (2017) 110–122. ISSN: 0920-5489, https://doi.org/10.1016/j.csi.2017.10.004.

[36] A. Shatnawi, H. Mili, G. El Boussaidi, A. Boubaker, Y.-G. Guéhéneuc, N. Moha, J. Privat, M. Abdellatif, Analyzing program dependencies in java EE applications, in: 2017 IEEE/ACM 14th International Conference on Mining Software Repositories (MSR), IEEE, 2017, pp. 64–74.

[37] A.S. Bozkir, E.A. Sezer, Layout-based computation of web page similarity ranks. Int. J. Hum. Comput. Stud. 110 (Suppl. C) (2018) 95–114. ISSN: 1071-5819, https://doi.org/10.1016/j.ijhcs.2017.10.008.

[38] I.A. Salihu, R. Ibrahim, Comparative Analysis of GUI reverse engineering techniques, in: H.A. Sulaiman, M.A. Othman, M.F.I. Othman, Y.A. Rahim, N.C. Pee (Eds.), Advanced Computer and Communication Engineering Technology, Springer, 2016, pp. 295–305.

[39] M.E. Joorabchi, A. Mesbah, Reverse engineering iOS mobile applications, in: 2012 19th Working Conference on Reverse Engineering, ISSN 1095-1350, 2012, pp. 177–186, https://doi.org/10.1109/WCRE.2012.27.

[40] J.M. Conejero, R. Rodríguez-Echeverría, F. Sánchez-Figueroa, M. Linaje, J.C. Preciado, P.J. Clemente, Re-engineering legacy web applications into RIAs by aligning modernization requirements, patterns and RIA features, J. Syst. Soft. 86 (12) (2013) 2981–2994. ISSN: 0164-1212, https://doi.org/10.1016/j.jss.2013.04.053.

[41] A. Mesbah, A. van Deursen, S. Lenselink, Crawling Ajax-Based web applications through dynamic analysis of user interface state changes, ACM Trans. Web 6 (1) (2012) 3:1–3:30. ISSN: 1559-1131, https://doi.org/10.1145/2109205.2109208.

[42] N. Matthijssen, A. Zaidman, FireDetective: understanding Ajax client/server interactions. in: Proceedings of the 33rd International Conference on Software Engineering, ACM, New York, NY, USA, 2011, ISBN: 978-1-4503-0445-0, pp. 998–1000, https://doi.org/10.1145/1985793.1985973.

[43] R. de Kleijn, Learning Selenium. Hands-on tutorials to create a robust and maintainable test automation framework, 2014.

[44] M. Schur, A. Roth, A. Zeller, Mining workflow models from web applications, IEEE Trans. Soft. Eng. 41 (12) (2015) 1184–1201.
[45] N. Walkinshaw, R. Taylor, J. Derrick, Inferring extended finite state machine models from software executions, Empir. Softw. Eng. 21 (3) (2016) 811–853.
[46] A. Gimblett, H. Thimbleby, User interface model discovery: towards a generic approach, in: Proceedings of the 2nd ACM SIGCHI symposium on Engineering interactive computing systems, ACM, 2010, pp. 145–154.
[47] C.E. Silva, J.C. Campos, Combining static and dynamic analysis for the reverse engineering of web applications, in: Proceedings of the 5th ACM SIGCHI Symposium on Engineering Interactive Computing Systems, ACM, New York, NY, USA, 2013, ISBN: 978-1-4503-2138-9, pp. 107–112, https://doi.org/10.1145/2494603.2480324.
[48] C.E. Silva, J.C. Campos, Characterizing the control logic of web applications' user interfaces, in: International Conference on Computational Science and Its Applications, Springer, 2014, pp. 263–276.
[49] F. Ricca, P. Tonella, Web site analysis: structure and evolution, in: Proceedings of the International Conference on Software Maintenance (ICSM 2000), 2000, pp. 76–86, San Jose, CA, USA.
[50] A. Shatnawi, A.-D. Seriai, H. Sahraoui, Z. Alshara, Reverse engineering reusable software components from object-oriented APIs, J. Syst. Soft. 131 (2017) 442–460.
[51] J. Cloutier, S. Kpodjedo, G. El Boussaidi, WAVI: a reverse engineering tool for web applications, in: 2016 IEEE 24th International Conference on Program Comprehension (ICPC), IEEE, 2016, pp. 1–3.
[52] J.-Y. Pan, S.-H. Ma, Advertisement removal of Android applications by reverse engineering, in: 2017 International Conference on Computing, Networking and Communications (ICNC), IEEE, 2017, pp. 695–700.
[53] S. Tilley, S. Huang, Evaluating the reverse engineering capabilities of web tools for understanding site content and structure: a case study, in: Proceedings of the 23rd International Conference on Software Engineering (ICSE 2001), 2001, pp. 514–523. Toronto, Ontario, Canada.
[54] H.T. Le, D.C. Nguyen, L. Briand, ReACP: a semi-automated framework for reverse-engineering and testing of access control policies of web applications, Université du Luxembourg, 2016. Tech. rep.
[55] D. Amalfitano, A.R. Fasolino, A. Polcaro, P. Tramontana, The DynaRIA tool for the comprehension of Ajax web applications by dynamic analysis, Innov. Syst. Softw. Eng. 10 (1) (2014) 41–57.
[56] G. Antoniol, M. Di Penta, M. Zazzara, Understanding web applications through dynamic analysis, in: Proceedings of the 12th IEEE International Workshop on Program Comprehension (IWPC 2004), 2004, pp. 120–129. Bari, Italy.
[57] G.A. Di Lucca, A.R. Fasolino, P. Tramontana, Reverse engineering Web applications: the WARE approach, J. Softw. Maint. Evol. Res. Pract. 16 (1–2) (2004) 71–101.
[58] A. Swearngin, A.J. Ko, J. Fogarty, Genie: input retargeting on the web through command reverse engineering, in: Proceedings of the 2017 CHI Conference on Human Factors in Computing Systems, ACM, 2017, pp. 4703–4714.
[59] P. Tonella, F. Ricca, Dynamic model extraction and statistical analysis of web applications, in: Proceedings of the Fourth International Workshop on Web Site Evolution (WSE 2002), 2002, pp. 43–52. Montreal, Quebec, Canada.
[60] I. Rožanc, M. Poženel, Reconstruction of the web application hypertext model using web logs, in: Proceedings of the 13th IASTED International Conference on Software Engineering (SE2013), ACTA Press, Inssbruck, Austria, 2014, pp. 148–155.
[61] J. Offutt, Y. Wu, Modeling presentation layers of web applications for testing, Soft. Syst. Model. 9 (2) (2010) 257–280.

[62] M. Spiliopoulou, B. Mobasher, B. Berendt, M. Nakagawa, A framework for the evaluation of session reconstruction heuristics in web-usage analysis, INFORMS J. Comput. 15 (2) (2003) 171–190.

[63] R. Kohavi, Mining e-commerce data: the good, the bad, and the ugly, in: F. Provost, R. Srikant (Eds.), Proceedings of the Seventh ACM SIGKDD International Conference on Knowledge Discovery and Data Mining, ACM, 2001, pp. 8–13. https://citeseer.ist. psu.edu/kohavi01mining.html.

[64] M. Poženel, V. Mahnič, M. Kukar, Separation of interleaved web sessions with heuristic search, in: Proceedings of the IEEE International Conference on Data Mining (ICDM), IEEE Computer Society Press, 2010, pp. 411–420.

[65] M. Levene, G. Loizou, Why is the snowflake schema a good data warehouse design? Inf. Syst. 28 (3) (2003) 225–240. ISSN: 0306-4379. https://doi.org/10.1016/S0306-4379 (02)00021-2.

[66] R. Kimball, R. Merz, The Data Webhouse Toolkit: Building the Web-Enabled Data Warehouse, vol. 78, John Wiley & Sons, Chichester, 2000.

[67] R. Kimball, M. Ross, The Data Warehouse Toolkit: The Complete Guide to Dimensional Modeling, John Wiley & Sons, Inc., New York, NY, USA, 2011. ISBN: 978-1-118-08214-0.

[68] X. Hu, N. Cercone, A data warehouse/online analytic processing framework for web usage mining and business intelligence reporting, Int. J. Intell. Syst. 19 (7) (2004) 585–606.

[69] M. Zorrilla, S. Millan, E. Menasalvas, Data web house to support web intelligence in e-learning environments, in: 2005 IEEE International Conference on Granular Computing, vol. 2, IEEE, 2005, pp. 722–727.

[70] B. Berendt, B. Mobasher, M. Nakagawa, M. Spiliopoulou, The impact of site structure and user environment on session reconstruction in web usage analysis, in: International Workshop on Mining Web Data for Discovering Usage Patterns and Profiles, Springer, 2002, pp. 159–179.

[71] Y. Fu, M.-Y. Shih, A framework for personal web usage mining, in: International Conference on Internet Computing, 2002, pp. 595–600.

[72] C. Shahabi, F. Banaei-Kashani, A framework for efficient and anonymous web usage mining based on client-side tracking, in: Lecture Notes in Computer Science, International Workshop on Mining Web Log Data Across All Customers Touch Points, vol. 2356, Springer, 2002, pp. 113–144.

[73] R. Cooley, B. Mobasher, J. Srivastava, Data preparation for mining world wide web browsing patterns. Knowl. Inf. Syst. 1 (1) (1999) 5–32, https://doi.org/10.1007/ BF03325089.

[74] R. Cooley, P.-N. Tan, J. Srivastava, Discovery of interesting usage patterns from web data, in: Web Usage Analysis and User Profiling, Springer, 2000, pp. 163–182.

[75] M. Munk, J. Kapusta, P. Švec, Data preprocessing evaluation for web log mining: reconstruction of activities of a web visitor, Procedia Comput. Sci. 1 (1) (2010) 2273–2280.

[76] R.F. Dell, P.E. Román, J.D. Velásquez, Web user session reconstruction with back button browsing, in: International Conference on Knowledge-Based and Intelligent Information and Engineering Systems, Springer, 2009, pp. 326–332.

[77] J. Huang, R.W. White, Parallel browsing behavior on the web, in: Proceedings of the 21st ACM conference on Hypertext and hypermedia, ACM, 2010, pp. 13–18.

[78] A. Paranjape, R. West, L. Zia, J. Leskovec, Improving website hyperlink structure using server logs, in: Proceedings of the Ninth ACM International Conference on Web Search and Data Mining, ACM, 2016, pp. 615–624.

[79] J.R. Mayer, J.C. Mitchell, Third-party web tracking: policy and technology, in: 2012 IEEE Symposium on Security and Privacy, IEEE, 2012, pp. 413–427.

[80] A. Sen, P.A. Dacin, C. Pattichis, Current trends in web data analysis. Commun. ACM 49 (11) (2006) 85–91. ISSN: 0001-0782, https://doi.org/10.1145/1167838.1167842.
[81] G. Antoniol, Y.-G. Gueheneuc, E. Merlo, P. Tonella, Mining the lexicon used by programmers during software evolution. in: 2007 IEEE International Conference on Software Maintenance, ISSN 1063-6773, 2007, pp. 14–23, https://doi.org/10.1109/ICSM.2007.4362614.
[82] N.R. Carvalho, J.J. Almeida, P.R. Henriques, M.J. Varanda, From source code identifiers to natural language terms. J. Syst. Soft. 100 (C) (2015) 117–128. ISSN: 0164-1212, https://doi.org/10.1016/j.jss.2014.10.013.
[83] T. Kamada, S. Kawai, An algorithm for drawing general undirected graphs, Inf. Process. Lett. 31 (1) (1989) 7–15.
[84] I. Borg, P.J.F. Groenen, P. Mair, Applied Multidimensional Scaling, Springer Verlag, 2013.

About the authors

Marko Poženel is a Teaching Assistant at the Faculty of Computer and Information Science at the University of Ljubljana, Ljubljana, Slovenia. His teaching and research interests include agile software development methods, empirical software engineering as well as Web data mining and user behavior analysis. He received his Ph.D. in Computer Science from the University of Ljubljana in 2010.

Boštjan Slivnik is an Assistant Professor of Computer Science at the University of Ljubljana, Faculty of Computer and Information Science, where he received the M.Sc. and Ph.D. degrees in Computer Science in 1996 and 2003, respectively. His research interests include compilers and programming languages with the special focus on parsing algorithms and formal languages, scheduling and distributed algorithms, and software engineering. He has been a member of the ACM since 1996.

Printed in the United States
By Bookmasters